T0224742

Numerical Problems in Crystallography

M. A. Wahab

Numerical Problems
in Crystallography

Springer

M. A. Wahab
Department of Physics
Jamia Millia Islamia
Delhi, India

ISBN 978-981-15-9756-5 ISBN 978-981-15-9754-1 (eBook)
https://doi.org/10.1007/978-981-15-9754-1

This Springer imprint is published by the registered company Springer Nature Singapore Pte Ltd.
The registered company address is: 152 Beach Road, #21-01/04 Gateway East, Singapore 189721,
Singapore

Preface

Crystallography, an interdisciplinary subject, plays an important role in a wide range of subjects of science and technology. Its rudimentary knowledge is essential for beginners in physics, chemistry, mathematics, molecular biology, geology, metallurgy, and particularly to materials science and mineralogy. Its relationship with physics can be understood in terms of the following examples:

1. All conservation laws in physics are essentially interrelated with symmetry operations.
2. Symmetry considerations allow us to establish Pauli's exclusion principle.
3. Periodic potential of the lattice is responsible for the creation of allowed and forbidden energy bands. Accordingly, materials can be classified as conductor, semi-conductor and insulator depending on the amount of band gap.
4. Relationships between the magnitudes of measurable parameters give rise to the tensor components of different physical properties in crystals. This list can be elaborated further.

In a similar manner, its relationship with other subjects can also be listed.

The topic "numerical problems in crystallography" therefore attaches even greater importance as the experience of solving numerical problems is extremely helpful in understanding different topics of crystallography. This book aims at making people understand different topics of crystallography which are conceptual in nature through solving numerical problems.

The author felt the need to write this book in view of the following reasons:

(i) the difficult nature of topics and the non-availability of any other book of this type in India (and perhaps in the international market),
(ii) the growing number of competitive examinations at various levels conducted by universities, where questions are generally of numerical in nature,
(iii) experience of teaching and research suggests that crystallography is a conceptual subject and needs a fair amount of imagination for understanding 3D images.

Crystallography is the science of crystals and has undergone innumerable changes in the course of its development. In fact, crystals have puzzled mankind since ancient times and crystallography as an independent branch of science began to take shape only during seventeenth to eighteenth centuries. In some way or the other, different topics considered in the book are taught under different courses/subjects such as solid-state physics, solid-state chemistry, mathematics, molecular biology, geology, metallurgy, and particularly the materials science and mineralogy at different universities and institutions. However, the introductory books covering these topics in a broader sense do not cater to the needs of students in real sense as they do not deal with the solved numerical problems. Similarly, students do not find the exhaustive treatises on crystallography written by the well-known authors very useful as they are difficult to comprehend and do not deal with numerical problems.

This book includes nine chapters based on different topics of crystallography, with each chapter containing a different number of intimately related subtopics. Each chapter begins with a brief but relevant description of the first subtopic (as provided in the content) explaining its basic concept, then followed by a number of solved numerical examples based on it for further clarity on the subject. This is how all other subtopics are completed in each chapter. This book will be extremely helpful to faculty members associated with the field and students of physics, chemistry, biology, geology, metallurgy and engineering-related subjects, particularly at graduate and research levels where they have to appear for various competitive examinations.

The author has taken utmost care in selecting various topics and presenting them in a simplified form as far as possible for the benefit of students and faculties. At the end of each chapter, the students will find multiple-choice questions (based on the solved examples) along with their answers for practice and gaining self-confidence.

The main features of this book are:

1. Carefully selected important and relevant topics of crystallography
2. Brief but relevant description of each subtopic explaining its basic concept
3. Solved examples under each conceptual subtopic
4. Multiple choice questions, along with the key at the end of each chapter

Although proper care has been taken during the preparation of the manuscript, few errors might creep in. Any omission, errors, suggestions brought to my knowledge will be appreciated and thankfully acknowledged.

I sincerely admit and acknowledge that I learnt the art of systematic presentation of the research work or other related scientific matters from my supervisor, Prof. G. C. Trigunayat (Late), Department. of Physics, University of Delhi. I also sincerely acknowledge the work of my two sons Mr. Khurram Mujtaba Wahab and

Mr. Shad Mustafa Wahab, for making the diagrams and formatting the manuscript. My special thanks to other members of my family for their continuous support and encouragement during the preparation of the manuscript.

Delhi, India   M. A. Wahab

Units of Measurements

Base Units

Quantity	Unit	Symbol
Length, L	meter	m
Mass, M	kilogram	kg
Time, t	second	s
Electric current, I	ampere	A
Temperature, T	Kelvin	K
Amount of substance, n	mole	mol
Luminous Intensity	candela	cd

Supplementary Units

Plane angle, θ	radian	rad
Solid angle, Ω	steradian	sr

Derived Units

Quantity	Special name	Symbol	Other derived units	Base units
Wavelength	-	λ	-	Å
Frequency	hertz	Hz	-	s^{-1}
Force, weight	newton	N	-	$kgms^{-2}$
Stress, strength, Pressure	pascal	Pa	-	$kgm^{-1}s^{-2}$
Energy, work, quantity of heat	joule	J	Nm	kgm^2s^{-2}
Power	watt	W	Js^{-1}	kgm^2s^{-3}
Electric charge	coulomb	C	-	As
Electric potential	volt	V	WA^{-1}	$kgm^2s^{-3}A^{-2}$
Resistance	ohm	W*	VA^{-1}	$kgm^2s^{-3}A^{-2}$
Capacitance	farad	F	CV^{-1}	$kg^{-1}m^{-2}s^4A^2$
Inductance	henry	H	WbA^{-1}	$kgm^2s^{-2}A^{-2}$

Prefix, Multiples and Submultiples

Factors by which the unit is multiplied	Name	Symbol
10^{12}	Tetra	T
10^9	Giga	G
10^6	Mega	M
10^3	Kilo	k
10^2	Hector	h
10^1	Deka	da
10^{-1}	Deci	d
10^{-2}	Centi	c
10^{-3}	Milli	m
10^{-6}	Micro	m*
10^{-9}	Nano	n
10^{-12}	Pico	p
10^{-15}	Femto	f
10^{-18}	Atto	a

Physical Constants

Avogadro's number	$N = 6.023 \times 10^{23} \text{ mol}^{-1} = 6.023 \times 10^{26} \text{ kmol}^{-1}$
Boltzmann's constant	$K = 1.380 \times 10^{-23} \text{ JK}^{-1} = 8.614 \times 10^{-5} \text{ eVK}^{-1}$
Gas constant	$R = 8.314 \text{ Jmol}^{-1} \text{ K}^{-1}$
Planck's constant	$h = 6.626 \times 10^{-34} \text{ Js} = 6.626 \times 10^{-27} \text{ ergs}$
Electronic charge	$e = 1.602 \times 10^{-19} \text{ C} = 4.8 \times 10^{10} \text{ esu}$
Electron rest mass	$m_0 = 9.11 \times 10^{-31} \text{ kg} = 9.11 \times 10^{-28} \text{ g}$
Proton rest mass	$m_p = 1.673 \times 10^{-27} \text{ kg} = 1.673 \times 10^{-24} \text{ g}$
Neutron rest mass	$M_n = 1.675 \times 10^{-27} \text{ kg} = 1.675 \times 10^{-24} \text{ g}$
Velocity of light	$c = 3 \times 10^8 \text{ ms}^{-1} = 3 \times 10^{10} \text{ cms}^{-1}$
Permitivity of free space	$\varepsilon_0 = 8.854 \times 10^{-12} \text{ Fm}^{-1}$
Coulomb force constant	$1/4\pi \, \varepsilon_0 = 9 \times 10^9 \text{ Nm}^2 \text{ C}^{-2}$
Atomic mass unit (amu)	$(1/10^3 N) = 1.660 \times 10^{-27} \text{ kg} = 9.11 \times 10^{-28} \text{ g}$
Acceleration due to gravity	$g = 9.81 \text{ ms}^{-2}$
Ice point	$0° \text{ C} = 273.15 \text{ K}$

Conversion Factors

1 micron $= 10^{-6}$ m
1 nm $= 10^{-9}$ m
1 Å $= 10^{-10}$ m
1° $= 1/57.3$ rad
1 eV $= 1.602 \times 10^{-19}$ J
1 erg $= 10^{-7}$ J
1 eV/entity $= 96.49$ kJ mol^{-1} of (entities)
1 calorie $= 4.18$ J
1 atmosphere $= 0.101325$ MPa
1 torr ($= 1$ mm of Hg) $= 133.3$ Nm^{-2}
1 bar $= 105$ Nm$^{-2} = 10^{-1}$ MPa
1 Psi $= 6.90$ kNm^{-2}
1 dyne/cm$^2 = 0.1$ Pa s
1 mol/cm^2/sec $= 10^4$ mol m^{-2} s^{-1}
1 mol/cm$^3 = 10^6$ mol m^{-3}
1 mol/cm$^4 = 10^8$ mol m^{-4}
1 cm^2/sec $= 10^{-4}$ mol m^{-2} s^{-1}
T° C $= (T + 273.15)$ K
1 erg/cation $= 1.43933 \times 10^{13}$ kcal/mol

1 electronic charge = 1.6×10^{-19} C

1 Wb = 10^8 Maxwell

1 Tesla = 1 Wb/m^2 = 10^4 max/cm^2 = 10^4 Gauss

1 A/m = $4\pi \times 10^{-3}$ Oe = 1.26×10^{-2} Oe

Contents

About the Author

M. A. Wahab is Former Professor and Head of the Department of Physics, Jamia Millia Islamia, New Delhi, India. He completed his Ph.D. in Physics from the University of Delhi, India. Earlier, he served as Lecturer at the P. G. Department of Physics and Electronics, University of Jammu, Jammu and Kashmir, India, from 1981 and later, at the P. G. Department of Physics, Jamia Millia Islamia, from 1985. During these years, he taught electrodynamics, statistical mechanics, theory of relativity, advanced solid-state physics, crystallography, physics of materials, growth and imperfections of materials, and general solid-state physics. He has authored 3 books and over 100 research papers in national and international journals of repute and supervised 15 Ph.D. theses during his career at Jamia Millia Islamia. Professor Wahab has published a paper on the discovery of hexagonal close packing (HCP) and rhombohedral close packing (RCP) as two new space lattices, along with his son (Khurram Mujtaba Wahab), as their first joint paper after his retirement.

Chapter 1
Unit Cell Composition

1.1 Translation Vectors in Plane and Space Lattices

(a) When an object (say the number 7) is repeatedly translated through an interval "a," we obtain a one-dimensional array of the number 7, as shown in Fig. 1.1a. When each object in the array is replaced with a point, a collection of points is obtained as shown in Fig. 1.1b. This is known as a linear lattice. Since, a geometrical point has no (or zero) dimension, therefore the lattice point is an imaginary concept but the array of the number "7" is real.

(b) When we add another non-collinear translation "b" to the entire lattice array due to the translation "a," a two-dimensional array of objects is obtained (Fig. 1.2a). The corresponding collection of two-dimensional points in Fig. 1.2b is called a plane lattice.

The characteristic feature of a plane lattice is that the environment around any one point is identical to the environment around any other point in the lattice. In a plane lattice, any two points connected through a translation vector $\vec{t}$ is given by

$$\vec{t} = n_1\vec{a} + n_2\vec{b} \tag{1.1}$$

where n_1 and n_2 are arbitrary integers, when putting together in the form $[n_1\ n_2]$ gives the direction of the translation vector (Fig. 1.3) in the given lattice, $\vec{a}$ and $\vec{b}$ are the primitive translation vectors along x and y axes, respectively.

The magnitude (say R) of the resultant translation vector between any two points (in a plane lattice) can be determined by the formula

$$R = \left(P^2 + Q^2 + 2PQ\cos\gamma\right)^{1/2} \tag{1.2}$$

© The Author(s), under exclusive license to Springer Nature Singapore Pte Ltd. 2021
M. A. Wahab, *Numerical Problems in Crystallography*,
https://doi.org/10.1007/978-981-15-9754-1_1

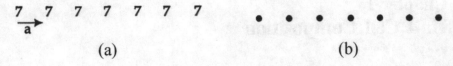

(a)   (b)

Fig. 1.1 One-dimensional array of: **a** objects, **b** points; a linear lattice

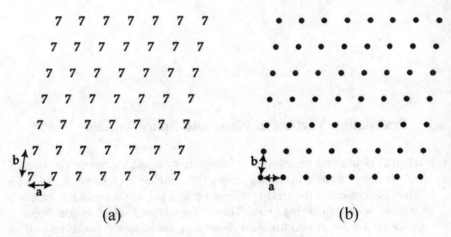

(a)   (b)

Fig. 1.2 Two-dimensional array of: **a** objects, **b** points; a plane lattice

Fig. 1.3 Translation vector, t_1 and t_2

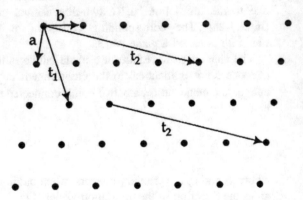

where $\vec{P} = n_1\vec{a}$ and $\vec{Q} = n_2\vec{b}$ are measured in terms of primitive vectors $\vec{a}, \vec{b}$ and γ is the angle between them.

(c) When we add a third non-coplanar translation "c" to the entire plane pattern due to translations "a" and "b," a three-dimensional lattice is obtained (Fig. 1.4). Like a plane lattice, a space lattice also exhibits a similar characteristic feature and hence any two points connected through a translation vector $\vec{T}$ is given by

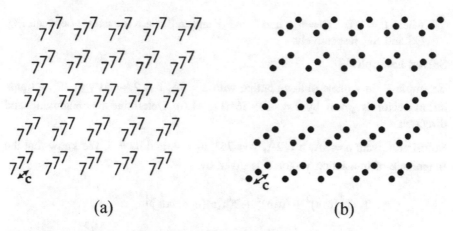

(a)   (b)

Fig. 1.4 Three-dimensional array of: **a** objects, **b** points; a space lattice

$$\vec{T} = n_1\vec{a} + n_2\vec{b} + n_3\vec{c} \tag{1.3}$$

where n_1, n_2 and n_3 are arbitrary integers, when putting together in the form [n_1 n_2 n_3] gives the direction of the translation vector $\vec{T}$ (Fig. 1.5) in the given lattice, and $\vec{a}$, $\vec{b}$ and $\vec{c}$ are the primitive translation vectors along x, y and z axes, respectively. The translation vectors $\vec{a}$, $\vec{b}$, $\vec{c}$ actually define the space (or the coordinate system) and hence are also known as basis vectors of the space.

The magnitude (say R) of the resultant translation vector between any two points (in a space lattice) can be determined by the formula

$$R = \left(P^2 + Q^2 + S^2 + 2PQ\cos\alpha + 2QS\cos\beta + 2SP\cos\gamma\right)^{1/2}$$

Fig. 1.5 Space lattice with basis vectors $\vec{a}$, $\vec{b}$, $\vec{c}$ and translation vector $\vec{T}$

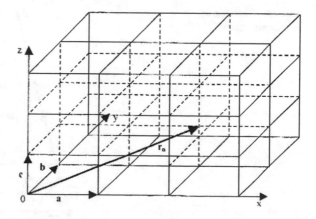

where $\vec{P} = n_1\vec{a}$ $\vec{Q} = n_2\vec{b}$ and $\vec{S} = n_3\vec{c}$ and α, β, γ are the angles between PQ, QS and SP, respectively.

Solved Examples

Example 1 In a plane oblique lattice with a = 3Å, b = 2Å and $\gamma = 75°$, a translation vector is given by $\vec{t} = 4\vec{a} + 5\vec{b}$ (Fig. 1.6). Determine its magnitude and direction.

Solution: Given: a = 3Å, b = 2Å, $\gamma = 75°$, $n_1 = 4$, and $n_2 = 5$. We know that the magnitude of translation vector $\vec{T}$ is given by.

$$
\begin{aligned}
T &= \left[(n_1a)^2 + (n_2b)^2 + 2(n_1a)(n_2b) \cos\gamma \right]^{1/2} \\
&= \left[(4 \times 3)^2 + (5 \times 2)^2 + 2(4 \times 3)(5 \times 2) \cos\gamma \right]^{1/2} \\
&= [(144 + 100 + 240 \times 0.2588)]^{1/2} \\
&= (306.1165)^{1/2} \\
&= 17.496Å
\end{aligned}
$$

Direction of the translation vector is $[n_1\ n_2] \equiv [45]$.

Example 2 In a plane rectangular lattice with a = 3Å, b = 2Å, a translation vector is given by $\vec{t} = 5\vec{a} + 7\vec{b}$ (Fig. 1.7). Determine its magnitude and direction.

Solution: Given: a = 3Å, b = 2Å, $n_1 = 5$, and $n_2 = 7$. We know that for a plane rectangular lattice, $\gamma = 90°$ and $\cos90° = 0$, therefore the magnitude of translation vector $\vec{T}$ is given by

Fig. 1.6 Direction of translation vector

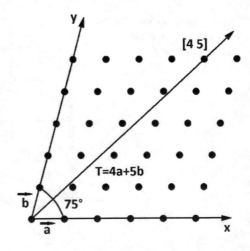

Fig. 1.7 Direction of
translation vector

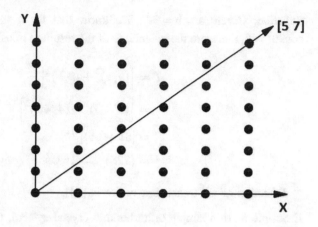

$$T = \left[(n_1 a)^2 + (n_2 b)^2 \right]^{1/2}$$

$$= \left[(5 \times 3)^2 + (7 \times 2)^2 \right]^{1/2}$$

$$= [225 + 196]^{1/2}$$

$$= (421)^{1/2}$$

$$= 20.52 \text{Å}$$

Direction of the translation vector is [57].

Example 3 In a plane hexagonal lattice, the translation vector is given by $\vec{t} = 2\vec{a} + 3\vec{b}$, where $a = b = 3\text{Å}$ and $\gamma = 120°$. Determine the magnitude and direction of the resultant translation vector.

Solution: Given: $a = b = 3\text{Å}$, $\gamma = 120°$, $n_1 = 2$, and $n_2 = 3$. We know that the magnitude of the resultant translation vector $\vec{T}$ is given by

$$T = \left[(n_1 a)^2 + (n_2 b)^2 + 2(n_1 a)(n_2 b) \cos \gamma \right]^{1/2}$$

$$= \left[\left((2 \times 3)^2 + (3 \times 3)^2 + 2(2 \times 3)(3 \times 3) \cos 120 \right) \right]^{1/2}$$

$$= \left[\left(36 + 81 + 2 \times 54 \times \left(\frac{-1}{2} \right) \right) \right]^{1/2} = 7.94 \text{Å}$$

Direction of the translation vector is [23].

Example 4 In a square lattice, the translation vector is given by $\vec{t} = 3\vec{a} + 4\vec{b}$, where $a = b = 3\text{Å}$. Determine the magnitude and direction of the resultant translation vector.

Solution: Given: a = b = 3Å. We know that for a square lattice, $\gamma = 90°$ and $\cos 90° = 0$, therefore the magnitude of the resultant translation vector $\vec{T}$ is given by

$$T = \left[(n_1 a)^2 + (n_2 b)^2\right]^{1/2}$$
$$= \left[(3 \times 3)^2 + (4 \times 3)^2\right]^{1/2}$$
$$= [81 + 144]^{1/2}$$
$$= (225)^{1/2} = 15.0\text{Å}$$

Direction of the translation vector is [34].

Example 5 In a simple orthorhombic crystal system, two lattice points are connected through a translation vector. $\vec{t} = 2\vec{a} + 3\vec{b} + 4\vec{c}$, where a = 2Å, b = 3Å and c = 4Å. Determine the magnitude and direction of the resultant translation vector.

Solution: Given: a = 2Å, b = 3Å and c = 4Å, $n_1 = 2$, $n_2 = 3$ and $n_3 = 4$. We know that for an orthorhombic crystal system, $\alpha = \beta = \gamma = 90°$ and $\cos 90° = 0$, therefore the magnitude of the resultant translation vector $\vec{T}$ is given by

$$T = \left[(n_1 a)^2 + (n_2 b)^2 + (n_3 c)^2\right]^{1/2}$$
$$= \left[\left(2 \times 2\right)^2 + (3 \times 3)^2\right) + (4 \times 4)^2\right]^{1/2}$$
$$= [(16 + 81 + 256)]^{1/2} = 18.79\text{Å}$$

Direction of the translation vector is [234].

Example 6 In a tetragonal crystal system, two lattice points are connected through a translation vector $\vec{t} = 3\vec{a} + 4\vec{b} + 5\vec{c}$, where a = b = 2Å and c = 3Å. Determine the magnitude and direction of the resultant translation vector.

Solution: Given: a = b = 2Å, c = 3Å, $n_1 = 3$, $n_2 = 4$ and $n_3 = 5$. We know that for an orthorhombic crystal system, $\alpha = \beta = \gamma = 90°$ and $\cos 90° = 0$, therefore the magnitude of the resultant translation vector $\vec{T}$ is given by

$$T = \left[(n_1 a)^2 + (n_2 b)^2 + (n_3 c)^2\right]^{1/2}$$
$$= \left[\left(3 \times 2\right)^2 + (4 \times 2)^2\right) + (3 \times 5)^2\right]^{1/2}$$
$$= [(36 + 64 + 225)]^{1/2}$$
$$= 18.03\,\text{Å}$$

Direction of the translation vector is [345].

Example 7 Determine the magnitude and direction of translation vector in a hexagonal crystal system with a = b = 3Å and c = 4Å. Further, the integers n_1 = 4 units, n_2 = 3 units and n_3 = 1, respectively.

Solution: Given: a = b = 3Å and c = 4Å, n_1 = 4, n_2 = 3 and n_3 = 1. We know that in a hexagonal crystal system, $\alpha = \beta = 90°$ and $\gamma = 120°$, therefore the magnitude of the resultant translation vector $\vec{T}$ is given by

$$
\begin{aligned}
T &= \left[(n_1 a)^2 + (n_2 b)^2 + 2(n_1 a)(n_2 b)\cos\gamma \right]^{1/2} \\
&= \left[(4 \times 3)^2 + (3 \times 3)^2 + (4 \times 1)^2 + 2(4 \times 3)(3 \times 3)\left(-\frac{1}{2}\right) \right]^{1/2} \\
&= [(144 + 81 + 16 - 108)]^{1/2} = 11.53\text{Å}
\end{aligned}
$$

Direction of the translation vector is [431].

1.2 Choice of Axes and Unit Cells

(a) In general, for a plane lattice, the choice of axes is infinite and so the basis vectors. Consequently, the choice of the unit cell is also infinite. However, the axial systems or the basis vectors $\vec{a}$ and $\vec{b}$ can be either left-handed or right-handed (Fig. 1.8). In a given lattice, this can be represented as shown in Fig. 1.9. The shapes and sizes of such unit cells are determined by the magnitude of the basis vectors and the interaxial angle γ. The area of such a unit cell is given by

$$\vec{a} \times \vec{b} = ab \sin\gamma \tag{1.4}$$

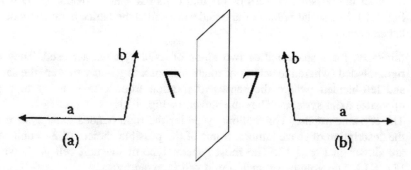

Fig. 1.8 **a** Left-handed, **b** Right-handed system, separated by the vertical mirror

(a)   (b)

Fig. 1.9 Lattice with basis vectors $\vec{a}$ and $\vec{b}$: **a** left-handed, **b** right-handed

Fig. 1.10 Some unit cells of a plane lattice obtained from different combinations of two unit translation vectors

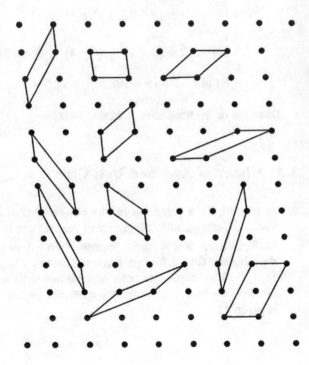

Some of the possible choices of the unit cells that can be made are shown in Fig. 1.10. The scalar values of a, b and γ are called the lattice parameters or the lattice constants.

(b) Similarly, for a space lattice two kinds of axial systems are used. They are right-handed (where the senses of rotation from x to y and y to z are the same) and left-handed (where the senses of rotation from x to y and y to z are opposite) axial systems. They are shown in Fig. 1.11.

The crystallographers conventionally prefer the right-handed axial system for the description of space lattices. Some of the possible choices of 3-D unit cells are shown in Fig. 1.12. The most general type of the unit cell is shown in Fig. 1.13. The volume of such a unit cell is given by

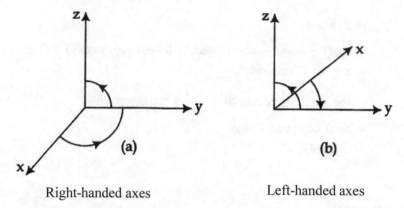

Right-handed axes   Left-handed axes

Fig. 1.11 Two axial systems **a** the senses of rotation from x to y and y to z are same **b** the senses of rotation from x to y and y to z are opposite

Fig. 1.12 Some choices of primitive unit cells in a space lattice

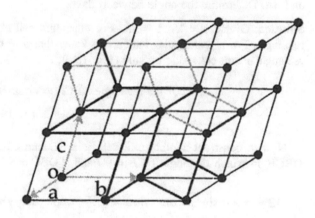

Fig. 1.13 Most general unit cell and its axes

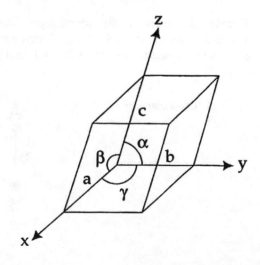

$$V_{cell} = \vec{a} \cdot \vec{b} \times \vec{c}$$
$$= abc \left(1 - \cos^2 \alpha - \cos^2 \beta - \cos^2 \gamma + 2\cos \alpha \cos \beta \cos \gamma\right)^{1/2} (\text{Triclinic})$$
$$= abc \, \sin \beta (\text{Monoclinic})$$
$$= abc \, \sin \gamma = a^2 c \sin 120° = \frac{\sqrt{3}}{2} a^2 c (\text{Hexagonal})$$
$$= abc (\text{Orthogonal Cells})$$
$$= a^3 (\text{Cubic})$$

$$(1.5)$$

Solved Examples

Example 1 Determine the area of a primitive rectangle whose sides are 4Å and 3Å, respectively. Construct another primitive cell of equal area whose sides are 4Å and 5Å. Determine the angle between them.

Solution: Given: a = 4Å, b = 3Å. For other unit cell a′ = a, b′ = diagonal of the rectangle. For a rectangular lattice, we know that a ≠ b, γ = 90°. Construct two rectangular unit cells side by side (Fig. 1.14).

$$\text{Area of the rectangle ABCD} = ab \sin 90°$$
$$= 4 \times 3 = 12 \text{ Å}^2$$

Now to construct another unit cell, join BD and CE. The required unit cell is DBCE, in which the angle DCB = 90°and < DBC = γ. Therefore,

$$12 = 5 \times 4 \sin \gamma \quad \text{or} \quad \sin \gamma = \frac{12}{5 \times 4} = \frac{3}{5} \quad \text{or} \quad \gamma = \sin^{-1}\left(\frac{3}{5}\right) = 36.87°$$

Example 2 Determine the area of a primitive square unit cell whose side is 2Å. Construct another unit cell of equal area whose one side is the diagonal of the square. Determine the length of this side and the angle it makes with x-axis.

Fig. 1.14 Two rectangular unit cells

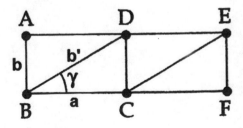

Fig. 1.15 Two square unit cells

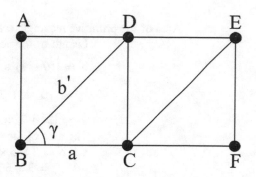

Solution: Given: a = b = 2Å. For other unit cell a′ = a, b′ = diagonal of the square. For a square lattice, we know that a = b, γ = 90°. Construct two square unit cells side by side (Fig. 1.15).

$$\text{Area of the square ABCD} = a^2 = 2^2 = 4 \text{ Å}^2$$

Now to construct another unit cell, join BD and CE. The required unit cell is DBCE, in which the angle DCB = 90° and < DBC = γ. Therefore,

$$BD = \left(2^2 + 2^2\right)^{1/2} = 2\sqrt{2} \text{ Å}$$

Further, according to the question, the areas of the two-unit cells are equal. Therefore,

$$4 = 2 \times 2\sqrt{2} \sin \gamma \quad \text{or} \quad \sin \gamma = \frac{1}{\sqrt{2}} \quad \text{or} \quad \gamma = \sin^{-1}\left(\frac{1}{\sqrt{2}}\right) = 45°$$

Example 3 The sides of a primitive rectangle are 3Å and 2Å, respectively. Construct a new unit cell with the edges defined by the vectors from the origin to the points with coordinates 1, 0 and 1, 2. Determine (i) the side b′ (ii) area of the new unit cell and (iii) the number of the lattice points in the new unit cell.

Solution: Given: Sides of primitive rectangle, a = 3Å, b = 2Å, coordinates of new cell edges: 1, 0 and 1, 2. Construct four rectangular unit cells side by side (Fig. 1.16).

Fig. 1.16 Four rectangular unit cells

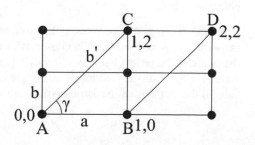

Area of the primitive rectangle cell $= a \times b = 3 \times 2 = 6 \, \text{Å}^2$

Length of the new cell edge

$$b' = \left[(0-1)^2 a^2 + (0-2)^2 b^2\right]^{1/2}$$

$$= \left[(0-1)^2 3^2 + (0-2)^2 2^2\right]^{1/2}$$

$$= \left[3^2 + 2^2 \times 2^2\right]^{1/2} = 5\text{Å}$$

The area of the new unit cell

$a \times b' = 3 \times 5 \sin\gamma$

$= 3 \times 5 \times \frac{4}{5}$

$= 12 \, \text{Å}^2$

Ratio of the two areas $= \frac{12}{6} = 2$

This implies that the new unit cell is non-primitive and the number of lattice points in it is 2.

Example 4 The side of a primitive square lattice is 2Å. Construct a new unit cell with edges defined by the vectors from the origin to the points with coordinates 1, 0 and 1, 2. Determine (i) the side b' (ii) area of the new unit cell and (iii) the number of the lattice points in the new unit cell.

Solution: Given: Sides of the primitive square lattice, a = 2Å, coordinates of new cell edges: 1, 0 and 1, 2. Area of the primitive cell $= a^2 = 2^2 = 4 \, \text{Å}^2$

Length of the new cell edge

$$b' = \left[(0-1)^2 a^2 + (0-2)^2 b^2\right]^{1/2}$$

$$= \left[(0-1)^2 2^2 + (0-2)^2 2^2\right]^{1/2}$$

$$= \left[2^2 + 2^2 \times 2^2\right]^{1/2}$$

$$= 4.47\text{Å}$$

The area of the new unit cell

$A = 4.47 \times 2 \times \sin\gamma$

$= 4.47 \times 2 \times \frac{4}{4.47}$

$= 8 \, \text{Å}^2$

Ratio of the two areas $= \frac{8}{4} = 2$

This implies that the new unit cell is non-primitive and the number of lattice points in it is 2.

Example 5 A primitive rectangular unit cell has a = 2Å, b = 3Å and $\gamma = 90°$, respectively. A new unit cell is chosen with the edges defined by the vectors from the origin to the points with coordinates 2, 0 and 0, 3. Determine (a) area of the unit cell (b) lengths of the two edges and angle between them (c) area of the new unit cell (d) the number of the lattice points in the new unit cell.

Solution: Given: Sides of primitive rectangle, a = 2Å, b = 3Å, coordinates of new cell edges: 2, 0 and 0, 3. (a) Area of the primitive rectangle cell = a × b = 2 3 = 6 Å².

(a) Lengths of the edges from the origin:
    (i) For coordinates 0, 0 and 2, 0

$$a' = \left[(0-2)^2 \times 2^2 + 0\right]^{1/2} = [16]^{1/2} = 4\text{Å}$$

(i) For coordinates 0, 0 and 0, 3

$$b' = [0 + (0-3)^2 \times 3^2]^{1/2} = [81]^{1/2} = 9\text{Å}$$

Angle between the lines with the end coordinates 2, 0 and 0, 3

$$\cos\gamma = \frac{0}{\sqrt{2^2}\ \sqrt{3^2}} = 0 \Rightarrow \gamma = 90°$$

(b) Area of the new unit cell a' × b' = 4 × 9 = 36 Å²
(c) Ratio of the two areas = $\frac{36}{6}$ = 6
(d) The new unit cell is non-primitive and the number of lattice points in it is 6.

Example 6   The side of a primitive square unit cell is, a = 2Å. A new unit cell is chosen with the edges defined by the vectors from the origin to the points with coordinates 1, 0 and 0, 3. Determine (a) area of the original unit cell (b) lengths of the two edges and angle between them (c) area of the new unit cell (d) the number of the lattice points in the new unit cell.

Solution: Given: Sides of primitive square unit cell, a = 2Å, coordinates of new cell edges: 1, 0 and 0, 3. We know that for a square lattice, γ = 90°. Therefore,

(a) Area of the primitive square unit cell = a^2 = 4 Å²
(b) Lengths of the edges from the origin:
    (i) For coordinates 0, 0 and 1, 0

$$a' = \left[(0-1)^2 \times 2^2 + 0\right]^{1/2} = [4]^{1/2} = 2\text{Å}$$

(ii) For coordinates 0, 0 and 0, 3

$$b' = \left[0 + (0-3)^2 \times 2^2\right]^{1/2} = [36]^{1/2} = 6\text{Å}$$

Angle between the lines with the end coordinates 1, 0 and 0, 3

$$\cos\gamma = \frac{0}{\sqrt{1^2}\ \sqrt{3^2}} = 0 \Rightarrow \gamma = 90°$$

(c) Area of the new unit cell   a′ x b′ = 2 × 6 = 12 Å^2
(d) Ratio of the two areas = $\frac{12}{4}$ = 3

This implies that the new unit cell is non-primitive and the number of lattice points in it is 3.

Example 7 A primitive orthorhombic unit cell has a = 5Å, b = 6Å and c = 7Å; α = β = γ = 90°. A new unit cell is chosen with the edges defined by the vectors from the origin to the points with coordinates 3, 1, 0; 1, 2, 0 and 0, 0, 1. Determine (a) volume of the original unit cell (b) lengths of the three edges and angles between them (c) volume of the new unit cell (d) the number of the lattice points in the new unit cell.

Solution: Given: Orthorhombic unit cell parameters, a = 5Å, b = 6Å and c = 7Å; α = β = γ = 90°, coordinates of the new cell edges: 3, 1, 0; 1, 2, 0 and 0, 0, 1.

(a) Volume of the primitive orthogonal unit cell,

V_0 = abc

$= 5 \times 6 \times 7 = 210 Å^3$

(b) Length of the vector a′ between the coordinates: 0, 0, 0 and 3, 1, 0

$$a' = \left[(0-3)^2 \times 5^2 + (0-1)^2 \times 6^2 + 0\right]^{1/2} = [9 \times 25 + 36]^{1/2} = 16.16 Å$$

Similarly, the length of the vector b′ between the coordinates; 0, 0, 0 and 1, 2, 0

$$b' = \left[(0-1)^2 \times 5^2 + (0-2)^2 \times 6^2 + 0\right]^{1/2} = [25 + 4 \times 36]^{1/2} = 13 Å$$

and the length of the vector c′ between the coordinates: 0, 0, 0 and 0, 0,1

$$c' = \left[0 + 0 + (0-1)^2 \times 7^2\right]^{1/2} = [7]^{1/2} = 7 Å$$

Now the angle between the edges with the end coordinates 3, 1, 0 and 1, 2, 0 is given by

$$\cos\gamma' = \frac{3+2}{(9+1+0)^{1/2}(1+4+0)^{1/2}} = \frac{5}{(10)^{1/2}(5)^{1/2}} = \frac{1}{\sqrt{2}}$$

$$\Rightarrow \gamma' = 45°$$

Similarly, the angle between the edges with the end coordinates 1, 2, 0 and 0, 0, 1 is given by

$$\cos \alpha' = \frac{0}{(1+4+0)^{1/2} (0+0+1)^{1/2}} = 0$$

$$\Rightarrow \alpha' = 90°$$

And the angle between the edges with the end coordinates 3, 1, 0 and 0, 0, 1 is given by

$$\cos \beta' = \frac{0}{(9+1+0)^{1/2}(0+0+1)^{1/2}} = 0$$

$$\Rightarrow \beta' = 90°$$

(c) The new unit cell with $a' = 16.16$Å, $b' = 13$Å, $c' = 7$Å and $\alpha' = \beta' = 90°$, $\gamma' = 45°$ is a monoclinic lattice and its volume is given by

$$V_m = a' \, b' \, c' \sin 45°$$
$$= 16.16 \times 13 \times 7 \times \sin 45°$$
$$= 1039.84 \text{Å}^3$$

(d) Ratio of the two volumes $= \dfrac{1039.84}{210} = 4.95 \cong 5$

This implies that the new unit cell is non-primitive and the number of lattice points in it is 5.

Example 8 A rhombohedral unit cell has $a_r = 5$Å and $\alpha = 75°$. Calculate: (a) the volume of the unit cell (b) the dimension of the triply primitive hexagonal unit cell that may be chosen (c) the volume of the hexagonal unit cell (d) the number of lattice points in the hexagonal unit cell.

Solution: Given: $a_r = 5$Å and $\alpha = 75°$, triply primitive hexagonal unit cell.

(a) The volume of a rhombohedral unit cell is given by

$$V_R = a_R^3 \sqrt{1 - 3\cos^2 \alpha + 2\cos^3 \alpha}$$
$$= 125\sqrt{1 - 3\cos^2 75 + 2\cos^3 75}$$
$$= 125\sqrt{1 - 0.2 + 0.035} = 114.2 \text{ Å}^3$$

(b) Take a plane projection of the rhombohedral unit cell and choose a triply primitive hexagonal unit cell (Fig. 1.17) whose unit cell parameters are a = b ≠ c, $\alpha = \beta = 90°$ and $\gamma = 120°$. In order to calculate a_H and c_H, take projection of a_R on c-axis. Therefore,

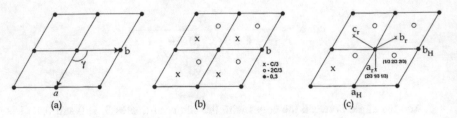

Fig. 1.17 Showing hexagonal unit cells and projected rhombohedral positions

$$\frac{c_H}{3} = (a_R)_p = a_R \cos 45° = 5 \cos 45° = 3.535$$

$$\text{or} \quad c_H = 3 \times 3.353 = 10.61 \text{ Å}$$

Again taking projection of a_R on ab-plane, we have

$$(a_R)_p = 5 \cos 45° = 3.535$$

From Fig. 1.17, we can obtain

$$a_H = \left[(3.535)^2 + (3.535)^2 + 2(3.535) \cos 60°\right]^{1/2} = 3.535 \times \sqrt{3} = 6.12 \text{ Å}$$

(c) Volume of the hexagonal unit cell,

$$V_H = 0.866 \, a^2 c$$
$$= 0.866 \times (6.12)^2 \times 10.61$$
$$= 344.14 \text{ Å}^3$$

(d) Ratio of the two volumes $= \dfrac{V_H}{V_R} = \dfrac{344.14}{114.2} \cong 3$

$\Rightarrow$ There are three lattice points in the hexagonal unit cell.

1.3 Crystal Systems in 2-D and 3-D Lattices

Arbitrary values of lattice parameters a, b and γ (in a plane lattice) and a, b, c and α, β, γ (in a space lattice) produce unlimited number of 2-D and 3-D lattices, respectively. However, when suitable restrictions on the lattice parameters are imposed, the numbers of possible primitive lattices drastically reduce to four in 2-D as shown in Fig. 1.18. They also represent the number of crystal systems in their respective category.

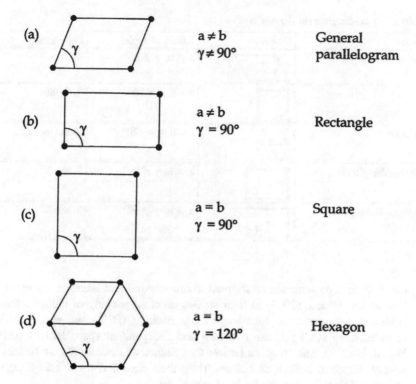

Fig.1.18 Two-dimensional primitive unit cells

(a) **Crystal Systems in 2-D**

In order to obtain the number of possible crystal systems in 2-D, we need to impose suitable restrictions on the lattice parameters a, b and γ in a given oblique plane lattice. The resulting 2-D lattices and the corresponding unit cell axes and angles are provided in Table 1.1. Based on the recently proposed mirror combination scheme (Wahab 2020), the revised point group distribution is provided in the last column of the Table.

(b) **Crystal Systems in 3-D**

Consider two identical layers of a given plane lattice and place one layer on the top of the other. The height of the upper layer gives the c-dimension, while its orientation with respect to the lower layer provides the angular relationships of the resulting 3-D lattices. Based on this principle, it is easy to derive eight 3-D primitive lattices from five primitive plane lattices (where the primitive form of the centered rectangular lattice is taken as primitive rhombic). The complete derivation scheme of 3-D primitive lattices is illustrated in Fig. 1.19.

It is well known that based on the geometrical shapes of the unit cells and lattice translation vectors, Bravais in 1848 proposed 14 space lattices and classified them into 7 crystal systems. However, due to persistent ambiguity in

Table 1.1 Two-dimensional Bravais lattices

Lattice	Unit cell	Axes and angles	Point group
Oblique		$a \neq b, \gamma \neq 90°$	m, 1
Square		$a = b, \gamma = 90°$	4, 4mm
Hexagonal		$a = b, \gamma = 120°$	3, 3m, 6, 6mm
Rectangular primitive		$a \neq b, \gamma \neq 90°$	2, mm2
Centered rectangular		$a \neq b, \gamma \neq 90°$	2, mm2

the Bravais representation of trigonal, rhombohedral and hexagonal structures, Wahab and Wahab (2015) in their studies on close packing of identical atoms (spheres) discovered the hexagonal close packing (HCP) and rhombohedral close packing (RCP) as the two new and independent space lattices (called Wahab lattices). The 16 space lattices they named as Bravais-Wahab lattices or simply as space lattices as before. They then classified them into 8 crystal systems (Table 1.2) on the basis of symmetry.

Further, a recent study on symmetry made by Wahab (2020) has shown that mirror is the only fundamental symmetry in crystals and all other symmetries such as rotation, inversion, rotoreflection, rotoinversion and translational periodicity are derivable from different combinations of mirrors. According to the mirror combination scheme, it is necessary and inevitable to make some changes in the earlier assigned point groups in some low symmetry crystal systems such as triclinic, monoclinic and orthorhombic. The resulting changes and some other information are provided in Table 1.2.

1.4 Centering in 2-D and 3-D Crystal Lattices

The basic criterion to verify the existence of non-primitive (or centered) lattice in crystals is to add a lattice point (or lattice points) at appropriate positions without disturbing the essential symmetry of the original lattice (or unit cell) and then examining whether a smaller primitive cell may be chosen with the same general shape or not. Two possibilities will arise:

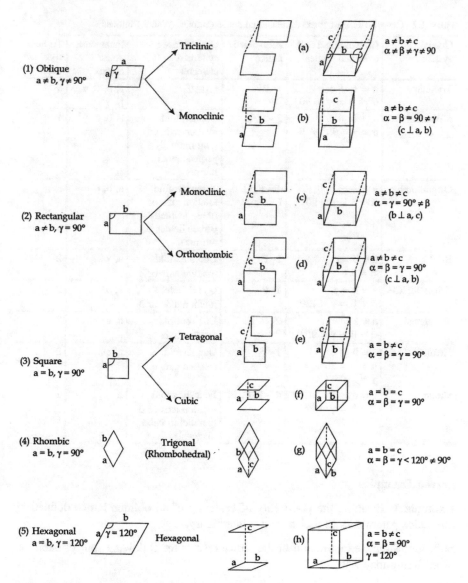

Fig. 1.19 Derivation of 3D-lattices from 2D-lattices

1. If the new primitive cell has the same general shape, then it represents the same general lattice.
2. If no new primitive cell is found, then the resulting non-primitive lattice must be recognized as a new lattice type.

Table 1.2 Classification of three-dimensional space (Bravais-Wahab) lattices

Crystal system	Conventional cell, axes and angles	Associated lattice	Characteristic symmetry elements	Parameters Axes/ angles	To be specified Total
Triclinic	$a \neq b \neq c$, $\alpha \neq \beta \neq \gamma \neq 90°$	1- P	$m(\equiv \bar{2})$	a, b, c; α, β, γ	6
Monoclinic	$a \neq b \neq c$, $\alpha = \gamma = 90° \neq \beta$	2 - P, C	One twofold rotation axis or (two mutually perpendicular mirrors)	a, b, c; γ	4
Orthorhombic	$a \neq b \neq c$, $\alpha = \beta = \gamma = 90°$	4 – P, C, F, I	Three twofold rotation axis or (three mutually perpendicular mirrors)	a, b, c	3
RCP	$a = b = c$, $\alpha = \beta = \gamma$	1- P	One threefold rotation axis or $\bar{3}$	a, γ	2
Trigonal/HCP	$a = b = c$, $\alpha = \beta = \gamma < 120°$	1- P	One threefold rotation axis or $\bar{3}$	a; α	2
Tetragonal	$a = b \neq c$, $\alpha = \beta = \gamma = 90°$	2- P, I	One fourfold rotation axis or $\bar{4}$	a, c	2
Hexagonal	$a = b \neq c$, $\alpha = \beta = 90°$, $\gamma = 120°$	1- P	One sixfold rotation axis or $\bar{6}$	a, c	2
Cubic	$a = b = c$, $\alpha = \beta = \gamma = 90°$	4 – P, C, F, I	Four threefold rotation axis or $\bar{3}$ (parallel to cube diagonal)	a	1

Solved Examples

Example 1 Examine the possibility of centering of an oblique lattice defined by the lattice parameters as $a \neq b$, and γ is arbitrary.

Solution: Simple examination of the lattice reveals the following four possibilities of centering may arise:

(a) An extra lattice point can be added to the center of an edge (say) at A to give a doubly primitive cell as shown in Fig. 1.20a. The resulting cell is simply another primitive cell with the same general shape but with the lattice parameters a, b/2 and γ. A similar argument will apply if the additional lattice point is added at the mid points of the other cell edges.

(b) Two additional lattice points can be added to centers of both the cell edges A and B, to give a triply primitive lattice as shown in Fig. 1.20b. However, this does not constitute a lattice, because the environment of A and B are now different and does not fulfill the criterion of a lattice.

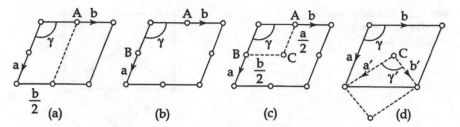

Fig. 1.20 The addition of extra lattice points to a primitive oblique cell. Alternative primitive cells are shown with broken lines

(c) Three additional lattice points can be added, one each at A, B and C to produce a quadruple primitive cell as shown in Fig. 1.20c, where the point C is at the center of the unit cell. The resulting cell is a new primitive cell with the same general shape and may be defined as a/2, b/2 and γ.

(d) An additional lattice is added to C, the center of the unit cell to produce a doubly primitive unit cell as shown in Fig. 1.20d. At first sight, it appears that this represents a new non-primitive lattice type. However, a careful look suggests that a smaller primitive oblique cell of new dimensions a′, b′ and γ′ (Fig. 1.20d) can be selected to describe this lattice arrangement.

The above examinations show that no centering is possible in an oblique lattice.

Example 2 Examine the possibility of centering of a rectangular lattice defined as a $\neq$ b and γ = 90°.

Solution: Similar to an oblique lattice, there exist four possibilities of centering of a rectangular lattice. In this case too, we obtain similar results in first three cases, (a)–(c). However, the result in case of (d) is different. In the rectangular lattice, the addition of an extra lattice point at the center of the unit cell does produce a new non-primitive (or centered) lattice. The centered rectangular lattice may also be described as a primitive rhombic (or primitive diamond) lattice as shown in Fig. 1.21. Thus the same lattice can have two alternative descriptions but a more useful centered lattice is preferred because of the simple geometry and greater symmetry.

This implies that in a rectangular lattice, a body centering is possible.

Example 3 Is any centering possible in a square lattice or a plane hexagonal lattice?

Solution: Examining the above-mentioned four possibilities of centering in a square lattice or a plane hexagonal lattice, we observe that neither of them exhibits any new lattice. Hence, no centering is possible either in a square lattice or in plane hexagonal lattice.

Example 4 Examine the possibility of centering in a monoclinic crystal system defined as a $\neq$ b $\neq$ c, α = β = 90° $\neq$ γ.

Fig. 1.21 Alternative way to describe the centered rectangular lattice

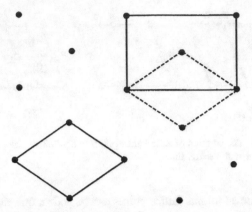

Solution: We know that in a monoclinic crystal system, there exists a unique twofold axis, normally taken as the c-axis. This is called the monoclinic system of first setting. Now, let us examine the following four possibilities of centering, viz. the body-centering, the base-centering and face-centering in the primitive monoclinic one by one.

(a) **Body-centering**

Let us draw two monoclinic lattices side by side. Join the points as shown by the dotted lines (Fig. 1.22a) This can equally well be described as a base-centered (B-centering on ac-plane) lattice by appropriate choice of a and b axes. Conversely, a B-centered lattice shown in Fig. 1.22b can equally well be described as a body-centered monoclinic lattice. Thus, I ≡ B.

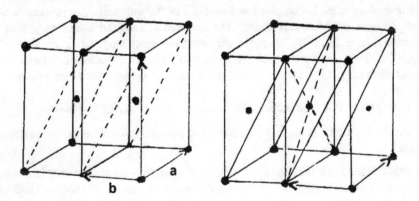

Fig. 1.22 a Body-centered (I) unit cells, **b** B-centered unit cells

(b) A-, B- and C-face centering

Let us consider A-, B- and C-face centering separately as shown in Fig. 1.23. We observe that A- and B-face centering provides a primitive lattice that does not follow the fundamental monoclinic conditions. Further, these two lattices can be made equivalent by simply interchanging the axes a ↔ b. Thus A ≡ B.

The C-face centering on the other hand does not provide any new lattice because the resulting lattice can be described as a primitive lattice (Fig. 1.23)c with a different value of lattice parameter a and γ while retaining the fundamental monoclinic conditions.

(c) F-centering

The F-centering is nothing but the A-, B- and C-centering taken together. The above discussion suggests that the F-centering will also be equivalent to a B-lattice. Thus F ≡ B. This finally gives us B ≡ I ≡ A ≡ F. It is only a matter of convention to use B-centering in preference to others. Therefore, there are two unique monoclinic lattices, that is, a primitive lattice and a B-centered lattice.

It is to be mentioned that the monoclinic system of the second setting is conventionally preferred by crystallographers, where the b-axis is the unique axis. The above discussion is still valid except for the interchange of axes. Thus, in this setting, the primitive lattice and the C-lattice are the two monoclinic lattices.

Example 5 Examine the possibility of centering in an orthorhombic crystal system defined as a $\neq$ b $\neq$ c, $\alpha = \beta = \gamma = 90°$.

Solution: Orthorhombic conditions imply that the axes are unequal but perpendicular to each other. To make the problem simple, let us compare the possibilities of centering in orthorhombic system with the 1st setting of monoclinic system discussed above.

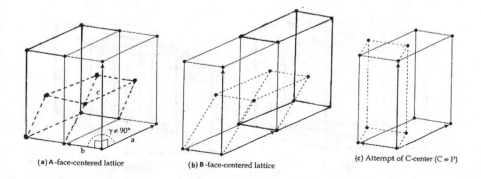

(a) A-face-centered lattice   (b) B-face-centered lattice   (c) Attempt of C-center (C ≡ P)

Fig. 1.23 Several aspects of centering a monoclinic lattice

According to monoclinic condition, $\alpha = \beta = 90° \neq \gamma$, b-c and c-a faces are perpendicular to each other and the corresponding A- and B- centering give rise to a new centered lattice. Applying this criterion to the orthorhombic system (where all faces are perpendicular to each other), all A-, B- and C- centering will give rise to a new centered lattice. However, they can be described in terms of any one simply by interchanging the orthogonal axes. Normally, the C-centering is conventionally preferred over others.

On a similar ground, one can verify that all the face-centering (F-centering) and body centering (I-centering) will give rise to distinct possible centered lattices in orthorhombic crystal system. Hence, all the four lattices, that is, P, C, I and F are possible in this crystal system.

Example 6 Examine the possibility of centering in a tetragonal crystal system defined as $a = b \neq c$, $\alpha = \beta = \gamma = 90°$.

Solution: Tetragonal conditions imply that there exists one fourfold symmetry axis and all the axes are perpendicular to each other.

Body centering in tetragonal system provides a new and still complies with the tetrahedral condition. Therefore, an I-lattice is a new lattice in tetragonal system. On the other hand, in an F-centering we observe that a body-centered lattice could be constructed out of it in such a way that the new axes are rotated through an angle of 45° with respect to the old axes as shown in Fig. 1.24a. Thus $F \equiv I$. Since the body-centered unit cell is smaller than the face-centered, the I-cell is usually preferred.

As per the tetragonal condition, three axes are mutually perpendicular to each other, also $a = b$. As a result of these, the A- and B-centering does not form any proper lattice (because the environments at all lattice points are not the same). However, only the C-centering does represent a lattice as shown in Fig. 1.24b. But this lattice can also be represented by a primitive lattice rotated through an angle of 45° with respect to c-axis. Thus $C \equiv P$. Further, since the primitive is smaller of the two, the P-cell is usually preferred. Accordingly, in tetragonal system there exist only two lattices, that is, the primitive and the body-centered.

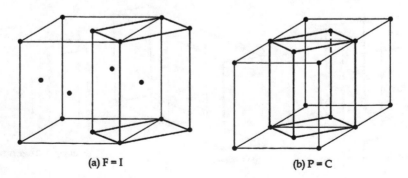

(a) F = I   (b) P = C

Fig. 1.24 Several aspects of centering a tetragonal lattice

Fig. 1.25 An impossible way
to center a lattice

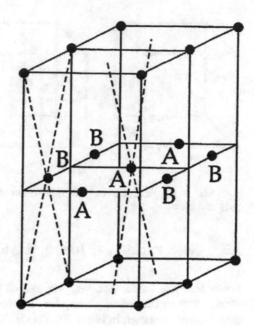

Example 7 Why two independent face centering are not possible in tetragonal system?

Solution: Draw two primitive tetragonal unit cells side by side. Add extra points at A and B positions as shown in Fig. 1.25. A close examination reveals that the environments (as indicated at A and B by dashed lines) of all the points are not identical, no matter how one chooses the translation vectors. Hence, the centering of two independent faces of a given lattice cannot form a proper lattice and is not possible.

Example 8 Examine the possibility of centering in a cubic crystal system defined as $a = b = c$, $\alpha = \beta = \gamma = 90°$.

Solution: Cubic conditions imply that all the axes are equal and perpendicular to each other. In this case, both body centering and the face centering are found to produce new lattices without disturbing the cubic conditions. However, a close examination reveals that a base centering destroys the basic cubic condition of threefold symmetry (along the body diagonals) and hence is not possible. Thus a cubic system has three lattices, the primitive, the body-centered and the face-centered. They are shown in Fig. 1.26.

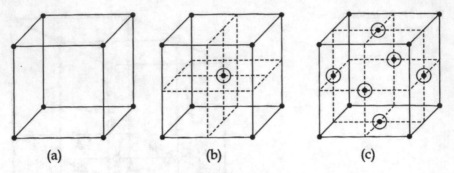

Fig. 1.26 a Simple cubic (sc), **b** Body-centered cubic (bcc) and **c** Face-centered cubic (fcc) structures

1.5   Close Packing of Identical Atoms (Spheres)

When identical atoms (spheres) are packed in a plane such that they touch each other, they automatically acquire a close-packed hexagonally coordinated arrangement as shown in Fig. 1.27. This is the only way of packing identical atoms (spheres) in most efficient way in 2-D. Coordination in such an arrangement is 6 and its maximum symmetry is 6mm. Let us call this as A-layer. This contains two types of voids, one with the apex of the triangle up (Δ) and labelled B, and the other with apex down (∇) and labelled C (Fig. 1.27).

In a 3-D packing, the hexagonally close-packed layer can occupy either B or C position, but not both at the same time. Similarly, above a B-layer, it can be either C or A and above a C-layer, it can be either A or B and so on without violating the rule of close packing (where no two adjacent layers can be in the same orientation). Coordination number in such a 3-D arrangement of atoms or spheres is 12, where 6 spheres are in the reference layer and 3 each in the layer just above and just below it as shown in Fig. 1.28 for two different packing arrangements AB/A and ABC/A, respectively.

Addition of each layer perpendicular to the principal axis [0001] will form a new structure, NH or 3NR depending upon the nature of the resulting ABC...... sequence (where N is the number of layer/layers in the hexagonal unit cell). NH and 3NR, respectively, represent the hexagonal close-packed (HCP) and rhombohedral close-packed (RCP) structures, they have been discovered as two new and independent lattices by Wahab and Wahab (2015) and called Wahab lattices. The number of possible structures increases rapidly with the size of the unit cell. Many elements (such as Si, Ge, etc.) and compounds (such as SiC, ZnS, CdI_2, PbI_2, etc.) are found to exhibit such structures known as polytypes.

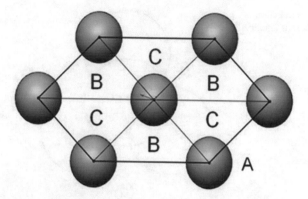

Fig. 1.27   A close packed layer showing B and C voids

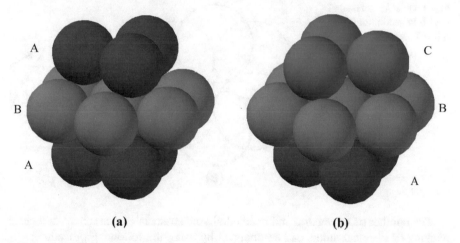

(a)   **(b)**

Fig. 1.28   a Hexagonal close packing **b** Cubic close packing

Solved Examples

Example 1 In close packing of identical spheres, show that: (i) there are as many octahedral voids as there are spheres, and (ii) there are twice as many tetrahedral voids as there are spheres.

Solution: A single close-packed layer of identical spheres contains two types of triangular voids. Stacking of additional close-packed layers produce tetrahedral and octahedral voids. If the triangular void has a sphere directly above it (Fig. 1.29), the resulting void is termed as tetrahedral void. On the other hand, if a triangular void in one layer is covered by another triangular void in the next layer (Fig. 1.30), the resulting void is termed as octahedral void.

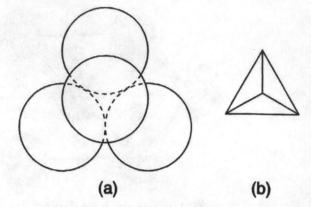

(a)   **(b)**

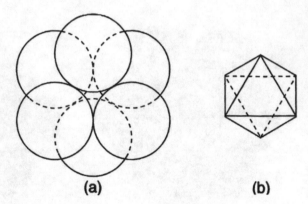

(a)   **(b)**

The number of tetrahedral and octahedral voids associated per sphere in a close packing of identical atoms can be obtained by using the following procedure:

(i)   Consider a reference sphere, around which there are six triangular voids (three alternate voids are of the same kind).

(ii) When the next layer is placed, only three voids of one kind are occupied. They become tetrahedral voids and other three become octahedral voids. This gives us three tetrahedral and three octahedral voids.

(iii) Similar is the situation when a layer below the reference layer is considered. This also gives us three tetrahedral and three octahedral voids.

(iv) The reference layer also covers a triangular void in the layer above it and another in the layer below it. This gives us two tetrahedral voids.
      Therefore, each sphere is surrounded by

$$3 + 3 = 6 \text{ octahedral voids and } 3 + 3 + 2 = 8 \text{ tetrahedral voids}$$

(v) Each octahedral void is surrounded by six spheres while each tetrahedral void is surrounded by four spheres.

Therefore, the number of octahedral voids (n_o) and the number of tetrahedral voids (n_t) belonging to a sphere, respectively, are:

$$n_o = \frac{\text{Number of octahedral voids around a sphere}}{\text{Number of spheres around an octhedral void}} = \frac{6}{6} = 1$$

$$n_t = \frac{\text{Number of tetrahedral voids around a sphere}}{\text{Number of spheres around a tetrahedral void}} = \frac{8}{4} = 2$$

This follows that:

(i) There are as many octahedral voids as there are spheres, and

(ii) There are twice as many tetrahedral voids as there are spheres.

Example 2 Show that the critical radius ratio for a triangular coordination is 0.155.

Solution: Let us consider a triangular void surrounded by three spheres of radius R touching each other. Let a small sphere of radius r is placed within the triangular void such that the central sphere just touches the three coordinating spheres as shown in Fig. 1.31. From this simple geometrical construction, we have

$$\frac{LM}{LO} = \frac{R}{R+r} = \cos 30°$$

$$\text{or} \quad r = \frac{R}{\cos 30°} - R = \frac{2}{\sqrt{3}}R - R = 1.155R - R$$

$$\text{or} \quad r = 0.155R$$

$$\frac{r}{R} = 0.155$$

Fig. 1.31 A planar void

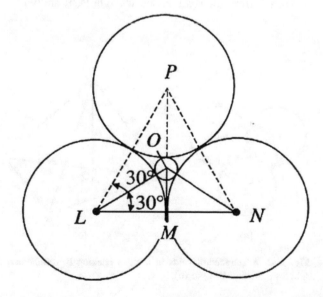

This gives the critical radius ratio representing the triangular void.

Example 3 Show that the critical radius ratio for a tetrahedral coordination is 0.225.

Solution: Let us consider a tetrahedral void surrounded by four spheres of radius R (Fig. 1.32a). The centers of these spheres lie at corners of a regular tetrahedron of side a = 2R and height h (Fig. 1.32b). For a regular tetrahedron of side "a," a sin60° is the median of any of the four bounding faces. Let BP is the median of a side face, drawn from the apex B of the tetrahedron and BQ be the perpendicular from B to the base of the tetrahedron. The triangle BPQ is shown in Fig. 1.32c. As the point P is the midpoint of one side of the base and Q is its centroid, then

$$BP = a\ \sin 60°, \quad PQ = \tfrac{1}{3}\ a\ \sin 60°$$
$$\text{and} \quad (BP)^2 = (BQ)^2 + (QP)^2$$
$$\text{or} \quad a^2 \sin^2 60° = h^2 + \tfrac{a^2}{9} a\ \sin^2 60°$$
$$\text{or} \quad \tfrac{h}{a} = \tfrac{\sqrt{2}}{\sqrt{3}} = 0.8165$$
$$\text{or} \quad \tfrac{c}{a} = \tfrac{nh}{a} = 0.8165 \times n$$

Let r be the radius of the sphere that jut fits into this void whose center is at P and is equidistant from all corners of the tetrahedron (Fig. 1.33a). Let the bond length p = r + R. As shown above, the height BQ = h is related to the side of the tetrahedron "a" as

$$h = \frac{\sqrt{2}}{\sqrt{3}}\ a$$

Hence, from the right angled triangle PQR shown in Fig. 1.33b , we have

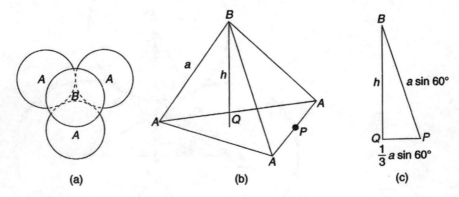

Fig. 1.32 a Tetrahedral voids in a close packing **b** Tetrahedron formed by the centers of the spheres **c** Section BPQ of (**b**)

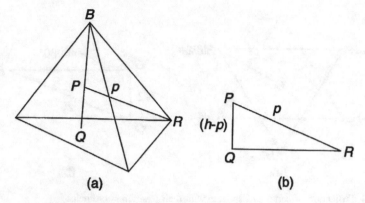

Fig. 1.33 a A tetrahedron **b** Section PQR of **(a)**

$$PQ = BQ - BP = h - p$$
here, $PR = p$ and $QR = \frac{2}{3} a \sin 60° = \frac{h}{\sqrt{2}}$
therefore, $(PR)^2 = (h - p)^2 + (QR)^2$ $\qquad$ (i)
or $p^2 = (h - p)^2 + \frac{h^2}{2}$
so that $p = \frac{3}{4} h = 0.75h$

Substituting the values for p and h in terms of r and R in Eq. (i), we get

$$r + R = \frac{3\sqrt{2}}{4\sqrt{3}} 2R = \frac{\sqrt{3}}{\sqrt{2}} \ R$$

or $r = \frac{\sqrt{3}}{\sqrt{2}} R - \frac{\sqrt{2}}{\sqrt{2}} R = \frac{\sqrt{3} - \sqrt{2}}{\sqrt{2}} = 0.225R$ or $\frac{r}{R} = 0.225$ $\qquad$ (ii)

This gives the critical radius ratio representing the tetrahedral void.

Example 4 Show that the critical radius ratio for an octahedral coordination is 0.414.

Solution: Let us consider an octahedral void of radius r surrounded by six spheres (three in the layer below and three in the layer above) of radius R, the centers of these spheres lie at the corners of a regular octahedron of side a = 2R. The projection of the centers of six spheres is shown in Fig. 1.34a . The point O represents the center of the sphere fitted within the void. Again the bond length OR = p = R + r. Now from the right angled triangle OO′R shown in Fig. 1.34b (where O′ lie in the plane of the upper layer), we have

$$(OR)^2 = (OO')^2 + (O'R)^2$$
or $p^2 = \left(\frac{h}{2}\right) + \left(\frac{2}{3} a \sin 60°\right)$ $\qquad$ (i)
therefore, $p = \frac{\sqrt{3}}{4} h = 0.88h$

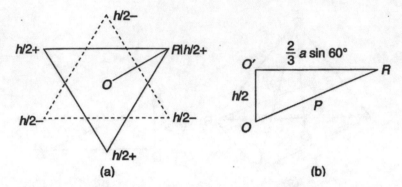

Fig. 1.34 Projection of centers of spheres with their position coordinates

The bond length for an atom placed in an octahedral void in a close packing is thus 0.88 times the layer separation. Substituting $p = r + R, h = (\sqrt{2}/\sqrt{3})$ a and a = 2R we have,

$$r + R = \sqrt{2}\,R$$
$$\text{therefore, } r = (\sqrt{2} - 1)\,R = 0.414R \tag{ii}$$
$$\text{or} \quad \tfrac{r}{R} = 0.414$$

This gives the critical radius ratio representing the octahedral void.

Example 5 Show that the critical radius ratio for a simple cubic coordination is 0.732.

Solution: Let us consider a void surrounded by eight spheres of radius R, the centers of the spheres lie at the corners of a simple cube of side a = 2R as shown in Fig. 1.35. Let a small sphere of radius r just fits into the void. The body diagonal and side of the cube are related through the equation

$$2(r + R) = \sqrt{3}a = \sqrt{3} \times 2R$$
$$\text{or} \quad r + R = \sqrt{3}\,R$$
$$\text{or} \quad r = (\sqrt{3} - 1)\,R = 0.732R$$
$$\text{or} \quad \tfrac{r}{R} = 0.732$$

This gives the critical radius ratio representing the simple cubic void.

Example 6 Calculate the void space in a close packing of n spheres of radius 1.000, n spheres of radius 0.414 and 2n spheres of radius 0.225.

Solution: Let us consider the close packing of identical spheres of radius R in a face-centered cubic unit along [111] direction. In this case, side of the unit cell and the radius of the sphere are related (Fig. 1.36) through the equation

Fig. 1.35 Critical radius ratio
in sc

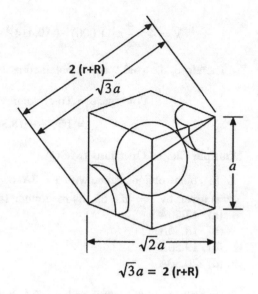

$$\sqrt{3}a = 2(r+R)$$

Fig. 1.36 Critical radius ratio
in fcc

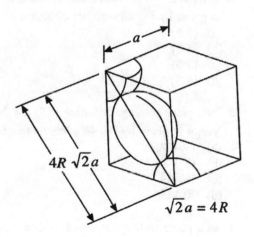

$$\sqrt{2}a = 4R$$

$$\sqrt{2}\,a = 4R \quad \text{or} \quad a = 2\sqrt{2}\,R$$

Therefore, the volume of the unit cell is

$$V = a^3 = 16\sqrt{2}\,R^3 = 16\sqrt{2} \quad \text{(for } R = 1\text{)}$$

Further for fcc, the number of spheres (atoms) in the unit cell is 4, then the volume occupied is

$$V = 4 \times \frac{4}{3}\pi\left[(1.000)^3 + (0.414)^3 + 2 \times (0.225)^3\right] = 18.319$$

Therefore, the void space is obtained as

$$\text{Void space} = \text{Total volume} - \text{occupied volume}$$
$$= 16\sqrt{2} - 18.319 = 4.308 = 19\%$$

Multiple Choice Questions (MCQ)

1. In a plane oblique lattice with a = 3Å, b = 2Å and φ = 75°, a translation vector is given by $\vec{t} = 4\vec{a} + 5\vec{b}$, its magnitude is:
   (a) 17.50Å
   (b) 18.50Å
   (c) 19.50Å
   (d) 20.50Å

2. In a plane oblique lattice with a = 3Å, b = 2Å and φ = 75°, a translation vector is given by $\vec{t} = 4\vec{a} + 5\vec{b}$, its direction is
   (a) [3 5]
   (b) [4 5]
   (c) [5 5]
   (d) [6 5]

3. In a plane rectangular lattice with a = 3Å, b = 2Å and φ = 90°, a translation vector is given by $\vec{t} = 4\vec{a} + 7\vec{b}$, its magnitude is:
   (a) 17.44Å
   (b) 18.44Å
   (c) 19.44Å
   (d) 20.44Å

4. In a plane rectangular lattice with a = 3Å, b = 2Å and φ = 90°, a translation vector is given by $\vec{t} = 4\vec{a} + 7\vec{b}$, its direction is:
   (a) [3 5]
   (b) [4 5]
   (c) [4 6]
   (d) [4 7]

5. In a plane hexagonal lattice with a = b = 3Å and φ = 120°, a translation vector
   is given by $\vec{t} = 2\vec{a} + 3\vec{b}$, its magnitude is:
   (a) 6.94Å
   (b) 7.94Å
   (c) 8.94Å
   (d) 9.94Å

6. In a plane hexagonal lattice with a = b = 3Å and φ = 120°, a translation vector
   is given by $\vec{t} = 2\vec{a} + 3\vec{b}$, its direction is:
   (a) [1 2]
   (b) [1 3]
   (c) [2 2]
   (d) [2 3]

7. In a square lattice with a = b = 3Å and φ = 90°, a translation vector is given by
   $\vec{t} = 3\vec{a} + 4\vec{b}$, its magnitude is:
   (a) 12.0Å
   (b) 13.0Å
   (c) 14.0Å
   (d) 15.0Å

8. In a square lattice with a = b = 3Å and φ = 90°, a translation vector is given by
   $\vec{t} = 3\vec{a} + 4\vec{b}$, its direction is:
   (a) [3 1]
   (b) [3 2]
   (c) [3 3]
   (d) [3 4]

9. In a simple orthorhombic crystal system, a = 2Å, b = 3Å and c = 4Å, two
   lattice points are connected through a translation vector $\vec{T} = 2\vec{a} + 3\vec{b} + 4\vec{c}$, its
   magnitude is:
   (a) 16.80Å
   (b) 17.80Å
   (c) 18.80Å
   (d) 19.80Å

10. In a simple orthorhombic crystal system, a = 2Å, b = 3Å and c = 4Å, two lattice points are connected through a translation vector $\vec{T} = 2\vec{a} + 3\vec{b} + 4\vec{c}$, its direction is:
   (a) [231]
   (b) [232]
   (c) [233]
   (d) [234]

11. In a tetragonal crystal system, a = b = 2Å and c = 3Å, two lattice points are connected through a translation vector $\vec{T} = 3\vec{a} + 4\vec{b} + 5\vec{c}$, its magnitude is:
   (a) 16.0Å
   (b) 17.0Å
   (c) 18.0Å
   (d) 19.0Å

12. In a tetragonal crystal system, a = b = 2Å, and c = 3Å, two lattice points are connected through a translation vector $\vec{T} = 2\vec{a} + 3\vec{b} + 4\vec{c}$, its direction is:
   (a) [123]
   (b) [234]
   (c) [345]
   (d) [456]

13. In a hexagonal crystal system, a = b = 3Å and c = 4Å, two lattice points are connected through a translation vector $\vec{T} = 4\vec{a} + 3\vec{b} + 1\vec{c}$, its magnitude is:
   (a) 11.50Å
   (b) 12.50Å
   (c) 13.50Å
   (d) 14.50Å

14. In a hexagonal crystal system, a = b = 3Å, and c = 4Å, two lattice points are connected through a translation vector $\vec{T} = 4\vec{a} + 3\vec{b} + 1\vec{c}$, its direction is:
   (a) [431]
   (b) [341]
   (c) [134]
   (d) [314]

15. Sides of a primitive rectangle are 4Å and 3Å, respectively. Construct another primitive cell of equal area whose sides are 4Å and 5Å. The angle between the new sides is:
   (a) 36.90°
   (b) 35.90°
   (c) 34.90°
   (d) 33.90°

16. Side of a primitive square unit cell is 2Å. Construct another unit cell of equal area whose one side is the diagonal of the square. The length of this side is:
    (a) $\sqrt{2}$ Å
    (b) $\sqrt{4}$ Å
    (c) $\sqrt{6}$ Å
    (d) $\sqrt{8}$ Å

17. Side of a primitive square unit cell is 2Å. Construct another unit cell of equal area whose one side is the diagonal of the square. The diagonal makes an angle with x-axis is:
    (a) 40°
    (b) 45°
    (c) 50°
    (d) 55°

18. The sides of a primitive rectangle are 3Å and 2Å, respectively. Construct a new unit cell with the edges defined by the vectors from the origin to the points with coordinates 1, 0 and 1, 2. Area of the new unit cell is:
    (a) 6 Å^2
    (b) 8 Å^2
    (c) 10 Å^2
    (d) 12 Å^2

19. The sides of a primitive rectangle are 3Å and 2Å, respectively. Construct a new unit cell with the edges defined by the vectors from the origin to the points with coordinates 1, 0 and 1, 2. The number of lattice points in the new unit cell is:
    (a) 4
    (b) 3
    (c) 2
    (d) 1

20. A primitive rectangular unit cell has a = 2Å, b = 3Å, and γ = 90°, respectively. A new unit cell is chosen with the edges defined by the vectors from the origin to the points with coordinates 2, 0 and 0, 3. Area of the new unit cell is:
    (a) 36 Å^2
    (b) 27 Å^2
    (c) 18 Å^2
    (d) 9 Å^2

21. A primitive rectangular unit cell has a = 2Å, b = 3Å, and γ = 90°, respectively. A new unit cell is chosen with the edges defined by the vectors from the origin to the points with coordinates 2, 0 and 0, 3. The number of lattice points in the new unit cell is:
    (a) 2
    (b) 4
    (c) 6
    (d) 8

22. The side of a primitive square unit cell is, a = 2Å. A new unit cell is chosen with the edges defined by the vectors from the origin to the points with coordinates 1, 0 and 0, 3. Area of the new unit cell is:
    (a) 12 Å^2
    (b) 10 Å^2
    (c) 8 Å^2
    (d) 6 Å^2

23. The side of a primitive square unit cell is, a = 2Å. A new unit cell is chosen with the edges defined by the vectors from the origin to the points with coordinates 1, 0 and 0, 3. The number of lattice points in the new unit cell is:
    (a) 2
    (b) 3
    (c) 4
    (d) 5

24. A primitive orthorhombic unit cell has a = 5Å, b = 6Å and c = 7Å; $\alpha = \beta =$ = 90°. A new unit cell is chosen with the edges defined by the vectors from the origin to the points with coordinates 3, 1, 0; 1, 2, 0 and 0, 0, 1. Volume of the new unit cell is:
    (a) 1020 Å^2
    (b) 1030 Å^2
    (c) 1040 Å^2
    (d) 1050 Å^2

25. A primitive orthorhombic unit cell has a = 5Å, b = 6Å and c = 7Å; $\alpha = \beta =$ = 90°. A new unit cell is chosen with the edges defined by the vectors from the origin to the points with coordinates 3, 1, 0; 1, 2, 0 and 0, 0, 1. The number of lattice points in the new unit cell is:
    (a) 2
    (b) 3
    (c) 4
    (d) 5

26. A rhombohedral unit cell has $a_r = 5\text{Å}$ and $\alpha = 75°$. Volume of the unit cell is:
    (a) 120 Å^3
    (b) 118 Å^3
    (c) 161 Å^3
    (d) 114 Å^3

27. A rhombohedral unit cell has $a_r = 5\text{Å}$ and $\alpha = 75°$. Volume of the hexagonal unit cell is:
    (a) 340 Å^3
    (b) 342 Å^3
    (c) 344 Å^3
    (d) 346 Å^3

28. In a close packing of identical spheres, the ratio of number of octahedral voids around a sphere and the number of spheres around an octahedral void is:
    (a) 1
    (b) 2
    (c) 3
    (d) 4

29. In a close packing of identical spheres, the ratio of number of tetrahedral voids around a sphere and the number of spheres a tetrahedral void is:
    (a) 1
    (b) 2
    (c) 3
    (d) 4

30. The critical radius ratio for a triangular coordination is:
    (a) 0.155
    (b) 0.225
    (c) 0.414
    (d) 0.732

31. The critical radius ratio for a tetrahedral coordination is:
    (a) 0.155
    (b) 0.225
    (c) 0.414
    (d) 0.732

32. The critical radius ratio for an octahedral coordination is:
    (a) 0.155
    (b) 0.225
    (c) 0.414
    (d) 0.732

33. The critical radius ratio for a simple cubic coordination is:
    (a) 0.155
    (b) 0.225
    (c) 0.414
    (d) 0.732

Answers

 1. (a)
 2. (b)
 3. (b)
 4. (d)
 5. (b)
 6. (d)
 7. (d)
 8. (d)
 9. (c)
10. (d)
11. (c)
12. (c)
13. (a)
14. (a)
15. (a)
16. (d)
17. (b)
18. (d)
19. (c)
20. (a)
21. (c)
22. (a)
23. (b)
24. (c)
25. (d)
26. (d)
27. (c)
28. (a)
29. (b)
30. (a)
31. (b)
32. (c)
33. (d)

Chapter 2
Unit Cell Construction

2.1 Construction of Wigner–Seitz Unit Cells

A Wigner–Seitz unit cell is an alternative way of selecting a primitive unit cell of area $(\vec{a} \times \vec{b} = ab \, \sin \gamma)$

or volume $(\vec{a}.\vec{b} \times \vec{c} = abc \, (1 - \cos^2\alpha - \cos^2\beta - \cos^2\gamma + 2 \cos \alpha \cos \beta \cos \gamma)^{1/2})$ equal to other conventional unit cell in a given (2-D or 3-D) lattice. This unit cell is constructed around a lattice point according to the following procedure:

1. Select a reference point in a given lattice and draw lines to connect this point with all other nearest lattice points.
2. At mid points of each line segment between the reference and neighboring lattice points, draw a line (in 2-D) and a plane (in 3-D) as perpendicular bisector.
3. Smallest area (in 2-D) and volume (in 3-D) enclosed around the reference lattice point gives the required Wigner–Seitz primitive unit cell.

The characteristic feature of a Wigner–Seitz unit cell is that it retains the symmetry of the original lattice.

Solved Examples

Example 1 Draw Wigner–Seitz unit cell for each of the five plane lattices.

Solution: Consider each lattice one by one. In each case, take the central atom as the reference atom and proceed according to the above said procedure. Mark the midpoint of each line. Draw new lines through their midpoints. Join the new lines with each other to get the required Wigner–Seitz unit cells for five lattices as shown in Fig. 2.1 in terms of three-step process.

Example 2 Construct a Wigner–Seitz unit cell for a simple cubic lattice.

Solution: Ideally eight simple cubes are needed to be drawn side by side to get the central atom and all other six neighboring atoms as shown in Fig. 2.2a. However, to

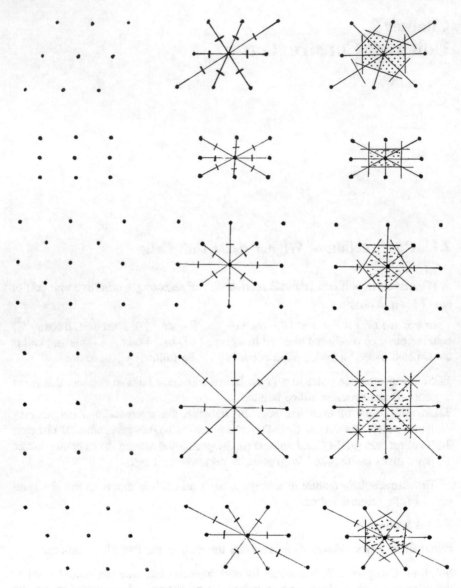

Fig. 2.1 Three-step process of Wigner–Seitz unit cells in 2-D

simplify the problem, one can draw only four cubes side by side to get the central atom along with five other nearest atoms, only one atom will lie outside.

Now, connect the central point with all other neighboring atoms as shown in the figure. Mark their midpoints and draw planes through these bisectors as mentioned in the procedure. Volume enclosed by the intersection of these planes (the resulting shape is a simple cube) is the required Wigner–Seitz unit cell as shown in Fig. 2.2c.

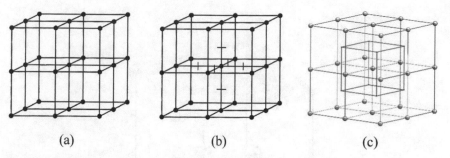

Fig. 2.2 Three-step process of Wigner–Seitz unit cell in sc

Example 3   Construct a Wigner–Seitz unit cell for a bcc lattice.

Solution: Draw a normal bcc unit cell and consider the body-centered atom as the reference atom. Connect this with all other neighboring atoms. Mark midpoints on these lines. Draw planes through these midpoints. Volume enclosed by the inter-section of these planes (the resulting shape is a truncated octahedron) is the required Wigner–Seitz unit cell. A three-step process for its construction is shown in Fig. 2.3.

Example 4   Show that every edge (side) of the polyhedron (square or hexagon) bounding the Wigner–Seitz unit cell of the body-centered cubic lattice is $\left(\sqrt{2}/4\right)$ times the length of the conventional unit cell.

Proof: Consider the Wigner–Seitz unit cell constructed for bcc in Fig. 2.3. Now, let us determine the angle between the directions $[\bar{1}10]$ and $[\bar{1}1\bar{1}]$ as shown separately (Fig. 2.4)

$$\cos\theta = \frac{1+1+0}{\sqrt{1^2+1^2}\sqrt{1^2+1^2+1^2}} = \frac{2}{\sqrt{2}\sqrt{3}} = \frac{\sqrt{2}}{\sqrt{3}} \qquad (i)$$

Further, consider the top diagonal towards $[\bar{1}10]$ (in Figs. 2.3c and 2.4) where x is supposed to be the edge (side) of the polyhedron and a is the side of the conventional cube, we can write

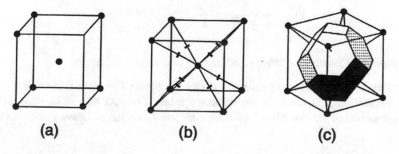

Fig. 2.3 Three-step process of Wigner–Seitz unit cell in bcc

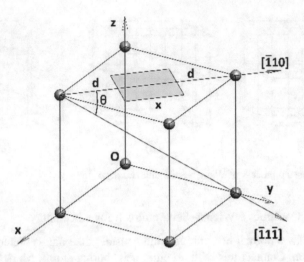

Fig. 2.4 Showing $[\bar{1}10]$ and $[\bar{1}1\bar{1}]$

$$\sqrt{2}a = 2d + x \quad \text{or} \quad d = \frac{\sqrt{2}\,a - x}{2}$$

Again, consider a triangle (Fig. 2.3c) whose normal lies in the hexagonal plane and intersects the top diagonal, its base is one-fourth of the body diagonal of the cube. From this triangle, we can write

$$\cos\theta = \frac{\sqrt{3}a/4}{(\sqrt{2}a - x)/2} \qquad \qquad \text{(ii)}$$

From Eqs. (i) and (ii), we have

$$\frac{\sqrt{2}}{\sqrt{3}} = \frac{\frac{\sqrt{3}a}{4}}{\frac{\sqrt{2}a - x}{2}} = \frac{\sqrt{3}a}{4} \times \frac{2}{(\sqrt{2}a - x)}$$

$$\text{or} \quad 3a = 2\sqrt{2}\left(\sqrt{2}\,a - x\right) = 4a - 2\sqrt{2}\,x$$

$$\text{or} \quad x = \frac{a}{2\sqrt{2}} = \frac{\sqrt{2}}{4}$$

Example 5 Construct a Wigner–Seitz unit cell for an fcc lattice.

Solution: Draw two fcc unit cells one above the other. The atom lying at the center of a square plane is taken as the reference atom. Connect this atom with all first nearest neighbor atoms. Mark the midpoints on these lines. Draw planes through these midpoints. Volume enclosed by the intersection of these planes (the resulting shape is a rhombic dodecahedron) is the required Wigner–Seitz unit cell. The three-step process for its construction is shown in Fig. 2.5.

Example 6 Show that the ratio of the lengths of the diagonals of each parallelo-gram face of the Wigner–Seitz unit cell for fcc lattice is $\sqrt{2} : 1$.

Proof: Consider the Wigner–Seitz unit cell constructed for fcc in Fig. 2.5c. Further, consider triangular region lying over a parallelogram face, shown separately (Fig. 2.6) whose base is one of the diagonals of the parallelogram. Other dimen-sions are known through its construction. To determine the diagonal, we can write

$$x^2 = \left(\frac{a}{2}\right)^2 - \left(\frac{\sqrt{2}a}{4}\right)^2 = \frac{a^2}{4} - \frac{a^2}{8}$$

$$\text{or} \quad x = \frac{a}{2\sqrt{2}} \quad \text{and} \quad 2x = d_1 (\text{say}) = \frac{a}{\sqrt{2}}$$

Further, the side of the small cube is $a/2$. Therefore, its body diagonal is $\sqrt{3}(a/2)$. However, the side of the parallelogram is half the body diagonal of the cube, which is $\sqrt{3}(a/4)$ as shown in Fig. 2.6. To determine the other diagonal say (d_2) we can write

$$\left(\frac{d_2}{2}\right)^2 = \left(\frac{\sqrt{3}a}{4}\right)^2 - \left(\frac{a}{2\sqrt{2}}\right)^2$$

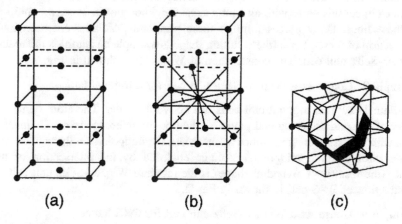

(a)   **(b)**   **(c)**

Fig. 2.5 Three-step process of Wigner–Seitz unit cell in fcc

Fig. 2.6 Showing top view
of the parallelogram in
Fig. 2.5c

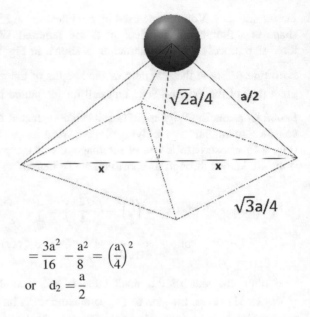

$$= \frac{3a^2}{16} - \frac{a^2}{8} = \left(\frac{a}{4}\right)^2$$
$$\text{or} \quad d_2 = \frac{a}{2}$$

Therefore, the ratio of

$$\frac{d_1}{d_2} = \frac{a/\sqrt{2}}{a/2} = \frac{\sqrt{2}}{1}$$

Example 7 Construct a Wigner–Seitz unit cell for a simple hexagonal lattice.

Solution: Draw two simple hexagonal unit cells one above the other as done in fcc. The atom lying at the center of the middle hexagonal plane is taken as the reference atom. Connect this atom with all first nearest neighbor atoms. Mark the midpoints on these lines. Draw planes through these midpoints. Volume enclosed by the intersection of these planes (the resulting shape is a simple hexagon) is the required Wigner–Seitz unit cell. The construction of W–S cell is shown in Fig. 2.7.

Example 8 Construct a Wigner–Seitz unit cell for a trigonal lattice.

Solution: Draw two trigonal unit cells one above the other. The atom lying at the center of the middle hexagonal plane is taken as the reference atom. Connect this atom with all first nearest neighbor atoms. Mark the midpoints on these lines. Draw planes through these midpoints. Volume enclosed by the intersection of these planes (the resulting polyhedral shape) is the required Wigner–Seitz unit cell. The construction of W–S cell is shown in Fig. 2.8.

Example 9 Construct a Wigner–Seitz unit cell for CsCl lattice.

Solution: Draw 27 CsCl unit cells and consider the Cs^+ ion at the body-centered position of the central CsCl unit cell as the reference ion. Connect this with six other neighboring Cs^+ ions. Mark midpoints on these lines. Draw planes through

Fig. 2.7 Three-step process
of Wigner–Seitz unit cell in
SH

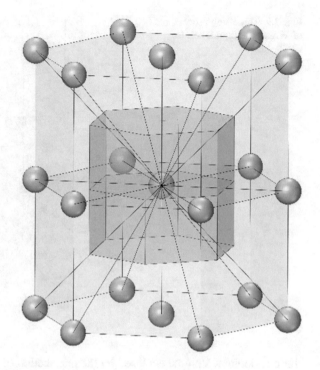

Fig. 2.8 Three-step process
of Wigner–Seitz unit cell in
trigon

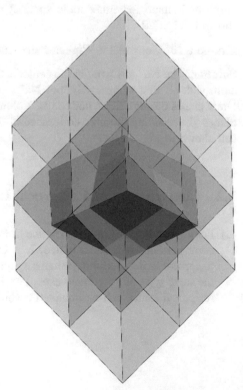

Fig. 2.9 Three-step process of Wigner–Seitz unit cell in CsCl

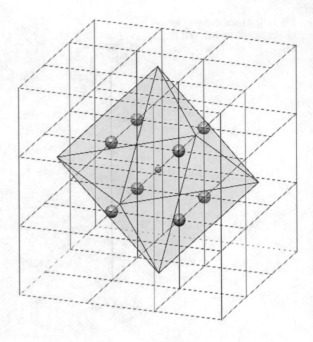

these midpoints. Volume enclosed by the intersection of these planes (the resulting polyhedral shape for either ionic species) is the required Wigner–Seitz unit cell shown in Fig. 2.9.

Example 10 Construct a Wigner–Seitz unit cell for NaCl lattice.

Solution: The Na^+ ion lying at the center is taken as the reference ion. Connect this atom with all first nearest neighbor Na^+ ions. Mark the midpoints on these lines. Draw planes through these midpoints. Volume enclosed by the intersection of these planes (the resulting rhombic dodecahedral shape) is the required Wigner–Seitz unit cell shown in Fig. 2.10.

2.2   Construction of Reciprocal Lattice

(a) For a given direct (primitive) lattice (irrespective of its dimension) there exists a corresponding reciprocal lattice. It is interesting to note that both direct and reciprocal lattices exhibit the same symmetries and hence can be represented by the same type of coordinate axes. For a given direct lattice, the corresponding reciprocal lattice can be obtained according to the following procedure.

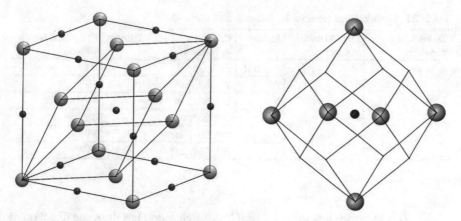

Fig. 2.10 Three-step process of Wigner–Seitz unit cell in NaCl

1. From a common origin, draw a normal to each crystal plane (in 3-D) or line (in 2-D).
2. Set the length of each normal equal to 2π times the reciprocal of interplanar spacing, d_{hkl} (separation between two consecutive planes in 3-D, lines in 2-D and points in 1-D).
3. Mark a point at the end of each normal, called the reciprocal lattice point.
4. A collection of points obtained in this way is known as reciprocal lattice.

If the direct lattice of the conventional unit cell is non-primitive, then the above procedure can be applied to obtain the reciprocal lattice after making necessary modifications as provided in Table 2.1.

(b) The coordinates of reciprocal lattice point is denoted by the symbol hkl or hk0 (without brackets) represents (hkl) plane or (hk0) line of the direct lattice of 3-D or 2-D, respectively. The reciprocal lattice preserves all the important characteristics of the plane in 3-D (or lines in 2-D) they represent:

1. The direction from the origin preserves the orientation of the plane (or line).
2. The distance of the reciprocal lattice point from the origin preserves the interplanar spacing (or separation between two consecutive planes in 3-D) of the set of planes (or lines in 2-D) it represents in the direct lattice.

(c) The dimension wise relationships between the direct lattice and reciprocal lattice are:

1. The reciprocal lattice translation for a given one-dimensional direct (primitive) lattice with translation vector "a" is given by

$$a^* . a = a . a^* = 2\pi$$

$$\text{or} \quad a^* = \frac{2\pi}{a}$$

Table 2.1 Modifications required for non-primitive unit cells

Direct lattice parameter	Direct lattice volume	Lattice type	Reciprocal lattice - Lattice parameters	Reciprocal lattice volume
a, b, c	V	P and R	a*, b*, c*	V*
		A	a*, 2b*, 2c*	4V*
		B	2a*, b*, 2 c*	4V*
		C	2a*, 2b*, c*	4V*
		F	2a*, 2b*, 2c*	8V*
		I	2a*, 2b*, 2c*	8V*

This indicates that the reciprocal lattice translation has the same direction as that of the direct lattice but the unit is in inverse mode. The two lattices are shown in Fig. 2.11a.

2. For a plane rectangular lattice with $a \neq b$ and $\gamma = 90°$, the following relationships exist:

$$a^* . a = a . a^* = 2\pi$$
$$b^* . b = b . b^* = 2\pi$$
$$a^* . b = b^* . a = 0$$

so that $$a^* = \frac{2\pi}{a}$$

and $$b^* = \frac{2\pi}{b}$$

These relationships suggest that

$$a^* \text{is } \perp \text{ to b or } a^* \text{is} \parallel \text{to a}$$
$$\text{and } b^* \text{is } \perp \text{ to a or } b^* \text{is} \parallel \text{to b}$$

They are shown in Fig. 2.11b.

However, for a general plane lattice with $a \neq b$ and $\gamma = \gamma°$ the relationships between the direct and reciprocal lattices are not straightforward as for orthogonal axes. Here, the angle between a* and b* will be $180° - \gamma$ as shown in Fig. 2.11c and their magnitudes are:

$$a^* = \frac{2\pi}{a\cos(90° - \gamma)} = \frac{2\pi}{a \sin \gamma}$$

and $$b^* = \frac{2\pi}{b\cos(90° - \gamma)} = \frac{2\pi}{b \sin \gamma}$$

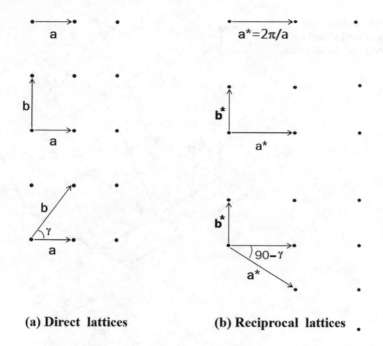

(a) Direct lattices   **(b) Reciprocal lattices**

Fig. 2.11   Direct and reciprocal of 1-D and 2-D lattices

3. For a 3-D space lattice, the relationships between the direct and reciprocal lattices, in general, are obtained as follows:

Consider a unit cell of general lattice (Fig. 2.12) whose b-c plane has an area A ($= b \times c = bc \sin \alpha$), then its volume is

$$V = \text{Area of the } b - c \text{ plane} \times \text{height} = A \times d_{100}$$

Similarly,

$$V = \text{Area of the } a - c \text{ plane} \times \text{height} = B \times d_{010}$$
$$= \text{Area of the } a - b \text{ plane} \times \text{height} = C \times d_{001}$$

So that, the magnitudes of the reciprocal lattice parameters are:

$$a^* = \frac{2\pi}{a} = \frac{2\pi}{d_{100}} = \frac{2\pi A}{V} = 2\pi \frac{b \times c}{a.b \times c}$$

Similarly,

Fig. 2.12 Relationship
between crystal planes and
their reciprocal lattice points

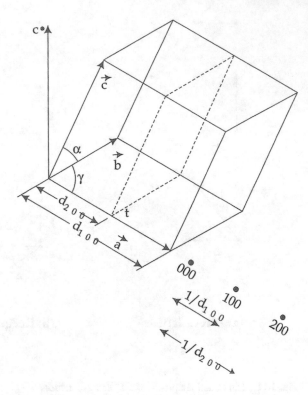

$$b^* = \frac{2\pi}{b} = \frac{2\pi}{d_{010}} = \frac{2\pi B}{V} = 2\pi \frac{c \times a}{a.b \times c}$$

$$c^* = \frac{2\pi}{c} = \frac{2\pi}{d_{001}} = \frac{2\pi C}{V} = 2\pi \frac{a \times b}{a.b \times c}$$

The exact relationships between the direct and reciprocal lattice parameters and interaxial angles of primitive unit cells of various crystal systems are provided in Table 2.2.

Solved Examples

Example 1 Using graphical construction, demonstrate the relation between a two-dimensional direct lattice and its corresponding reciprocal lattice net.

Proof: Let us consider a monoclinic crystal system of 1st setting with the unique 2-fold axis along c-axis defined as a ≠ b ≠ c; α = β = 90° ≠ γ.

For simplicity, let us orient the unit cell in such a way that the c-axis is perpendicular to the plane of the paper. Hence, a- and b- axes lie in the plane of the paper as shown in Fig. 2.13. Further, let us consider some of the (hk0) planes, namely (100), (010), (110), (120) and (210) in the direct lattice unit cell which are parallel to the c-axis and therefore their normal lie in the plane of the paper. Also, we know that a crystal plane in 2-D is a line.

Table 2.2 Relationship between direct and reciprocal lattice parameters and interaxial angles

Triclinic

$$a^* = \frac{Kbc \sin\alpha}{V}; \quad b^* = \frac{Kca \sin\beta}{V}; \quad c^* = \frac{Kab \sin\gamma}{V}$$

$$V = abc\left(1 - \cos^2\alpha - \cos^2\beta - \cos^2\gamma + 2\cos\alpha\cos\beta\cos\gamma\right)^{1/2}$$

$$= 2abc\{\sin s.\sin(s - \alpha).\sin(s - \beta).\sin(s - \gamma)\}^{1/2}; V^* = \frac{1}{V}$$

$$2s = \alpha + \beta + \gamma, \quad K = 2\pi$$

$$\cos\alpha^* = \frac{\cos\beta\cos\gamma - \cos\alpha}{\sin\beta\sin\gamma}; \quad \cos\beta^* = \frac{\cos\gamma\cos\alpha - \cos\beta}{\sin\gamma\sin\alpha}; \quad \cos\gamma^* = \frac{\cos\alpha\cos\beta - \cos\gamma}{\sin\alpha\sin\beta}$$

Monoclinic

1st setting $\quad a^* = \frac{K}{a\sin\gamma}; \quad b^* = \frac{K}{b\sin\gamma}; \quad c^* = \frac{K}{c}; \quad \alpha^* = \beta^* = 90°; \quad \gamma^* = 180° - \gamma$

2nd setting $\quad a^* = \frac{K}{a\sin\beta}; \quad b^* = \frac{K}{b}; \quad c^* = \frac{K}{c\sin\beta}; \quad \gamma^* = \gamma^* = 90°; \quad \beta^* = 180° - \beta$

Orthorhombic

$$a^* = \frac{K}{a}; b^* = \frac{K}{b}; c^* = \frac{K}{c}; \alpha^* = \beta^* = \gamma^* = 90°$$

Tetragonal

$$a^* = b^* = \frac{K}{a}; c^* = \frac{K}{c}; \alpha^* = \beta^* = \gamma^* = 90°$$

Cubic

$$a^* = b^* = c^* = \frac{K}{a}; \alpha^* = \beta^* = \gamma^* = 90°$$

Hexagonal

$$a^* = b^* = \frac{2K}{a\sqrt{3}}; c^* = \frac{K}{c}; \alpha^* = \beta^* = 90°; \gamma^* = 60°$$

Rhombohedral

$$a^* = b^* = c^* = \frac{K.a^2\sin\alpha}{V}, \quad \text{where} \quad V = a^3(1 - 3\cos^2\alpha + 2\cos^3\alpha)^{1/2}$$

$$\cos\alpha^* = \cos\beta^* = \cos\gamma^* = \frac{\cos^2\alpha - \cos\alpha}{\sin^2\alpha} = -\frac{\cos\alpha}{(1+\cos\alpha)}$$

Now, following the above-mentioned procedure, let us draw normal to each plane (line in 2D) mentioned above and mark a point on each normal at a distance $1/d_{hk0}$, where d_{hk0} is the interplanar spacing for the planes {hk0}. Here, the normal OA to the (100) plane defines a*-axis and the normal OB to the (010) plane defines b*-axis, respectively. Further, in general, to reach any reciprocal lattice point hk0, we have to travel h units along a*-axis and k-units along b*-axis. This way, we can generate the reciprocal lattice net as shown in Fig. 2.13.

Example 2 The primitive translation vectors of a two-dimensional lattice are $a = 2\hat{i} + \hat{j}$ and $b = 2\hat{j}$. Determine the primitive translation vectors of its reciprocal lattice.

Solution: Given: $a = 2\hat{i} + \hat{j}$ and $b = 2\hat{j}$. Let us assume the third translation vector c of the given lattice lies along the z-axis and it has a unit magnitude, that is, $c = \hat{k}$. Therefore, the volume of the unit cell is

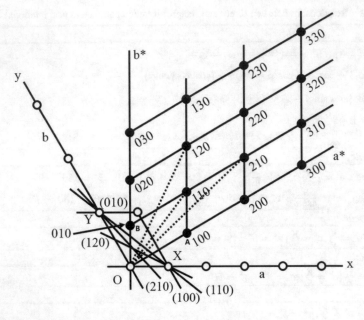

Fig. 2.13 Graphical construction of a 2-D reciprocal lattice

$$a.b \times c = (2\hat{i} + \hat{j}).(2\hat{j} \times \hat{k})$$
$$= 2(2\hat{i} + \hat{j}).\hat{i} \quad (\text{where } \hat{i}.\hat{i} = 1, \hat{i}.\hat{j} = 0)$$
$$= 2(2 + 0) = 4$$

Further, we know that the reciprocal lattice vectors are given by

$$a^* = 2\pi \frac{b \times c}{a.b \times c} \quad , \quad b^* = 2\pi \frac{c \times a}{a.b \times c}$$

Since the reciprocal lattice vectors a* and b* lie in the same plane, therefore

$$a^* = 2\pi \frac{b \times c}{a.b \times c} = \frac{2\pi}{4}(2\hat{j} \times \hat{k}) = \pi \hat{i}$$

and

$$b^* = 2\pi \frac{c \times a}{a.b \times c} = \frac{2\pi}{4}\left[\hat{k} \times (2\hat{i} + \hat{j})\right] = \frac{\pi}{2}\left[2\hat{j} - \hat{i}\right] = \frac{\pi}{2}\left[-\hat{i} + 2\hat{j}\right]$$

Example 3 A two-dimensional lattice has its basis vectors as a = $2\hat{i}$ and b = $\hat{i} + 2\hat{j}$.
Determine its corresponding reciprocal lattice vectors.

Solution: Given: a = $2\hat{i}$ and b = $\hat{i} + 2\hat{j}$. Let us assume the third translation vector c of the given lattice lies along the z-axis and it has a unit magnitude, that is, c = $\hat{k}$. Therefore, the volume of the unit cell is

$$a.b \times c = 2\hat{i}.(\hat{i} + 2\hat{j}) \times \hat{k} = 2\hat{i}.(-\hat{j} + 2\hat{i}) = 2(0 + 2) = 4$$

Further, we know that the reciprocal lattice vectors are given by

$$a^* = 2\pi \frac{b \times c}{a.b \times c} \quad , \quad b^* = 2\pi \frac{c \times a}{a.b \times c}$$

Since the reciprocal lattice vectors a* and b* lie in the same plane, therefore

$$a^* = 2\pi \frac{b \times c}{a.b \times c} = \frac{2\pi}{4}(\hat{i} + 2\hat{j}) \times \hat{k} = \frac{2\pi}{4}(2\hat{i} - \hat{j}) = \frac{\pi}{2}\left[2\hat{i} - \hat{j}\right]$$

and

$$b^* = 2\pi \frac{c \times a}{a.b \times c} = \frac{2\pi}{4}\left[\hat{k} \times 2\hat{i}\right] = \pi\hat{j}$$

Example 4   A two-dimensional direct lattice is formed from a repetition of points ABCD (Fig. 2.14) in which AB = CD = 3Å, AD = BC = 5Å and the angle BAD (say γ) = 60°. Draw a small area (about four unit cells) of the direct lattice. Calculate the basis vectors and draw the corresponding reciprocal cells.

Solution: From Fig. 2.14, we suppose that a and b, respectively, are the lattice vectors representing AB and AD in the direct lattice. Similarly, a* and b*are the corresponding reciprocal lattice vectors. Then,

$$a^* . a = 2\pi \text{ and } a^* . b = 0$$

This indicates that a* is perpendicular to b and is given by

$$a^* = \frac{2\pi}{a\sin\gamma} = \frac{2\pi}{3\sin60°} = \frac{2\pi \times 2}{3 \times \sqrt{3}} = \frac{4\pi}{3\sqrt{3}}\text{Å}^{-1}$$

Similarly, b* is perpendicular to a and is given by

$$b^* = \frac{2\pi}{b\sin\gamma} = \frac{2\pi}{5\sin60°} = \frac{2\pi \times 2}{5 \times \sqrt{3}} = \frac{4\pi}{5\sqrt{3}}\text{Å}^{-1}$$

The angle between a* and b* will be 120°. Four cells of each, the direct lattice and the reciprocal lattice are shown in Fig. 2.15.

Fig. 2.14 Showing lattice
vectors in the direct lattice

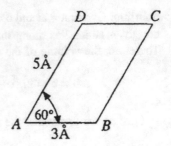

Example 5 A monoclinic unit cell of 1st setting has the following parameters: a =
5Å, b = 10Å, c = 15Å, $\alpha = \beta = 90°$ and $\gamma = 120°$. Determine its corresponding
reciprocal lattice parameters and the two cell volumes.

Solution: General equation of unit cell volume is given by

$$V = abc\left(1 - \cos{}^2\alpha - \cos{}^2\beta - \cos{}^2\gamma + 2\cos\alpha\,\cos\beta\,\cos\gamma\right)^{1/2}$$

$$= 5 \times 10 \times 15\left(1 - 0 - 0 - \frac{1}{4} + 0\right)^{1/2}$$

$$= 750\left(\frac{3}{4}\right)^{1/2} = 649.52\,\text{Å}^3$$

$$a^* = \frac{bc\sin\alpha}{V} = \frac{10 \times 15\sin90°}{649.52} = 0.23\,\text{Å}^{-1}$$

$$b^* = \frac{ac\sin\beta}{V} = \frac{5 \times 15\sin90°}{649.52} = 0.115\,\text{Å}^{-1}$$

$$c^* = \frac{ab\sin\gamma}{V} = \frac{5 \times 10\sin120°}{649.52} = 0.067\,\text{Å}^{-1}$$

and

$$\sin\alpha^* = \frac{V}{abc\,\sin\beta\,\sin\gamma} = \frac{649.52}{750\sin90° \times \sin120°} = 1; \alpha^* = 90°$$

$$\sin\beta^* = \frac{V}{abc\,\sin\alpha\,\sin\gamma} = \frac{649.52}{750\sin90° \times \sin120°} = 1; \beta^* = 90°$$

$$\sin\gamma^* = \frac{V}{abc\,\sin\alpha\,\sin\beta} = \frac{649.52}{750\sin90° \times \sin90°} = 0.866; \gamma^* = 60°$$

Finally,

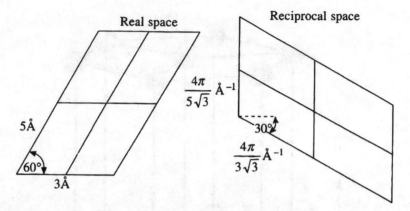

Fig. 2.15 Showing direct cells and corresponding reciprocal cells

$$V^* = a^*b^*c^* \left(1 - \cos{}^2\alpha^* - \cos{}^2\beta^* - \cos^2\gamma^* + 2\cos\alpha^* \cos\beta^* \cos\gamma^*\right)^{1/2}$$

$$= 0.23 \times 0.115 \times 0.67\left(1 - 0 - 0 - \frac{1}{4} + 0\right)^{1/2}$$

$$= 1.77 \times 10^{-3}\left(\frac{3}{4}\right)^{1/2}$$

$$= 1.53 \times 10^{-3}\text{Å}^3$$

Example 6 The primitive translation vectors of a simple hexagonal (Fig.2.16) space lattice may be taken as

$$A = \left(\frac{a}{2}\hat{i} + \frac{\sqrt{3}a}{2}\hat{j}\right), \quad B = \left(-\frac{a}{2}\hat{i} + \frac{\sqrt{3}a}{2}\hat{j}\right) \text{ and } C = c\hat{k}$$

where $\hat{i}$, $\hat{j}$ and $\hat{k}$ are unit vectors.

Determine the primitive translation vectors of the reciprocal lattice and the unit cell volume.

Solution: Given: Direct lattice parameters A, B and C; determine the reciprocal lattice parameters A*, B*, C* and volume V*. Volume of the direct unit cell is given by

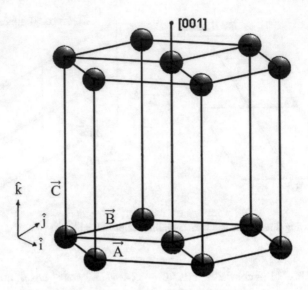

Fig. 2.16 Primitive translation vectors in a simple hexagonal lattice

$$V = A.B \times C$$

$$= \left(\frac{a}{2}\hat{i} + \frac{\sqrt{3}a}{2}\hat{j}\right) \cdot \left(-\frac{a}{2}\hat{i} + \frac{\sqrt{3}a}{2}\hat{j}\right) \times c\hat{k}$$

$$= \left(\frac{a}{2}\hat{i} + \frac{\sqrt{3}a}{2}\hat{j}\right) \cdot \left(\frac{ac}{2}\hat{j} + \frac{\sqrt{3}ac}{2}\hat{i}\right)$$

$$= \left(\frac{\sqrt{3}a^2c}{4} + \frac{\sqrt{3}a^2c}{4}\right)$$

$$= \frac{\sqrt{3}a^2c}{2}$$

Therefore,

$$A^* = 2\pi\frac{B \times C}{A.B \times C} = 2\pi\frac{\left(\frac{ac}{2}\hat{j} + \frac{\sqrt{3}ac}{2}\hat{i}\right)}{\frac{\sqrt{3}a^2c}{2}} = \left(\frac{2\pi}{a}\hat{i} + \frac{2\pi}{\sqrt{3}a}\hat{j}\right)$$

Similarly,

$$B^* = 2\pi \frac{C \times A}{A.B \times C} = 2\pi \frac{\left(\frac{ac}{2}\hat{j} - \frac{\sqrt{3}ac}{2}\hat{i}\right)}{\frac{\sqrt{3}a^2c}{2}} = \left(-\frac{2\pi}{a}\hat{i} + \frac{2\pi}{\sqrt{3}a}\hat{j}\right)$$

$$C^* = 2\pi \frac{A \times B}{A.B \times C} = 2\pi \frac{\left(\frac{\sqrt{3}a^2}{2}\hat{k}\right)}{\frac{\sqrt{3}a^2c}{2}} = \left(\frac{2\pi}{c}\hat{k}\right)$$

Further, the volume of the reciprocal unit cell is given by

$$V^* = A^*.B^* \times C^*$$

$$= \left(\frac{2\pi}{a}\hat{i} + \frac{2\pi}{\sqrt{3}a}\hat{j}\right).\left(-\frac{2\pi}{a}\hat{i} + \frac{2\pi}{\sqrt{3}a}\hat{j}\right) \times \frac{2\pi}{c}\hat{k}$$

$$= \left(\frac{2\pi}{a}\hat{i} + \frac{2\pi}{\sqrt{3}a}\hat{j}\right).\left(\frac{4\pi^2}{ac}\hat{j} + \frac{4\pi^2}{\sqrt{3}ac}\hat{i}\right)$$

$$= \left(\frac{8\pi^3}{\sqrt{3}a^2c} + \frac{8\pi^3}{\sqrt{3}a^2c}\right)$$

$$= \frac{2(2\pi)^3}{\sqrt{3}a^2c}$$

Example 7 The primitive translation vectors of a simple hexagonal space lattice may be taken as

$$A = a\hat{i}, \quad B = \left(-\frac{a}{2}\hat{i} + \frac{\sqrt{3}a}{2}\hat{j}\right) \text{and } C = c\hat{k}$$

where $\hat{i}, \hat{j}$ and $\hat{k}$ are unit vectors.

Determine the primitive translation vectors of the reciprocal lattice and show that they are rotated with respect to C or C* axis through an angle of 30°. Also, determine reciprocal unit cell volume.

Solution: Given: Direct lattice parameters A, B and C; determine the reciprocal lattice parameters A*, B*, C* and volume V*. Volume of the direct unit cell is given by

$$V = A.B \times C = a\hat{i}.\left(-\frac{a}{2}\hat{i} + \frac{\sqrt{3}a}{2}\hat{j}\right) \times c\hat{k}$$

$$= a\hat{i}.\left(\frac{ac}{2}\hat{j} + \frac{\sqrt{3}ac}{2}\hat{i}\right)$$

$$= \frac{\sqrt{3}a^2c}{2}$$

Therefore,

$$A^* = 2\pi \frac{B \times C}{A.B \times C} = 2\pi \frac{\left(\frac{\sqrt{3}ac}{2}\hat{i} + \frac{ac}{2}\hat{j}\right)}{\frac{\sqrt{3}a^2c}{2}} = \left(\frac{2\pi}{a}\hat{i} + \frac{2\pi}{\sqrt{3}a}\hat{j}\right)$$

Similarly,

$$B^* = 2\pi \frac{C \times A}{A.B \times C} = 2\pi \frac{ac}{\frac{\sqrt{3}a^2c}{2}}\hat{j} = \left(\frac{4\pi}{\sqrt{3}a}\hat{j}\right)$$

$$C^* = 2\pi \frac{A \times B}{A.B \times C} = 2\pi \frac{\left(\frac{\sqrt{3}a^2}{2}\hat{k}\right)}{\frac{\sqrt{3}a^2c}{2}} = \left(\frac{2\pi}{c}\hat{k}\right)$$

Comparing the direct and reciprocal lattice (Fig. 2.17), we observe that C and C*-axes are parallel but A*-axis has rotated through an angle of 30° with respect to A-axis in the hexagonal basal plane.

Further, the volume of the reciprocal unit cell is given by

$$V^* = A^*.B^* \times C^*$$
$$= \left(\frac{2\pi}{a}\hat{i} + \frac{2\pi}{\sqrt{3}a}\hat{j}\right) \cdot \left(\frac{4\pi}{\sqrt{3}a}\hat{j} \times \frac{2\pi}{c}\hat{k}\right)$$
$$= \left(\frac{2\pi}{a}\hat{i} + \frac{2\pi}{\sqrt{3}a}\hat{j}\right) \cdot \left(\frac{8\pi^2}{\sqrt{3}ac}\hat{i}\right)$$
$$= \frac{2(2\pi)^3}{\sqrt{3}a^2c}$$

Example 8　Show that the reciprocal lattice to a simple cubic lattice is also a simple cubic.

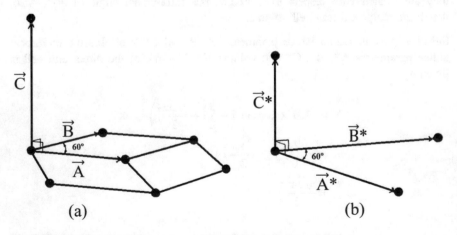

(a)　　　　　　　　　　　　　　　　　　　(b)

Fig. 2.17 Primitive translation vectors **a** Direct lattice **b** Reciprocal lattice

Proof: The primitive translation vectors of a simple cubic lattice (2.18a) may be written as:

$$a = a\hat{i}, \quad b = b\hat{j}, \quad c = c\hat{k}$$

where $a = b = c$ are lattice translations and $\hat{i}, \hat{j}, \hat{k}$ are unit vectors. Volume of the direct unit cell is given by

$$V = a.b \times c = a\hat{i}.a\hat{j} \times a\hat{k} = a\hat{i} \cdot a^2\hat{i} = a^3$$

Therefore,

$$a^* = 2\pi\frac{b \times c}{a.b \times c} = \frac{2\pi}{a}\hat{i}$$

$$b^* = 2\pi\frac{c \times a}{a.b \times c} = \frac{2\pi}{a}\hat{j}$$

$$c^* = 2\pi\frac{a \times b}{a.b \times c} = \frac{2\pi}{a}\hat{k}$$

The primitive reciprocal lattice translation vectors suggest that the reciprocal lattice to simple cubic lattice is itself a simple cubic shown as the first BZ (2.18b) with a lattice translation $2\pi/a$. The volume of the reciprocal unit cell is given by

$$V^* = a^*.b^* \times c^* = \frac{2\pi}{a}\hat{i}.\frac{2\pi}{a}\hat{j} \times \frac{2\pi}{a}\hat{k} = \frac{2\pi}{a}\hat{i}.\left(\frac{2\pi}{a}\right)^2\hat{i} = \left(\frac{2\pi}{a}\right)^3$$

Example 9 Show that the reciprocal lattice to a body-centered cubic is a face-centered cubic.

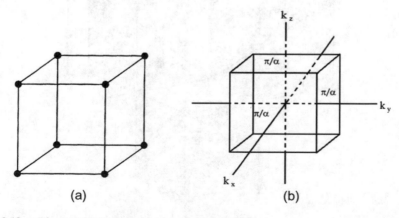

(a)   (b)

Fig. 2.18 **a** Direct and **b** reciprocal lattice of simple cubic

Proof: The primitive translation vectors of the bcc lattice shown in Fig. 2.19 are given by

$$a' = \frac{a}{2}\left(\hat{i}+\hat{j}-\hat{k}\right)$$

$$b' = \frac{a}{2}\left(-\hat{i}+\hat{j}+\hat{k}\right)$$

$$c' = \frac{a}{2}\left(\hat{i}-\hat{j}+\hat{k}\right)$$

where a is the side of the conventional unit cube and $\hat{i}, \hat{j}, \hat{k}$ are unit vectors. Volume of the direct (primitive) unit cell is given by

$$V = a'.b' \times c' = \frac{a}{2}\left(\hat{i}+\hat{j}-\hat{k}\right).\left[\frac{a}{2}\left(-\hat{i}+\hat{j}+\hat{k}\right) \times \frac{a}{2}\left(\hat{i}-\hat{j}+\hat{k}\right)\right]$$

$$= \frac{a}{2}(i+j-k).\frac{a^2}{4}\begin{vmatrix} \hat{i} & \hat{j} & \hat{k} \\ -1 & 1 & 1 \\ 1 & -1 & 1 \end{vmatrix}$$

$$= \frac{a^3}{8}(i+j-k).\begin{vmatrix} \hat{i} & \hat{j} & \hat{k} \\ -1 & 1 & 1 \\ 0 & 0 & 2 \end{vmatrix}$$

$$= \frac{a^3}{8}\left(\hat{i}+\hat{j}-\hat{k}\right).2\left(\hat{i}+\hat{j}\right)$$

$$= \frac{a^3}{8}(2+2)$$

$$= \frac{a^3}{2}$$

Therefore, the primitive translation vectors of the reciprocal lattice are:

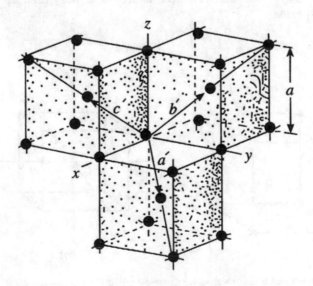

Fig. 2.19 Primitive translation vectors of bcc lattice

$$(a')^* = 2\pi \frac{b' \times c'}{a'.b' \times c'} = 2\pi \frac{\frac{a}{2}\left(-\hat{i}+\hat{j}+\hat{k}\right) \times \frac{a}{2}\left(\hat{i}-\hat{j}+\hat{k}\right)}{\frac{a^3}{2}}$$

$$= \frac{\pi}{a} \begin{vmatrix} \hat{i} & \hat{j} & \hat{k} \\ -1 & 1 & 1 \\ 1 & -1 & 1 \end{vmatrix}$$

$$= \frac{\pi}{a}.2\left(\hat{i}+\hat{j}\right)$$

$$= \frac{2\pi}{a}\left(\hat{i}+\hat{j}\right)$$

Similarly,

$$(b')^* = \frac{2\pi}{a}\left(\hat{j}+\hat{k}\right)$$

$$(c')^* = \frac{2\pi}{a}\left(\hat{k}+\hat{i}\right)$$

These reciprocal lattice vectors are found to be the primitive vectors of fcc lattice (Fig. 2.20).

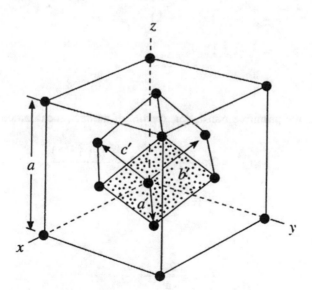

Fig. 2.20 Primitive translation vectors of fcc lattice

Example 10 Show that the reciprocal lattice to a face-centered cubic is a body-centered cubic.

Proof: The primitive translation vectors of the fcc lattice shown in Fig. 2.20 are given by

$$a' = \frac{a}{2}\left(\hat{i}+\hat{j}\right)$$

$$b' = \frac{a}{2}\left(\hat{j}+\hat{k}\right)$$

$$c' = \frac{a}{2}\left(\hat{k}+\hat{i}\right)$$

where a is the side of the conventional unit cube and $\hat{i}, \hat{j}, \hat{k}$ are unit vectors. Volume of the direct (primitive) unit cell is given by

$$V = a'.b' \times c' = \frac{a}{2}\left(\hat{i}+\hat{j}\right).\left[\frac{a}{2}\left(\hat{j}+\hat{k}\right) \times \frac{a}{2}\left(\hat{k}+\hat{i}\right)\right]$$

$$= \frac{a}{2}(i+j).\frac{a^2}{4}\begin{vmatrix} \hat{i} & \hat{j} & \hat{k} \\ 0 & 1 & 1 \\ 1 & 0 & 1 \end{vmatrix}$$

$$= \frac{a^3}{8}\left(\hat{i}+\hat{j}\right).\left(\hat{i}+\hat{j}-\hat{k}\right)$$

$$= \frac{a^3}{8}\left(1+1\right)$$

$$= \frac{a^3}{4}$$

Therefore, the primitive translation vectors of the reciprocal lattice are:

$$\left(a'\right)^* = 2\pi\frac{b' \times c'}{a'.b' \times c'} = 2\pi\frac{\frac{a^2}{4}\left(\hat{i}+\hat{j}-\hat{k}\right)}{\frac{a^3}{4}}$$

$$= \frac{2\pi}{a}\left(\hat{i}+\hat{j}-\hat{k}\right)$$

Similarly,

$$\left(b'\right)^* = \frac{2\pi}{a}\left(-\hat{i}+\hat{j}+\hat{k}\right)$$

$$\left(c'\right)^* = \frac{2\pi}{a}\left(\hat{i}-\hat{j}+\hat{k}\right)$$

These reciprocal lattice vectors are found to be the primitive vectors of bcc lattice (Fig. 2.19).

Example 11   Show that the reciprocal lattice vector G (hkl) is normal to the crystal plane (hkl).

Proof: This can be proved if we can show that the scalar product of the reciprocal lattice vector G (hkl) and any vector lying in the crystal plane (hkl) vanishes.

Let us consider the (hkl) plane as shown in Fig. 2.21. This plane intercepts the a-axis at a/h, b-axis at b/k and c-axis at c/l, respectively. The vectors A, B and C lie in the (hkl) plane. Now, let us take the scalar product of the vector C with G, we obtain

$$
\begin{aligned}
C.G &= \left(\frac{a}{h} - \frac{b}{k}\right).(ha^* + kb^* + lc^*) \\
&= \frac{a}{h}.(ha^* + kb^* + lc^*) - \frac{b}{k}(ha^* + kb^* + lc^*) \\
&= \left(\frac{h}{h} + 0 + 0\right) - \left(0 + \frac{k}{k} + 0\right) = 0
\end{aligned}
$$

Similarly, it can be shown that A.G = B.G = 0. This proves that the reciprocal lattice vector G (hkl) is normal to the crystal plane (hkl).

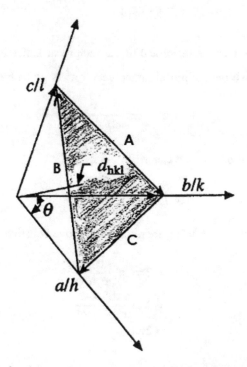

Fig. 2.21   Intercepts of a plane on three axes and the interplanar spacing

Example 12 Show that the interplanar spacing d_{hkl} in a real crystal is proportional to $|1/G(hkl)|$.

Proof: From the above proof, we know that G (hkl) is normal to the crystal plane (hkl). Accordingly, we can write

$$|G(hkl)|\hat{n} = (ha^* + kb^* + lc^*)$$

where $\hat{n}$ is a unit vector, so that

$$\hat{n} = \frac{(ha^* + kb^* + lc^*)}{|G(hkl)|}$$

In Fig. 2.21, the length of the interplanar spacing is

$$
\begin{aligned}
d_{hkl} &= \frac{a}{h}\cos\theta \\
&= \frac{a}{h}\hat{n} \\
&= \frac{a}{h} \cdot \frac{(ha^* + kb^* + lc^*)}{|G(hkl)|} \\
&= \frac{1}{|G(hkl)|}
\end{aligned}
$$

Example 13 Show that the reciprocal of the reciprocal lattice is a direct lattice.

Proof: We know that the reciprocal lattice vector a* is given by

$$a^* = 2\pi \frac{b \times c}{a.b \times c}$$

Taking reciprocal of this, we can write

$$(a^*)^* = 4\pi^2 \frac{b^* \times c^*}{a^*.b^* \times c^*}$$

Since a.a* = b.b* = c.c* = 2π, therefore, the above equation becomes

$$
\begin{aligned}
(a^*)^* &= 2\pi a . a^* \frac{b^* \times c^*}{a^*.b^* \times c^*} \\
&= 2\pi a \frac{a^*.b^* \times c^*}{a^*.b^* \times c^*} \\
&= 2\pi a
\end{aligned}
$$

Similarly,

$$(b^*)^* = 2\pi b$$
$$(c^*)^* = 2\pi c$$

Example 14 Show that the volume of the reciprocal unit cell is inversely proportional to that of corresponding direct unit cell.

Proof: We know that the volume of the primitive unit cell of a direct lattice is

$$V = a.b \times c$$

Similarly, the volume of the primitive cell of reciprocal lattice is

$$
\begin{aligned}
V^* &= a^*.b^* \times c^* \\
&= \left[\frac{b \times c}{a.b \times c}\right].\left[\frac{c \times a}{a.b \times c}\right] \times \left[\frac{a \times b}{a.b \times c}\right] \\
&= \frac{1}{(a.b \times c)^3}[b \times c.\{(c \times a) \times (a \times b)\}] \\
&= \frac{1}{(a.b \times c)^3}(b \times c)[\{c.(a \times b)a \quad a.(a \times b)\}c] \\
&= \frac{1}{(a.b \times c)^3}[c.(a \times b)\{(a.b \times c) - 0\}] \\
&= \frac{1}{(a.b \times c)^3}(a.b \times c)^2 \\
&= \frac{1}{(a.b \times c)}
\end{aligned}
$$

2.3 Construction of Brillouin Zones

A Brilloiun zone is defined as the region (in 1-D, 2-D or 3-D) in k-space (reciprocal space) that the electrons/phonons occupy to move freely without being diffracted. Because of its connection with k-space, the first Brillouin zone is equivalent to a Wigner–Seitz unit cell in reciprocal space, that is,

$$B - Z = W - S \text{ Cell of } R - L$$

The first B-Z and all other (higher) zones are symmetrical about k = 0. The construction of the first B-Z for any given (direct) lattice can be made according to the following procedure.

1. Obtain the reciprocal lattice net of the given (direct) lattice as per the procedure given in section 2.2.
2. Construct a Wigner–Seitz (W–S) unit cell in the resulting reciprocal lattice according to the set procedure given in section 2.1. This will represent the first B-Z corresponding to the given direct lattice.

Since the exact nature of the region in the form of line, plane or space within the first B-Z (according to lattice dimension) is governed by the diffraction condition, therefore let us start with Bragg's diffraction condition to understand the construction of B-Z in 1-D, 2-D and 3D lattices, one by one.

(a) One-Dimensional Lattice

We know that the general form of Bragg's diffraction condition is given by

$$G^2 + 2k \cdot G = 0 \tag{2.1}$$

However for a 1-D lattice, the components of the reciprocal lattice vector G and the wave vector k are given by

$$G = \left(\frac{2\pi}{a}\right)\hat{i}\, n_x \quad \text{and} \quad k = \hat{i}\,k_x$$

Substituting these values in Eq. 2.1, we obtain

$$\left[\left(\frac{2\pi}{a}\right)\hat{i}\, n_x\right]^2 + 2\hat{i}\,k_x \cdot \left(\frac{2\pi}{a}\right)\hat{i}\, n_x = 0$$

$$\text{or} \qquad \frac{4\pi^2}{a^2}\, n_x^2 + \frac{4\pi}{a} n_x\, k_x = 0 \tag{2.2}$$

$$\text{or} \qquad k_x = -\frac{\pi}{a} n_x$$

where $n_x = \pm 1, \pm 2, \ldots$ will provide us various 1-D Brillouin zones as shown in Fig. 2.22.

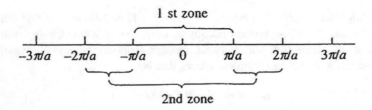

Fig. 2.22 First two Brillouin zones of 1-D lattice

(b) **Two-Dimensional (Square) Lattice**

For simplicity of the problem, let us take the case of a square lattice with a = b and $\gamma = 90°$. Corresponding components of the reciprocal lattice vector G and the wave vector k are given by

$$G = \frac{2\pi}{a}\left(\hat{i}\,n_x + \hat{j}\,n_y\right) \quad \text{and} \quad k = \left(\hat{i}\,k_x + \hat{j}\,k_y\right)$$

Substituting these values in Eq. 2.1, we obtain

$$\left[\frac{2\pi}{a}\left(\hat{i}\,n_x + \hat{j}\,n_y\right)\right]^2 + 2\left(\hat{i}\,k_x + \hat{j}\,k_y\right) \cdot \left[\frac{2\pi}{a}\left(\hat{i}\,n_x + \hat{j}\,n_y\right)\right] = 0$$

or $$\frac{4\pi^2}{a^2}\left(n_x^2 + n_y^2\right) + \frac{4\pi}{a}\left(n_x\,k_x + n_y\,k_y\right) = 0$$

where $\hat{i}.\hat{i} = \hat{j}.\hat{j} = 1$ and $\hat{i}.\hat{j} = 0$.
Further simplifying this, we obtain

$$n_x k_x + n_y k_y = -\frac{\pi}{a}\left(n_x^2 + n_y^2\right) \tag{2.3}$$

where n_x and n_y are integers for diffraction by the vertical columns and horizontal rows of atoms. For the first zone, one integer (say n_x) is ± 1 and other integer (say n_y) is zero. Therefore, the zone boundaries for the first zone are:

$$\text{For } n_x = \pm 1, \quad n_y = 0$$
$$\pm k_x = -\frac{\pi}{a}\,\text{giving} \quad k_x = \pm\frac{\pi}{a}$$

Similarly, for $n_x = 0$, $n_y = \pm 1$

$$\pm k_y = -\frac{\pi}{a}\,\text{giving} \quad k_y = \pm\frac{\pi}{a}$$

This is the required B.Z in 2-D and is illustrated in Fig. 2.23.
In the region $k < \pm\frac{\pi}{a}$, electrons/phonons move freely without being diffracted. However, when $k = \pm\frac{\pi}{a}$, they are prevented from moving in the x or y direction due to diffraction. As the value of k exceeds $\frac{\pi}{a}$ the number of possible directions of motion decreases gradually, until when $k = \frac{\pi}{a\sin 45°} = \frac{\sqrt{2}\pi}{a}$
This implies that the electrons/phonons are diffracted even when they move diagonally (at 45° with k_x and k_y axes) inside the square. It is here the first zone ends and the second zone begins.
For the second zone, both the integers (n_x and n_y) in Eq. 2.3 are equal to ± 1, that is,

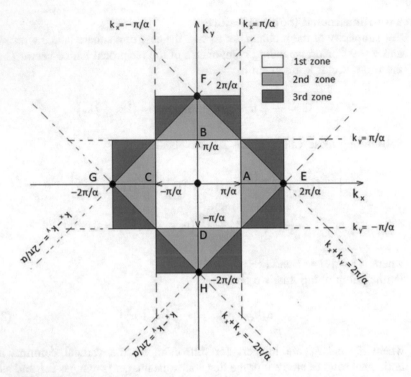

Fig. 2.23 First three Brillouin zones of a 2-D square lattice

$$\pm k_x \pm k_y = \frac{2\pi}{a}$$

This gives us four equations representing zone boundaries:

$$n_x = +1, n_y = +1 \ \text{ giving } \ k_x + k_y = \frac{2\pi}{a} (\text{line FE})$$

$$n_x = -1, n_y = +1 \ \text{ giving } \ -k_x + k_y = \frac{2\pi}{a} (\text{line FG})$$

$$n_x = -1, n_y = -1 \ \text{ giving } \ -k_x - k_y = \frac{2\pi}{a} (\text{line GH})$$

$$n_x = +1, n_y = -1 \ \text{ giving } \ k_x - k_y = \frac{2\pi}{a} (\text{line EH})$$

The second B-Z contains electrons/phonons with k values from π/a (that do not fit into the first zone) to $2\pi/a$, or between the square ABCD and EFGH as shown in Fig. 2.23.

(c) **Three-Dimensional (Simple cubic) Lattice**

Again for simplicity of the problem, let us take the case of a simple cubic lattice with a = b = c and $\alpha = \beta = \gamma = 90°$. The components of the reciprocal lattice vector G and the wave vector k are given by

$$G = \frac{2\pi}{a}\left(\hat{i}\,n_x + \hat{j}\,n_y + \hat{k}\,n_z\right) \quad \text{and} \quad k = \left(\hat{i}\,k_x + \hat{j}\,k_y + \hat{k}\,k_z\right)$$

Substituting these values in Eq. 2.1, we obtain

$$\left[\frac{2\pi}{a}\left(\hat{i}\,n_x + \hat{j}\,n_y + \hat{k}\,n_z\right)\right]^2 + 2\left(\hat{i}\,k_x + \hat{j}\,k_y + \hat{k}\,n_z\right)\cdot\left[\frac{2\pi}{a}\left(\hat{i}\,n_x + \hat{j}\,n_y + \hat{k}\,n_z\right)\right] = 0$$

or $\qquad \dfrac{4\pi^2}{a^2}\left(n_x^2 + n_y^2 + n_z^2\right) + \dfrac{4\pi}{a}\left(n_x k_x + n_y k_y + k k_z\right) = 0$

where $\hat{i}.\hat{i} = \hat{j}.\hat{j} = \hat{k}.\hat{k} = 1$ and $\hat{i}.\hat{j} = \hat{j}.\hat{k} = \hat{k}.\hat{i} = 0.$
Further simplifying this, we obtain

$$n_x\,k_x + n_y\,k_y + n_z\,k_z = -\frac{\pi}{a}\left(n_x^2 + n_y^2 + n_z^2\right) \qquad (2.4)$$

This equation could have been obtained by simply generalizing the Eq. 2.3 obtained in 2-D case. However, from Eq. 2.4, it follows that the first zone for a simple cubic lattice is a simple cube whose walls intersect the k_x, k_y and k_z axes at the $\pm\pi/a$ as shown in Fig. 2.24a. The second zone is obtained by adding a pyramid (like a triangle in 2-D) to each face of the cube of the first zone. The resulting diagram is shown in Fig. 2.24b.

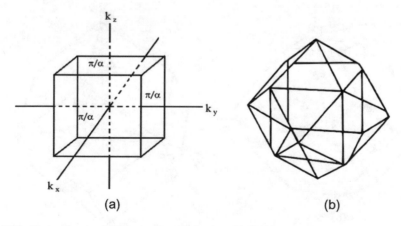

(a)   (b)

Fig. 2.24 First and second Brillouin zone of simple cubic lattice

(d) **Some Important Notations Used in B.Z**

The names of certain points and directions of a B-Z (Fig. 2.25), commonly used by experts are the following:

Γ—Origin
W—Corner
L—Center of hexagonal faces
X—Center of square faces
Γ ⟶ H denotes [100] direction and symbolized as Δ
Γ ⟶ N denotes [110] direction and symbolized as Σ
Γ ⟶ P denotes [111] direction and symbolized as ∧

Solved Examples

Example 1 Construct B-Z corresponding to a given plane oblique lattice with a ≠ b and $\gamma = \gamma°$.

Solution: *Given*: Plane oblique lattice with a ≠ b and $\gamma = \gamma°$. Let us follow the procedure given below:

(i)   Draw a primitive cell in the given lattice.
(ii) Draw a normal to each axis and obtain a* and b* as

$$a^* = \frac{2\pi}{a\sin\gamma} \quad \text{and} \quad b^* = \frac{2\pi}{a\sin\gamma}$$

(iii) Obtain the reciprocal lattice net and construct a W–S unit cell. This is the required B-Z for the given primitive oblique lattice.

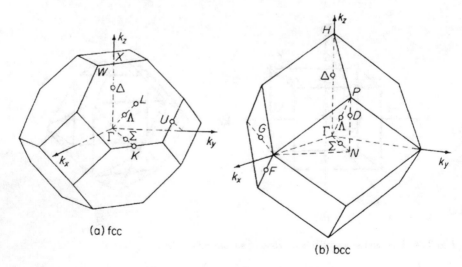

(a) fcc

(b) bcc

Fig. 2.25 Some important notations used in Brillouin zones

The geometric relationship between the direct lattice, the reciprocal lattice and the B-Z is shown in Table 2.3.

Table 2.3 Direct lattice and Brillouin zone in 2-D crystal system

Crystal system	Crystal lattice	Brillouin zone
Oblique		
Rectangular		
Centered rectangular		
Square		
Hexagon		

Example 2 Construct B-Z corresponding to a given plane hexagonal lattice with a = b and $\gamma = 120°$ and show that the shortest distance of a zone face from the center of the zone is $1.155\left(\frac{\pi}{a}\right)$.

Solution: *Given*: Plane hexagonal lattice with a = b and $\gamma = 120°$. Follow the given procedure, we can obtain:

(i)   A primitive cell in the given lattice.
(ii) Draw normal to each axis and obtain a* and b* as

$$a^* = b^* = \frac{2\pi}{a\sin\gamma} = \frac{2\pi}{a\left(\sqrt{3}/2\right)} = \frac{4}{\sqrt{3}}\left(\frac{\pi}{a}\right)$$

(iii) Obtain reciprocal lattice net and construct a W–S unit cell. This is the required B-Z for the given plane hexagonal lattice.

The geometric relationship between the direct lattice, the reciprocal lattice and the B-Z is shown in Table 2.3. From the figure, we observe that the shortest distance of a zone face from the center of the zone is

$$d = \frac{a^*}{2} = \frac{2}{\sqrt{3}}\left(\frac{\pi}{a}\right) = 1.155\left(\frac{\pi}{a}\right)$$

Example 3 Construct B-Z corresponding to a given plane rectangular lattice with a ≠ b and $\gamma = 90°$ and show that the zone faces are at π/a and π/b, respectively, from the center of the zone.

Solution: *Given*: Plane oblique lattice with a ≠ b and $\gamma = 90°$. Following the given procedure, we obtain:

(i)   A primitive cell in the given lattice.
(ii) Draw a normal to each axis and obtain a* and b* as

$$a^* = \frac{2\pi}{a} \quad \text{and} \quad b^* = \frac{2\pi}{b}$$

(iii) Obtain the reciprocal lattice net and construct a W–S unit cell. This is the required B-Z for the given primitive oblique lattice.

The geometric relationship between the direct lattice, the reciprocal lattice and the B-Z is shown in Table 2.3. From the figure, we observe that the two zone faces are at π/a and π/b, respectively, from the center of the zone.

Example 4 Construct B-Z corresponding to a simple cubic lattice with a = b = c and $\alpha = \beta = \gamma = 90°$. Show that the six equivalent zone faces are at a distance π/a from the center of the zone.

Solution: Given: Simple cubic lattice with a = b = c and $\alpha = \beta = \gamma = 90°$, the primitive translation vectors of a simple cubic lattice may be written as:

$$a = a\hat{i} \quad , \quad b = b\hat{j} \quad , \quad c = c\hat{k}$$

where a = b = c are lattice translations and $\hat{i}, \hat{j}, \hat{k}$ are unit vectors parallel to the cube edges. Volume of the direct unit cell is given by

$$V = |a.b \times c| = a^3$$

Therefore,

$$a^* = 2\pi \frac{b \times c}{a.b \times c} = \frac{2\pi}{a}\hat{i} = \frac{2\pi}{a}(100)$$

$$b^* = 2\pi \frac{c \times a}{a.b \times c} = \frac{2\pi}{a}\hat{j} = \frac{2\pi}{a}(010)$$

$$c^* = 2\pi \frac{a \times b}{a.b \times c} = \frac{2\pi}{a}\hat{k} = \frac{2\pi}{a}(001)$$

This shows that the reciprocal lattice is a simple cubic lattice whose lattice parameter is $2\pi/a$. The first B-Z is constructed by drawing six perpendicular bisector planes to the vectors a*, b* and c*. They are:

$$\pm\frac{a^*}{2} = \pm\frac{\pi}{a}\hat{i}, \pm\frac{b^*}{2} = \pm\frac{\pi}{a}\hat{j}, \pm\frac{c^*}{2} = \pm\frac{\pi}{a}\hat{k},$$

The space bounded by these six {100} planes (termed as zone faces) is also a cube of side $(2\pi/a)$ and is known as first B-Z of the simple cubic lattice (Fig. 2.24a).

Example 5 Construct B-Z corresponding to a bcc lattice with a = b = c and $\alpha = \beta = \gamma = 90°$. Show that the twelve equivalent zone faces are at a distance $\sqrt{2}$ (π/a) from the center of the zone.

Solution: *Given*: Body-centered cubic lattice with a = b = c and $\alpha = \beta = \gamma = 90°$. The primitive translation vectors of a bcc shown in Fig. 2.19 are given by:

$$a' = \frac{a}{2}\left(\hat{i}+\hat{j}-\hat{k}\right)$$

$$b' = \frac{a}{2}\left(-\hat{i}+\hat{j}+\hat{k}\right)$$

$$c' = \frac{a}{2}\left(\hat{i}-\hat{j}+\hat{k}\right)$$

where a is the side of the conventional unit cube and $\hat{i}, \hat{j}, \hat{k}$ are unit vectors parallel to the cube edges. Volume of the direct (primitive) unit cell is given by

$$V = a'.b' \times c' = \frac{a^3}{2}$$

Therefore, the primitive translation vectors of the reciprocal lattice are:

$$(a')^* = 2\pi \frac{b' \times c'}{a'.b' \times c'} = \frac{2\pi}{a}(\hat{i} + \hat{j}) = \frac{2\pi}{a}(110)$$

Similarly,

$$(b')^* = 2\pi \frac{c' \times a'}{a'.b' \times c'} = \frac{2\pi}{a}(\hat{j} + \hat{k}) = \frac{2\pi}{a}(011)$$

$$(c')^* = 2\pi \frac{a' \times b'}{a'.b' \times c'} = \frac{2\pi}{a}(\hat{k} + \hat{i}) = \frac{2\pi}{a}(101)$$

This shows that the reciprocal lattice is an fcc lattice. The reciprocal lattice vectors are shown in Fig. 2.20. Further, in terms of the Miller indices (hkl), the general form of the reciprocal lattice vector is written as:

$$G(hkl) = ha^* + kb^* + lc^*$$

$$= \frac{2\pi}{a}\left[h(\hat{i} + \hat{j}) + k(\hat{j} + \hat{k}) + l(\hat{k} + \hat{i})\right]$$

$$= \frac{2\pi}{a}\left[(h+l)\hat{i} + (h+k)\hat{j} + (k+l)\hat{k}\right]$$

However, there are twelve vectors (see W–S unit cell for direct fcc lattice, Fig. 2.4) of equal magnitude, all starting from the origin and terminating at the nearest reciprocal lattice points. They are given as:

$$\frac{2\pi}{a}(\pm\hat{i} \pm \hat{j}) \;;\; \frac{2\pi}{a}(\pm\hat{j} \pm \hat{k}) \;;\; \frac{2\pi}{a}(\pm\hat{k} \pm \hat{i})$$

where the choices of sign are independent.

Now, we can construct the first B-Z by drawing planes normal to each of the 12 reciprocal lattice vectors at their midpoints. The resulting polyhedron is a rhombic dodecahedron as shown in Fig. 2.26. This is bounded by twelve {110} planes at:

$$\frac{\pi}{a}(\pm\hat{i} \pm \hat{j}) \;;\; \frac{\pi}{a}(\pm\hat{j} \pm \hat{k}) \;;\; \frac{\pi}{a}(\pm\hat{k} \pm \hat{i})$$

Magnitude of each of the 12 vectors is $\sqrt{2}(\pi/a)$.

Example 6 Construct B-Z corresponding to an fcc lattice with $a = b = c$ and $\alpha = \beta = \gamma = 90°$. Show that there are 8 hexagonal and 6 square equivalent zone faces in it and they are at a distance of $\sqrt{3}\,(\pi/a)$ and $2(\pi/a)$, respectively, from the center of the zone.

Fig. 2.26 Brillouin zone of bcc lattice

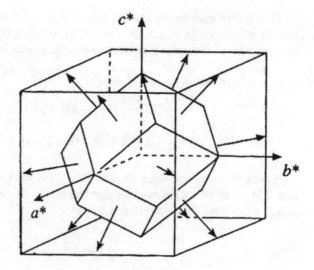

Solution: Given: Face-centered cubic lattice with a = b = c and α = β = γ = 90°. The primitive translation vectors of an fcc shown in Fig. 2.20 are given by:

$$a' = \frac{a}{2}\left(\hat{i}+\hat{j}\right)$$

$$b' = \frac{a}{2}\left(\hat{j}+\hat{k}\right)$$

$$c' = \frac{a}{2}\left(\hat{k}+\hat{i}\right)$$

where a is the side of the conventional unit cube and $\hat{i}, \hat{j}, \hat{k}$ are unit vectors. Volume of the direct (primitive) unit cell is given by

$$V = |a'.b' \times c'| = \frac{a^3}{4}$$

The primitive translation vectors of the reciprocal lattice are:

$$(a')^* = 2\pi \frac{b' \times c'}{a'.b' \times c'} = \frac{2\pi}{a}\left(\hat{i}+\hat{j}-\hat{k}\right) = \frac{2\pi}{a}(11\bar{1})$$

Similarly,

$$(b')^* = 2\pi \frac{c' \times a'}{a'.b' \times c'} = \frac{2\pi}{a}\left(-\hat{i}+\hat{j}+\hat{k}\right) = \frac{2\pi}{a}(\bar{1}11)$$

$$(c')^* = 2\pi \frac{a' \times b'}{a'.b' \times c'} = \frac{2\pi}{a}\left(\hat{i}-\hat{j}+\hat{k}\right) = \frac{2\pi}{a}(1\bar{1}1)$$

This shows that the reciprocal lattice is a bcc lattice. The reciprocal lattice vectors are shown in Fig. 2.19. Further, in terms of the Miller indices (hkl), the general form of the reciprocal lattice vector is written as:

$$G(hkl) = ha^* + kb^* + 1c^*$$
$$= \frac{2\pi}{a}\left[h(\hat{i}+\hat{j}-\hat{k}) + k(-\hat{i}+\hat{j}+\hat{k}) + 1(\hat{i}-\hat{j}+\hat{k})\right]$$
$$= \frac{2\pi}{a}\left[(h-k+1)\hat{i} + (h+k-1)\hat{j} + (-h+k+1)\hat{k}\right]$$

However, there are eight vectors (see W–S unit cell for direct bcc lattice, Fig. 2.3) of equal magnitude, joining the origin to the nearest reciprocal lattice points at corners. They are given as:

$$\frac{2\pi}{a}(\pm\hat{i}\pm\hat{j}\pm\hat{k})$$

where the choices of sign are independent.

Now, we can construct the first B-Z by drawing planes normal to each of the 8 reciprocal lattice vectors at their midpoints. Thus the bounding {111} planes are at:

$$\frac{\pi}{a}(\pm\hat{i}\pm\hat{j}\pm\hat{k}) \tag{i}$$

In this case, there are 6 other vectors of equal magnitude corresponding to the next nearest neighbor atoms lying at body centers of the neighboring bcc lattice. They are given as:

$$\frac{2\pi}{a}(\pm2\hat{i}) \; ; \; \frac{2\pi}{a}(\pm2\hat{j}) \; ; \; \frac{2\pi}{a}(\pm2\hat{k})$$

Again, drawing planes normal to each of the 6 reciprocal lattice vectors at their midpoints, we obtain the six bounding {200} planes at:

$$\frac{\pi}{a}(\pm2\hat{i}) \; ; \; \frac{\pi}{a}(\pm2\hat{j}) \; ; \; \frac{\pi}{a}(\pm2\hat{k}) \tag{ii}$$

The 14 bounding faces (8 hexagonal and 6 squares) give a truncated octahedron shape as shown in Fig. 2.27. Magnitude of each of the 8 vectors passing through the center of the hexagonal faces is obtained from Eq. (i) and is given by $\sqrt{3}(\pi/a)$.

Similarly, the magnitude of the six vectors passing through the center of square faces is obtained from Eq. (ii) and is given by $2(\pi/a)$.

Fig. 2.27 Brillouin zone of a fcc lattice

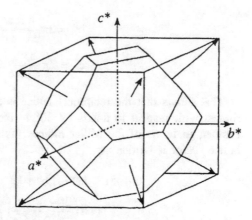

Example 7 Construct B-Z corresponding to a simple hexagonal lattice with a = b $\neq$ c and $\alpha = \beta = 90°$ and $\gamma = 120°$. Show that six vertical planes of B-Z are at a distance of $\frac{2}{\sqrt{3}}\left(\frac{\pi}{a}\right)$ and two horizontal planes are at a distance (π/c) from the center of the zone.

Solution: *Given*: Simple hexagonal lattice with a = b $\neq$ c and $\alpha = \beta = 90°$ and $\gamma = 120°$. The primitive translation vectors of simple hexagonal lattice shown in Fig. 2.16 are given by:

$$a = a\hat{i}, \quad b = \frac{a}{2}\hat{i} + \frac{\sqrt{3}a}{2}\hat{j}, \quad c = c\hat{k}$$

where $\hat{i}, \hat{j}, \hat{k}$ are the orthogonal unit vectors. Volume of the direct (primitive) unit cell is given by

$$V = a\hat{i}.\left(\frac{a}{2}\hat{i} + \frac{\sqrt{3}a}{2}\hat{j}\right) \times c\hat{k}$$

$$= a\hat{i}.\left(-\frac{ac}{2}\hat{j} + \frac{\sqrt{3}ac}{2}\hat{i}\right)$$

$$= \frac{\sqrt{3}a^2c}{2}$$

The primitive translation vectors of the reciprocal lattice are:

$$a^* = 2\pi\frac{b \times c}{a.b \times c} = \frac{2\pi}{a}\hat{i} - \frac{2\pi}{\sqrt{3}a}\hat{j}$$

Similarly,

$$b^* = 2\pi \frac{c \times a}{a.b \times c} = \frac{4\pi}{\sqrt{3}a} \hat{j}$$

$$c^* = 2\pi \frac{a \times b}{a.b \times c} = \frac{2\pi}{c} \hat{k}$$

This shows that the reciprocal lattice is a simple hexagonal where a* axis is rotated with respect to a-axis by 30° in the basal plane as shown in Fig. 2.17. Further, in terms of the Miller indices (hkl), the general form of the reciprocal lattice vector is written as:

$$G(hkl) = ha^* + kb^* + lc^*$$

$$= h\left(\frac{2\pi}{a}\hat{i} - \frac{2\pi}{\sqrt{3}a}\hat{j}\right) + k\frac{4\pi}{\sqrt{3}a}\hat{j} + l\frac{2\pi}{c}\hat{k}$$

$$= \frac{2\pi}{a}(h\hat{i}) + \frac{2\pi}{\sqrt{3}a}(2k - h)\hat{j} + \frac{2\pi}{c}l\hat{k}$$

However, there are six vectors of equal magnitude (lying in the basal plane), joining the origin to the nearest reciprocal lattice points at corners of the hexagon. They are given as:

$$\frac{4\pi}{\sqrt{3}a}(\pm 2\hat{i}); \quad \frac{4\pi}{\sqrt{3}a}(\pm 2\hat{j}); \quad \frac{4\pi}{\sqrt{3}a}\left[-(\hat{i}+\hat{j})\right]$$

Now, we can construct the first B-Z by drawing planes normal to each of the 6 reciprocal lattice vectors at their midpoints. Thus the bounding six vertical planes are at:

$$\frac{2\pi}{\sqrt{3}a}(\pm 2\hat{i}); \quad \frac{2\pi}{\sqrt{3}a}(\pm 2\hat{j}); \quad \frac{2\pi}{\sqrt{3}a}\left[-(\hat{i}+\hat{j})\right] \tag{i}$$

In this case, there are two other vectors of equal magnitude corresponding to the next nearest neighbor atoms lying at the centers of the hexagon. They are given as:

$$\frac{2\pi}{c}(\pm \hat{k}) \tag{ii}$$

Corresponding to this, two horizontal planes at their midpoints are obtained. The resulting B-Z with simple hexagonal shape is shown in Fig. 2.7.

Magnitude of each of the 6 vectors passing through 6 vertical planes is obtained from Eq. (i) and is given by $\frac{2}{\sqrt{3}}\left(\frac{\pi}{a}\right)$. Similarly, the magnitude of each of the two vectors passing through two neighboring horizontal planes is obtained from Eq. (ii) and is given by (π/c).

Fig. 2.28 The evolution of
FS with the increase in
concentration of valence
electrons

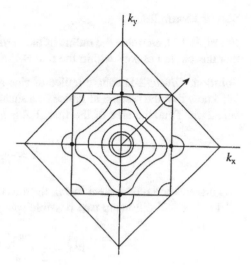

2.4   Fermi Surface and Brillouin Zone

The Fermi surface (or Fermi sphere) is defined as the surface of constant energy E_F
in k-space inside which all the energy states are filled by the valence electrons,
while all the energy states lying outside it are empty. The shape of the Fermi surface
(FS) is determined by the geometry of the contours in a Brillouin Zone (B-Z). Well
within the B-Z, the FS remains spherical in shape while its shape changes near the
B-Z boundaries as shown in Fig. 2.28.

Total number of free/valence electrons, n within the sphere of radius k_F is
equivalent to the number of possible filled energy states of electrons and can be
determined by the equation [where each particle in k-space occupies a volume $(2\pi/a)$]:

$$n = \frac{\text{Volume of FS}}{\text{Volume of BZ}} \times 2$$

$$= \frac{(4\pi/3)k_F^3}{(2\pi/a)^3} \times 2$$

This gives us

$$k_F = \left(\frac{3\pi^2 n}{a^3}\right)^{1/3}$$

where the factor 2 is included for two spin states in each energy level and a is a
direct unit cell parameter. For monovalent metals (e.g., alkali metals), n is the
number of atoms per unit cell.

Solved Examples

Example 1 Determine the radius of the Fermi circle for a monovalent metal. Show that this circle lies well within the first B-Z obtained from a square lattice of side a.

Solution: *Given*: 2-D square lattice of side a, monovalent metal, $k_F = ?$
We know that the reciprocal lattice of a square lattice of side a is a square lattice of side $2\pi/a$. Thus the area of the first B-Z is given by

$$A = \frac{4\pi^2}{a^2}$$

Further, for a monovalent metal, the area occupied by an electron is half the area of the first B-Z (allowing two possible spin states for each electron). Therefore,

$$\pi k_F^2 = \frac{1}{2} \times \frac{4\pi^2}{a^2}$$

$$\text{or} \quad k_F = \sqrt{\frac{2}{\pi}}\left(\frac{\pi}{a}\right) = 0.798\left(\frac{\pi}{a}\right)$$

where π/a is the distance of the zone boundary from the center of the zone. Also, the value of k_F lies between

$$0 < k_F < \frac{\pi}{a}$$

This indicates that the Fermi circle lies well within the first B-Z, as shown in Fig. 2.29.

Example 2 Determine the radius of the Fermi circle for a divalent metal. Show that this circle does not remain confined to the first B-Z obtained from a square lattice of side a.

Solution: *Given*: 2-D square lattice of side a, divalent metal, $k_F = ?$
We know that for a square lattice of side a, the corresponding area of the first B-Z is given by

$$A = \frac{4\pi^2}{a^2}$$

Further, for a divalent metal, the area occupied by two electrons is equal to the area of the first B-Z (allowing two possible spin states for each electron). Therefore,

$$\pi k_F^2 = \frac{2}{2} \times \frac{4\pi^2}{a^2}$$

Fig. 2.29 Free electron FS
within the first BZ

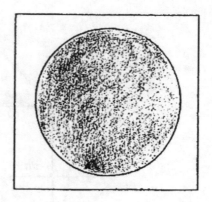

$$\text{or} \qquad k_F = \frac{2}{\sqrt{\pi}}\left(\frac{\pi}{a}\right) = 1.228\left(\frac{\pi}{a}\right)$$

But the distance of a corner (diagonal) of the first B-Z from the center is

$$d = \sqrt{2}\left(\frac{\pi}{a}\right) = 1.414\left(\frac{\pi}{a}\right)$$

The above calculations give us that the value of k_F lies between

$$\frac{\pi}{a} < k_F < 1.414\left(\frac{\pi}{a}\right)$$

This indicates that the radius k_F goes beyond the first B-Z near edges (boundaries) but remains within near corners (Fig. 2.30).

Example 3 Determine the radius of the Fermi sphere for a body-centered cubic crystal of side a. Show that (i) the Fermi sphere is entirely contained within the first B-Z (ii) it covers 88 % of the shortest distance from the center of the zone, and (iii) it is separated by a distance of $0.174\left(\frac{\pi}{a}\right)$ from the zone boundaries.

Solution: *Given*: A bcc crystal, number of atoms per unit cell, n = 2, k_F = ?
Radius of the Fermi sphere can be obtained by

$$k_F = \left(\frac{3\pi^2 n}{a^3}\right)^{1/3} = \left(\frac{3\pi^2 2}{a^3}\right)^{1/3} = \left(\frac{6}{\pi}\right)^{1/3}\frac{\pi}{a} = 1.24\left(\frac{\pi}{a}\right)$$

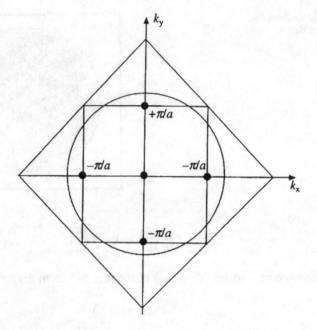

Fig. 2.30 FS partially on first and second BZ

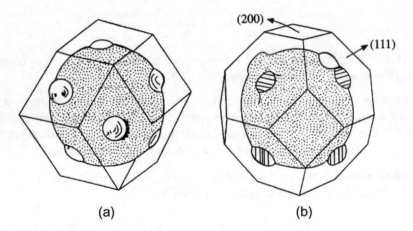

(a)   (b)

Fig. 2.31 BZ and FS of **a** bcc **b** fcc lattice

Further, the first B-Z for a bcc crystal lattice has the rhombic dodecahedron shape (Fig. 2.31a). It has twelve identical faces located at the distance $\sqrt{2}\left(\frac{\pi}{a}\right)$ from the center of the zone, where a is the periodicity of the lattice.

The shortest distance of zone face from the center of the zone is:

$$d = \frac{2\pi}{a}\left[\left(\frac{1}{2}\right)^2 + \left(\frac{1}{2}\right)^2 + (0)^2\right]^{1/2} = \sqrt{2}\left(\frac{\pi}{a}\right) = 1.414\left(\frac{\pi}{a}\right)$$

The above calculations, give us that the value of k_F lies between

$$0 < k_F < d$$

This indicates that the Fermi sphere is entirely contained within the first B-Z. Further, we have

$$\frac{k_F}{d} = \frac{1.24\left(\frac{\pi}{a}\right)}{1.414\left(\frac{\pi}{a}\right)} = 0.88 = 88\%$$

This shows that the Fermi sphere covers 88% of the shortest distance. Finally, we have

$$d - k_F = 1.414\left(\frac{\pi}{a}\right) - 1.24\left(\frac{\pi}{a}\right) = 0.174\left(\frac{\pi}{a}\right)$$

This confirms that the Fermi sphere is separated by a distance of $0.174\left(\frac{\pi}{a}\right)$ from the zone boundaries.

Example 4 Determine the radius of the Fermi sphere for a face-centered cubic crystal of side a. Show that (i) the Fermi sphere is entirely contained within the first B-Z (ii) it covers 90 % of the distance (of the hexagonal face) from the center of the zone, and (iii) it is separated by a distance of $0.169\left(\frac{\pi}{a}\right)$ from the zone boundaries.

Solution *Given*: An fcc crystal, number of atoms per unit cell, n = 4, k_F = ? The radius of the Fermi sphere can be obtained by

$$k_F = \left(\frac{3\pi^2 n}{a^3}\right)^{1/3} = \left(\frac{3\pi^2\,4}{a^3}\right)^{1/3} = \left(\frac{12}{\pi}\right)^{1/3}\frac{\pi}{a} = 1.563\left(\frac{\pi}{a}\right)$$

Further, the first B-Z for an fcc crystal lattice has the truncated octahedron shape (Fig. 2.31b). It has fourteen faces, eight hexagonal and six squares.

The shortest distance of a square face from the center of the zone is:

$$d_S = \frac{2\pi}{a}\left[(1)^2 + (0)^2 + (0)^2\right]^{1/2} = \frac{2\pi}{a}$$

Similarly, the shortest distance of the hexagonal face from the center of the zone is

$$d_h = \frac{2\pi}{a}\left[\left(\frac{1}{2}\right)^2 + \left(\frac{1}{2}\right)^2 + \left(\frac{1}{2}\right)^2\right]^{1/2} = \sqrt{3}\left(\frac{\pi}{a}\right) = 1.732\left(\frac{\pi}{a}\right)$$

The above calculations, give us that the value of k_F lies between

$$0 < k_F < d_h(d_S)$$

This indicates that the Fermi sphere is entirely contained within the first B-Z. Further, we have

$$\frac{k_F}{d_h} = \frac{1.563\left(\frac{\pi}{a}\right)}{1.732\left(\frac{\pi}{a}\right)} = 0.90 = 90\%$$

This shows that the Fermi sphere covers 90% of the distance (of the hexagonal face). Finally, we have

$$d_h - k_F = 1.732\left(\frac{\pi}{a}\right) - 1.563\left(\frac{\pi}{a}\right) = 0.169\left(\frac{\pi}{a}\right)$$

This confirms that the Fermi sphere is separated by a distance of $0.169\left(\frac{\pi}{a}\right)$ from the zone boundaries.

Example 5 Determine the radius of the Fermi sphere for bcc tetragonal crystal with $(c/a) = 2$. Also, determine the shortest distance of the tetragonal face from the center of the zone and show that the Fermi sphere does not remain confined to the first B-Z.

Solution *Given*: A bcc tetragonal crystal, n = 2, (c/a) = 2, k_F = ?, d_S = ?
We know that in a tetragonal crystal a = b ≠ c. If suppose a = 1Å, then c = 2Å. Therefore,

$$a^* = \frac{2\pi}{a} = 2\pi\,\text{Å}^{-1}$$

$$b^* = \frac{2\pi}{b} = 2\pi\,\text{Å}^{-1}$$

$$c^* = \frac{2\pi}{c} = \pi\,\text{Å}^{-1}$$

Reciprocal unit cell with a*, b* and c* will also be a tetragonal lattice. Now, the radius of the Fermi sphere for a body-centered tetragonal with 2 atoms per unit cell is

$$k_F = \left(\frac{3\pi^2 n}{a^3}\right)^{1/3} = \left(\frac{3\pi^2 2}{1 \times 1 \times 2}\right)^{1/3} = \left(\frac{3}{\pi}\right)^{1/3}\pi = 0.985\,\pi$$

Further, the shortest distance of the tetragonal face from the center of the zone is

$$d_S = \frac{\pi}{2}$$

The above calculations provide us

$$\frac{\pi}{2} < k_F < \pi$$

This shows that the Fermi sphere does not remain within the first B-Z.

Multiple Choice Questions (MCQ)

1. The shape of the Wigner–Seitz unit cell of a square lattice is:
   (a) Rectangle
   (b) Triangular
   (c) Square
   (d) Hexagonal

2. The shape of the Wigner–Seitz unit cell of a hexagonal lattice (in 2-D) is:
   (a) Rectangle
   (b) Triangular
   (c) Square
   (d) Hexagonal

3. The shape of the Wigner–Seitz unit cell of a simple cubic lattice is:
   (a) Tetrahedral
   (b) Octahedral
   (c) Cubic
   (d) Hexagonal

4. The shape of the Wigner–Seitz unit cell of a body-centered cubic lattice is:
   (a) Tetrahedral
   (b) Rhombic dodecahedral
   (c) Cubic
   (d) Truncated octahedral

5. The shape of the Wigner–Seitz unit cell of a face-centered cubic lattice is:
   (a) Tetrahedral
   (b) Rhombic dodecahedral
   (c) Cubic
   (d) Truncated octahedral

6. The shape of the Wigner–Seitz unit cell of a hexagonal lattice (in 3-D) is:
   (a) Hexagonal
   (b) Rhombic dodecahedral
   (c) Cubic
   (d) Truncated octahedral

7. Show that every edge (side) of the polyhedron (square or hexagon) bounding the Wigner–Seitz unit cell of the body-centered cubic lattice is x times the length of the conventional unit cell, where x is:
   (a) $\sqrt{2}/1$
   (b) $\sqrt{2}/2$
   (c) $\sqrt{2}/3$
   (d) $\sqrt{2}/4$

8. Show that the ratio of the lengths of the diagonals of each parallelogram face of the Wigner–Seitz unit cell for fcc lattice is:
   (a) $\sqrt{2}:1$
   (b) $\sqrt{2}:2$
   (c) $\sqrt{2}:3$
   (d) $\sqrt{2}:4$

9. The primitive translation vectors of a three-dimensional lattice are $a = 2\hat{i} + \hat{j}$, $b = 2\hat{j}$ and $c = \hat{k}$. The primitive reciprocal lattice vector a* is:
   (a) $\pi\hat{i}$
   (b) $2\pi\hat{i}$
   (c) $3\pi\hat{i}$
   (d) $4\pi\hat{i}$

10. The primitive translation vectors of a three-dimensional lattice are $a = 2\hat{i} + j$, $b = 2\hat{j}$ and $c = \hat{k}$. The primitive reciprocal lattice vector b* is:
   (a) $\pi/2(-\hat{i} - 2\hat{j})$
   (b) $\pi/2(-\hat{i} + 2\hat{j})$
   (c) $\pi/2(\hat{i} + 2\hat{j})$
   (d) $\pi/2(\hat{i} - 2\hat{j})$

11. A monoclinic unit cell of 1st setting has the following parameters: a = 5Å, b = 10Å, c = 15Å, $\alpha = \beta = 90°$ and $\gamma = 120°$. The corresponding reciprocal lattice parameters in the unit of (Å^{-1}) are:
    (a) 0.23, 0.12, 0.08
    (b) 0.20, 0.12, 0.08
    (c) 0.23, 0.10, 0.08
    (d) 0.23, 0.12, 0.07

12. A monoclinic unit cell of 1st setting has the following parameters: a = 5Å, b = 10Å, c = 15Å, $\alpha = \beta = 90°$ and $\gamma = 120°$. The reciprocal cell volume is:
    (a) 1.33×10^{-3} Å^3
    (b) 1.43×10^{-3} Å^3
    (c) 1.53×10^{-3} Å^3
    (d) 1.63×10^{-3} Å^3

13. The primitive translation vectors of a simple hexagonal space lattice may be taken as

$$A = \left(\frac{a}{2}\hat{i} + \frac{\sqrt{3}a}{2}\hat{j}\right), \quad B = \left(-\frac{a}{2}\hat{i} + \frac{\sqrt{3}a}{2}\hat{j}\right) \text{ and } C = c\,\hat{k}$$

where $\hat{i}, \hat{j}$ and $\hat{k}$ are unit vectors. The reciprocal unit cell volume is:

    (a) $\dfrac{(2\pi)^3}{\sqrt{3}a^2c}$

    (b) $\dfrac{2(2\pi)^3}{\sqrt{3}a^2c}$

    (c) $\dfrac{3(2\pi)^3}{\sqrt{3}a^2c}$

    (d) $\dfrac{4(2\pi)^3}{\sqrt{3}a^2c}$

14. The reciprocal lattice to a simple cubic (sc) lattice is:
    (a) sc
    (b) bcc
    (c) fcc
    (d) Dc

15. The reciprocal lattice to a body-centered cubic (bcc) lattice is:
    (a) sc
    (b) bcc
    (c) fcc
    (d) Dc

16. The reciprocal lattice to a face-centered cubic (fcc) lattice is:
    (a) sc
    (b) bcc
    (c) fcc
    (d) Dc

17. The reciprocal lattice vector G (hkl) to a crystal plane (hkl) is:
    (a) parallel
    (b) normal
    (c) diagonal
    (d) tortional

18. In a Brillouin Zone (B-Z) constructed from a given plane hexagonal lattice with
    a = b and $\gamma = 120°$, the shortest distance of the zone face from the center of the
    zone is:
    (a) $0.155\left(\frac{\pi}{a}\right)$
    (b) $1.155\left(\frac{\pi}{a}\right)$
    (c) $2.155\left(\frac{\pi}{a}\right)$
    (d) $3.155\left(\frac{\pi}{a}\right)$

19. In a B-Z constructed from a given simple cubic lattice with a = b = c and
    $\alpha = \beta = \gamma = 90°$, the six equivalent zone faces from the center of the zone are
    at a distance:
    (a) (π/a)
    (b) $2(\pi/a)$
    (c) $\sqrt{2}(\pi/a)$
    (d) $(\pi/2a)$

20. In a B-Z constructed from a given bcc lattice with a = b = c and $\alpha = \beta = \gamma =
    90°$, the twelve equivalent zone faces from the center of the zone are at a
    distance:
    (a) (π/a)
    (b) $\sqrt{2}(\pi/a)$
    (c) $\sqrt{3}(\pi/a)$
    (d) $2(\pi/a)$

21. In a B-Z constructed from a given fcc lattice with a = b = c and $\alpha = \beta = \gamma = 90°$. the eight hexagonal equivalent zone faces from the center of the zone are at a distance:
    (a) (π/a)
    (b) $\sqrt{2}(\pi/a)$
    (c) $\sqrt{3}(\pi/a)$
    (d) $2(\pi/a)$

22. In a B-Z constructed from a given fcc lattice with a = b = c and $\alpha = \beta = \gamma = 90°$. the six square equivalent zone faces from the center of the zone are at a distance:
    (a) (π/a)
    (b) $\sqrt{2}(\pi/a)$
    (c) $\sqrt{3}(\pi/a)$
    (d) $2(\pi/a)$

23. In a B-Z constructed from a given simple hexagonal lattice with a = b ≠ c and $\alpha = \beta = 90°$ and $\gamma = 120°$, the six vertical planes from the center of the zone are at a distance:
    (a) $\left(\frac{\pi}{a}\right)$
    (b) $\left(\frac{2\pi}{a}\right)$
    (c) $\frac{1}{\sqrt{3}}\left(\frac{\pi}{a}\right)$
    (d) $\frac{2}{\sqrt{3}}\left(\frac{\pi}{a}\right)$

24. In a square lattice, the distance of a corner of the first BZ from the center is:
    (a) (π/a)
    (b) $\sqrt{2}(\pi/a)$
    (c) $\sqrt{3}(\pi/a)$
    (d) $2(\pi/a)$

25. The radius of the Fermi circle for a monovalent metal is:
    (a) $0.798\left(\frac{\pi}{a}\right)$
    (b) $1.228\left(\frac{\pi}{a}\right)$
    (c) $1.382\left(\frac{\pi}{a}\right)$
    (d) $1.596\left(\frac{\pi}{a}\right)$

26. The radius of the Fermi circle for a divalent metal is:
    (a) $0.798\left(\frac{\pi}{a}\right)$
    (b) $1.228\left(\frac{\pi}{a}\right)$
    (c) $1.382\left(\frac{\pi}{a}\right)$
    (d) $1.596\left(\frac{\pi}{a}\right)$

27. The radius of the Fermi circle for a trivalent metal is:
    (a) $0.798\left(\frac{\pi}{a}\right)$
    (b) $1.228\left(\frac{\pi}{a}\right)$
    (c) $1.382\left(\frac{\pi}{a}\right)$
    (d) $1.596\left(\frac{\pi}{a}\right)$

28. The radius of the Fermi circle for a tetravalent metal is:
    (a) $0.798\left(\frac{\pi}{a}\right)$
    (b) $1.228\left(\frac{\pi}{a}\right)$
    (c) $1.382\left(\frac{\pi}{a}\right)$
    (d) $1.596\left(\frac{\pi}{a}\right)$

29. The radius of the Fermi sphere for a body-centered cubic crystal of side "a" is:
    (a) $\left(\frac{\pi}{a}\right)$
    (b) $1.120\left(\frac{\pi}{a}\right)$
    (c) $1.240\left(\frac{\pi}{a}\right)$
    (d) $1.563\left(\frac{\pi}{a}\right)$

30. Distance covered by the Fermi sphere of a body-centered cubic crystal of side' a' towards the center of the zone face is:
    (a) 68%
    (b) 78%
    (c) 88%
    (d) 98%

31. The radius of the Fermi sphere for a body-centered cubic crystal of side "a" is separated from the zone boundary by a distance of:
    (a) $0.174\left(\frac{\pi}{a}\right)$
    (b) $1.174\left(\frac{\pi}{a}\right)$
    (c) $2.174\left(\frac{\pi}{a}\right)$
    (d) $3.174\left(\frac{\pi}{a}\right)$

32. The radius of the Fermi sphere for a face-centered cubic crystal of side "a" is:
    (a) $\left(\frac{\pi}{a}\right)$
    (b) $1.120\left(\frac{\pi}{a}\right)$
    (c) $1.240\left(\frac{\pi}{a}\right)$
    (d) $1.563\left(\frac{\pi}{a}\right)$

33. Distance covered by the Fermi sphere of a face-centered cubic crystal of side'
    a' towards the center of the zone face is:
    (a) 90%
    (b) 80%
    (c) 70%
    (d) 60%

34. The radius of the Fermi sphere for a face-centered cubic crystal of side "a" is
    separated from the zone boundary by a distance of:
    (a) $3.169\left(\frac{\pi}{a}\right)$
    (b) $2.169\left(\frac{\pi}{a}\right)$
    (c) $1.169\left(\frac{\pi}{a}\right)$
    (d) $0.169\left(\frac{\pi}{a}\right)$

35. The radius of the Fermi sphere of bcc tetragonal crystal with $(c/a) = 2$ is:
    (a) 0.985π
    (b) 0.885π
    (c) 0.785π
    (d) 0.685π

Answers

1. (c)
2. (d)
3. (c)
4. (d)
5. (b)
6. (a)
7. (d)
8. (a)
9. (a)
10. (b)
11. (a)
12. (c)
13. (b)
14. (a)
15. (c)
16. (b)
17. (b)
18. (a)
19. (a)
20. (b)
21. (c)
22. (d)

23. (d)
24. (b)
25. (a)
26. (b)
27. (c)
28. (d)
29. (c)
30. (c)
31. (a)
32. (d)
33. (a)
34. (d)
35. (a)

Chapter 3
Unit Cell Calculations

3.1 Fractional Coordinates

The location of any point within a unit cell (oblique or orthogonal) by means of three coordinates (x, y, z) and three basis vectors $\vec{a}, \vec{b}, \vec{c}$ may be specified in terms of position vector $\vec{r}$ in 2-D and 3-D as

$$\vec{r} = x\vec{a} + y\vec{b} \tag{3.1}$$

$$\vec{r} = x\vec{a} + y\vec{b} + z\vec{c} \tag{3.2}$$

They are shown in Fig. 3.1. The coordinates x, y, z being fractional, they are unitless and the basis vectors $\vec{a}, \vec{b}, \vec{c}$ are usually measured in Å.

When an integer is added or subtracted from a given fractional coordinate, an equivalent (point) position is obtained in a neighboring unit cell. For example, a point with fractional coordinates (0.30, 0.25, 0.15) in the reference unit cell will have an equivalent position (1.30, 0.25, 0.15) in the unit cell just on right and (−0.70, 0.25, 0.15) in the unit cell just on left. Similar equivalent positions can be obtained by changing the y and z fractional coordinates. These are the example of translational symmetry.

The fractional coordinates of equivalent positions can also be obtained by applying a rotational, mirror, inversion or a compatible combination of these symmetries. For example, a twofold rotation along the z-axis will change the fractional coordinates (x, y, z) into $(\bar{x}, \bar{y}, z)$, a mirror plane perpendicular to the z-axis will change (x, y, z) into $(x, y, \bar{z})$ and similarly an inversion symmetry through the reference point will change (x, y, z) into $(\bar{x}, \bar{y}, \bar{z})$, respectively, as shown in Fig. 3.2.

The number of equivalent positions is found to increase as we move from low symmetry crystal system (e.g., triclinic) to high symmetry crystal system (e.g., cubic). As a result, the determination of fractional coordinates in higher symmetry

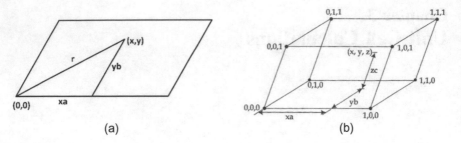

Fig. 3.1 Representation of position vectors in 2-D and 3-D

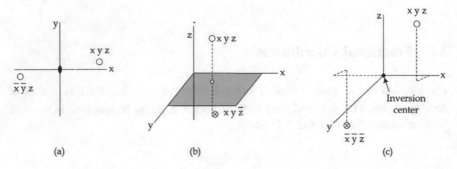

Fig. 3.2 Fractional coordinates corresponding to **a** twofold rotation, **b** mirror plane, and **c** inversion symmetry operations

crystal systems becomes increasingly cumbersome. However, they can be worked out by adopting the following procedure.

1. Select a coordinate system of axes appropriate to the crystal system, for example, orthogonal or crystallographic.
2. Consider a point (consisting of asymmetric group of atom/atoms), define its coordinates and then apply the given symmetries one by one to explore other possible equivalent positions.
3. Each symmetry operation transforms the axes and gives rise the corresponding matrix, with the help of which one can successively generate the fractional coordinates of other equivalent positions under a given point group.
4. Total number of equivalent positions (N) for a given point group is the product of number of equivalent points (N_{Ep}) of each independent symmetry element/ operation with the number of independent symmetry elements/operations (N_S) in the point group, that is,

$$N = N_{Ep} \times N_S$$

For example, the independent symmetry elements of the point group mm2 are two mirror planes m, m and the number of equivalent points associated with each mirror plane is 2. Therefore, the number of equivalent positions for the point group mm2 is $2 \times 2 = 4$.

A list of symmetry matrix and fractional coordinates of equivalent positions is provided for 2-D point groups in Table 3.1.

Making use of the concept of equivalent points, crystal structures can be entirely specified with fractional coordinates.

Solved Examples

Example 1 In a unit cell of plane hexagonal lattice with $a = b = 10$ Å and $\gamma = 120°$, an atom has its fractional coordinates $x = 0.5$ and $y = 0.5$. Draw four unit cells on its edges and write their fractional coordinates.

Solution: *Given*: $a = b = 10$ Å, $\gamma = 120°$, fractional coordinates of an atom in the given unit cell: $x = 0.5$ and $y = 0.5$.

Table. 3.1 Symmetry matrix and fractional coordinates of equivalent points in 2-D

S. No	Symmetry operation	No. of equivalent points	Symmetry matrix	Fractional coordinates
1	Identity	1	$\begin{pmatrix} 1 & 0 \\ 0 & 1 \end{pmatrix}$	(x, y)
2	2-fold rotation (∥ to z-axis)	2	$\begin{pmatrix} \bar{1} & 0 \\ 0 & \bar{1} \end{pmatrix}$	(x, y), ($\bar{x}$, $\bar{y}$)
3	Mirror line (∥ to y or x-axis)	2	$\begin{pmatrix} \bar{1} & 0 \\ 0 & 1 \end{pmatrix}$	(x, y), ($\bar{x}$, y) or (x, y), (x, $\bar{y}$)
4	4-fold rotation	4	$\begin{pmatrix} 0 & \bar{1} \\ 1 & 0 \end{pmatrix}$	(x, y), ($\bar{y}$, x), ($\bar{x}$, $\bar{y}$), (y, $\bar{x}$)
5	3-fold rotation	3	$\begin{pmatrix} 0 & \bar{1} \\ 1 & \bar{1} \end{pmatrix}$	(x, y), ($\bar{y}$, x-y), (y-x, $\bar{x}$)
6	6-fold rotation	6	$\begin{pmatrix} 1 & \bar{1} \\ 1 & 0 \end{pmatrix}$	(x, y), (x-y, x), ($\bar{y}$, x-y), ($\bar{x}$, $\bar{y}$), (y-x, $\bar{x}$), (y, y-x)
7	mm2	4	$\begin{pmatrix} 1 & 0 \\ 0 & \bar{1} \end{pmatrix}$	(x, y), ($\bar{x}$, y), ($\bar{x}$, $\bar{y}$), (x, $\bar{y}$)
8	4 mm	8	$\begin{pmatrix} 0 & \bar{1} \\ \bar{1} & 0 \end{pmatrix}$	(x, y), ($\bar{y}$, x), ($\bar{x}$, $\bar{y}$), (y, $\bar{x}$), (x, $\bar{y}$), ($\bar{x}$, y), (y, x), ($\bar{y}$, $\bar{x}$)
9	3 m (3m1)	6	$\begin{pmatrix} 1 & 0 \\ 1 & \bar{1} \end{pmatrix}$	(x, y), ($\bar{y}$, x-y), (y-x, $\bar{x}$), ($\bar{y}$, $\bar{x}$), (y-x, $\bar{x}$), (x, x-y)
10	6 mm	12	$\begin{pmatrix} 1 & \bar{1} \\ 0 & \bar{1} \end{pmatrix}$	(x, y), (x-y, x), ($\bar{y}$, x-y), ($\bar{x}$, $\bar{y}$), (y-x, $\bar{x}$), (y, y-x), (x-y, $\bar{y}$), ($\bar{x}$, y-x), (y, x), ($\bar{y}$, $\bar{x}$), (x, x-y), (y-x, y)

Now, draw a unit cell (A) as per the given lattice parameters. The position coordinates of the atom within the reference cell are:

$$xa = 0.5a = 0.5 \times 10 = 5\text{Å (length along the basis vector a)}$$
$$yb = 0.5b = 0.5 \times 10\text{Å} = 5\text{Å (length along the basis vector b)}$$

Locate the position of the atom in the unit cell (A). Now, draw other four unit cells on its edges as shown in Fig. 3.3. They are named as B, C, D and E, counterclockwise. Now, following the principle of addition or subtraction of an integer from the given fractional coordinates, we can obtain the equivalent points of the neighboring unit cells. They are:

When an integer is added in x, the fractional coordinates of the atom in the unit cell B are (1.5, 0.5). Similarly, the fractional coordinates of an equivalent atom in the unit cells C, D and E, respectively, are: (0.5, 1.5), (−0.5, 0.5) and (0.5, −0.5). In a similar manner, the fractional coordinates in other plane lattices can be obtained.

Example 2 In a Triclinic unit cell with a = 5 Å, b = 6 Å, c = 8 Å and α, β, γ not equal to 90°, an atom has its fractional coordinates x = 0.6 and y = 0.5 and z = 0.25. Draw six unit cells on its faces and write their fractional coordinates.

Solution: *Given*: a = 5 Å, b = 6 Å, c = 8 Å and α, β, γ not equal to 90°, fractional coordinates of an atom in the given unit cell: x = 0.6 and y = 0.5 and z = 0.25.

Therefore, the position coordinates of the atom within the reference unit cell are:

$$xa = 0.6a = 0.6 \times 5\text{Å} = 3\text{Å (length along the basis vector a)}$$
$$yb = 0.5b = 0.5 \times 6\text{Å} = 3\text{Å (length along the basis vector b)}$$
$$zc = 0.25b = 0.25 \times 8\text{Å} = 2\text{Å (length along the basis vector c)}$$

Locate the position of the atom in the unit cell (A). Now, draw other six unit cells on its edges and obtain the equivalent points based on the principle of addition or subtraction of an integer from the given fractional coordinates. They are (in the

Fig. 3.3 Position of the
atoms in the five unit cells

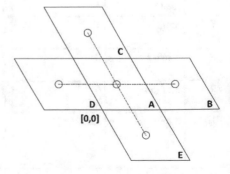

unit cells on the positive x, y, z axes and the unit cells on the negative x, y, z axes, respectively):

$$(1.6, 0.5, 0.25), (0.6, 1.5, 0.25), (0.6, 0.5, 1.25)$$

$$\text{and } (-0.4, 0.5, 0.25), (0.6, -0.5, 0.25), (0.6, 0.5, -0.75)$$

In a similar manner, the fractional coordinates in other 3-D lattices can be obtained.

Example 3 Determine the set of fractional coordinates of equivalent positions related to a 3-fold rotational symmetry in 2-D and 3-D.

Solution: *Given*: A 3-fold rotational symmetry (means 120° rotations).

Select crystallographic axes and consider a point with fractional coordinates (x, y, z) and then apply the given symmetry (Fig. 3.4).

The position vector of the point at (x, y, z) is

$$\vec{r} = x\vec{a} + y\vec{b} + z\vec{c}$$

Application of 120° rotation changes the position vector to

$$\vec{r'} = x\vec{a'} + y\vec{b'} + z\vec{c'}$$

where a′ = b, b′ = −a−b and c′= c. Therefore,

$$r' = xb + y(-a - b) + zc$$
$$= -ya + (x - y)\ b + zc$$

Fig. 3.4 Fractional coordinates of three equivalent positions

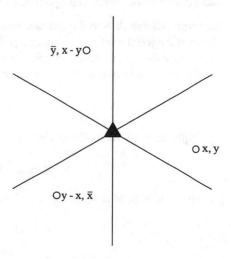

In the matrix notation, this becomes

$$\begin{pmatrix} a' \\ b' \\ c' \end{pmatrix} = \begin{pmatrix} 0 & -1 & 0 \\ 1 & -1 & 0 \\ 0 & 0 & 1 \end{pmatrix} \begin{pmatrix} a \\ b \\ c \end{pmatrix}$$

The corresponding fractional coordinates of the second point in 3-D are: $(\bar{y}, x-y, z)$.

However in 2-D, the matrix is $\begin{pmatrix} 0 & -1 \\ 1 & -1 \end{pmatrix}$ and the corresponding fractional coordinates of the second point are: $(\bar{y}, x-y)$. In order to find the third point, let us consider the product of two such matrices. This will provide us

$$\begin{pmatrix} 0 & -1 \\ 1 & -1 \end{pmatrix} \begin{pmatrix} 0 & -1 \\ 1 & -1 \end{pmatrix} = \begin{pmatrix} -1 & 1 \\ -1 & 0 \end{pmatrix}$$

Hence, the fractional coordinates of the third point are: $(y-x, \bar{x})$. Thus, the fractional coordinates of three equivalent positions in 2-D are:

$$(x, y), (\bar{y}, x - y), (y - x, \bar{x}), (z = 0).$$

They are shown in Fig. 3.4.

Similarly, the fractional coordinates of the three equivalent positions in 3-D are:

$$(x, y, z), (\bar{y}, x - y, z), (y - x, \bar{x}, z).$$

Example 4 Determine the set of fractional coordinates of equivalent positions related to a 4-fold rotational symmetry in 2-D and 3-D.

Solution: *Given*: A 4-fold rotational symmetry (means 90° rotations).
Select orthogonal axes and consider a point with fractional coordinates (x, y, z) and then apply the given symmetry (Fig. 3.5).

The position vector of the point at (x, y, z) is

$$\vec{r} = x\vec{a} + y\vec{b} + z\vec{c}$$

Application of 90° rotation changes the position vector to

$$\vec{r'} = x\vec{a'} + y\vec{b'} + z\vec{c'}$$

where $a' = b, b' = -a$ and $c' = c$. Therefore,

$$r' = xb - ya + zc$$
$$= -ya + xb + zc$$

Fig. 3.5 Fractional
coordinates of four equivalent
positions

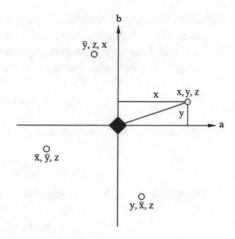

In the matrix notation, this becomes

$$\begin{pmatrix} a' \\ b' \\ c' \end{pmatrix} = \begin{pmatrix} 0 & -1 & 0 \\ 1 & 0 & 0 \\ 0 & 0 & 1 \end{pmatrix} \begin{pmatrix} a \\ b \\ c \end{pmatrix}$$

The corresponding fractional coordinates of the second point in 3-D are: $(\bar{y}, x, z)$.

However in 2-D, the matrix is $\begin{pmatrix} 0 & -1 \\ 1 & 0 \end{pmatrix}$ and the corresponding fractional

coordinates of the second point are: $(\bar{y}, x)$. Further, considering the following products of two matrices, we can obtain

$$\begin{pmatrix} 0 & -1 \\ 1 & 0 \end{pmatrix}\begin{pmatrix} 0 & -1 \\ 1 & 0 \end{pmatrix} = \begin{pmatrix} -1 & 0 \\ 0 & -1 \end{pmatrix} \equiv (\bar{x}, \bar{y})$$

and $\quad \begin{pmatrix} 0 & -1 \\ 1 & 0 \end{pmatrix}\begin{pmatrix} -1 & 0 \\ 0 & -1 \end{pmatrix} = \begin{pmatrix} 0 & 1 \\ -1 & 0 \end{pmatrix} \equiv (y, \bar{x})$

Hence, the fractional coordinates in 2-D are:

$$(x, y), (\bar{y}, x), (\bar{x}, \bar{y}), (y, \bar{x}).$$

Similarly, the fractional coordinates in 3-D are:

$$(x, y, z), (\bar{y}, z, x), (\bar{x}, \bar{y}, z), (y, \bar{x}, z).$$

They are shown in Fig. 3.5.

Example 5 Determine the set of fractional coordinates of equivalent positions related to a 6-fold rotational symmetry in 2-D and 3-D.

Solution: *Given*: A 6-fold rotational symmetry (means 60° rotations).

Select crystallographic axes and consider a point with fractional coordinates (x, y, z) and then apply the given symmetry (Fig. 3.6).

The position vector of the point at (x, y, z) is

$$\vec{r} = x\vec{a} + y\vec{b} + z\vec{c}$$

Application of 60° rotation changes the position vector to

$$\vec{r'} = x\vec{a'} + y\vec{b'} + z\vec{c'}$$

where a' = a + b, b' = –a and c' = c. Therefore,

$$r' = x(a+b) - ya + zc$$
$$= (x - y)a + xb + zc$$

In the matrix notation, this becomes

$$\begin{pmatrix} a' \\ b' \\ c' \end{pmatrix} = \begin{pmatrix} 1 & -1 & 0 \\ 1 & 0 & 0 \\ 0 & 0 & 1 \end{pmatrix} \begin{pmatrix} a \\ b \\ c \end{pmatrix}$$

The corresponding fractional coordinates of the second point in 3-D are: (x–y, x, z).

However in 2-D, the matrix is $\begin{pmatrix} 1 & -1 \\ 1 & 0 \end{pmatrix}$ and the corresponding fractional coordinates of the second point are: (x–y, x). In order to find the third point, let us consider the product of two such matrices. This will provide us

Fig. 3.6 Fractional coordinates of six equivalent positions

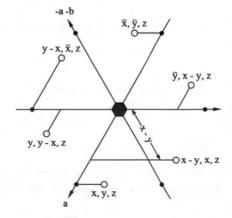

$$\begin{pmatrix} 1 & -1 \\ 1 & 0 \end{pmatrix} \begin{pmatrix} 1 & -1 \\ 1 & 0 \end{pmatrix} = \begin{pmatrix} 0 & -1 \\ 1 & -1 \end{pmatrix}$$

Hence, the fractional coordinates of the third point are: $(\bar{y}, x{-}y)$. Thus, the fractional coordinates of three equivalent positions in 2-D are: (x, y), $(x{-}y, x)$, $(\bar{y}, x{-}y)$. Similarly, considering the product of proper matrices other three fractional coordinates can also be obtained (actually they are simply inverses of the earlier three). Therefore all the six fractional coordinates in 2-D are:

$$(x, y), (x - y, x), (\bar{y}, x - y), (\bar{x}, \bar{y}), (y - x, \bar{x}), (y, y - x).$$

Similarly in 3-D, they are:

$$(x, y, z), (x - y, x, z), (\bar{y}, x - y, z), (\bar{x}, \bar{y}, z), (y - x, \bar{x}, z), (y, y - x, z).$$

3.2 Distance Between Two Lattice Points (Oblique System)

(a) **In Two Dimensions**

Let us consider an axial system OX and OY where the angle XOY = γ as shown in Fig. 3.7. Within the oblique unit cell, consider two points P (x_1, y_1) and Q (x_2, y_2) whose distance is to be determined.

Draw the lines PM and QN parallel to OY axis and PT parallel to OX axis, respectively. From this construction, we have

$$OM = ax_1, PM = by_1, ON = ax_2, QN = by_2$$

Now, PT = MN = a $(x_2{-}x_1)$ and QT = QN{-}TN = QN{-}PM = b $(y_2{-}y_1)$. Further, in the triangle PQT, the angle < PQT = $180°{-}\gamma$. Therefore, we can write

Fig. 3.7 Distance between two lattice points

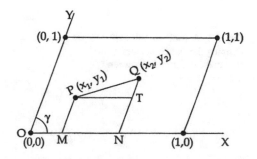

$$(PQ)^2 = (PT)^2 + (QT)^2 + 2\,PT.QT\,\cos\gamma$$
$$= (x_2 - x_1)^2 a^2 + (y_2 - y_1)^2 b^2 + 2(x_2 - x_1)(y_2 - y_1)\,ab\,\cos\gamma$$
$$\text{or}\quad PQ = \left[(x_2 - x_1)^2 a^2 + (y_2 - y_1)^2 b^2 + 2(x_2 - x_1)(y_2 - y_1)\,ab\,\cos\gamma\right]^{1/2}$$
$$(3.3)$$

This is a general equation expressing the distance between two lattice points in an oblique system. Equations for other lattices can be obtained by substituting the respective axial parameters (i.e., axes and the angle between them) in Eq. 3.3.

(b) In Three Dimensions

Let us consider Fig. 3.8, which illustrates the location of a point with coordinates (x, y, z) within a triclinic unit cell. Also, the coordinate vector $\vec{r}$ between the origin and the point (x, y, z) defined as

$$\vec{r} = xa\,\hat{i} + yb\,\hat{j} + zc\,\hat{k} \tag{3.4}$$

Then two similar points P (x_1, y_1, z_1) and Q (x_2, y_2, z_2) will produce the coordinate vectors as

$$\vec{r_1} = x_1 a\,\hat{i} + y_1 b\,\hat{j} + z_1 c\,\hat{k}$$
$$\text{and}\quad \vec{r_2} = x_2 a\,\hat{i} + y_2 b\,\hat{j} + z_2 c\,\hat{k} \tag{3.5}$$

Therefore, the distance between two coordinate vectors, d_{21} is

$$\vec{d_{21}} = \vec{r_2} - \vec{r_1} = \hat{i}(x_2 - x_1)a + \hat{j}(y_2 - y_1)b + \hat{k}(z_2 - z_1)c \tag{3.6}$$

Further, in terms of dot product, we can write

Fig. 3.8 A triclinic unit cell defining the lattice vectors a, b, c and the distance between the two atoms

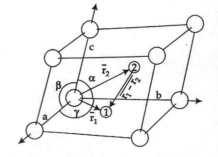

$$(r_2 - r_1).(r_2 - r_1) = \left[\hat{i}(x_2 - x_1)a + \hat{j}(y_2 - y_1)b + \hat{k}(z_2 - z_1)c \right].$$
$$\left[\hat{i}(x_2 - x_1)a + \hat{j}(y_2 - y_1)b + \hat{k}(z_2 - z_1)c \right]$$

$$\text{or } |r_2 - r_1|^2 = (x_2 - x_1)^2 a^2 + (y_2 - y_1)^2 b^2 + (z_2 - z_1)^2 c^2$$
$$+ 2ab \, \hat{i}.\hat{j}(x_2 - x_1)(y_2 - y_1)$$
$$+ 2bc \, \hat{j}.\hat{k}(y_2 - y_1)(z_2 - z_1)$$
$$+ 2ca \, \hat{k}.\hat{i}(z_2 - z_1)(x_2 - x_1)$$

(3.7)

where $\hat{i}.\hat{j} = \cos \gamma$, $\hat{j}.\hat{k} = \cos \alpha$ and $\hat{k}.\hat{i} = \cos \beta$. Substituting these values in Eq. 3.7, we obtain

$$|r_2 - r_1| = \left[\begin{array}{c} (x_2 - x_1)^2 a^2 + (y_2 - y_1)^2 b^2 + (z_2 - z_1)^2 c^2 + \\ 2ab(x_2 - x_1)(y_2 - y_1) \cos \gamma + \\ 2bc(y_2 - y_1)(z_2 - z_1) \cos \alpha + \\ 2ca(z_2 - z_1)(x_2 - x_1) \cos \beta \end{array} \right]^{1/2}$$

(3.8)

This is a general equation valid for triclinic unit cell. The equations for other unit cells can be obtained by substituting the respective axial parameters (i.e., axes and angle between them) in Eq. 3.8.

Solved Examples

Example 1 A unit cell has the following dimensions: a = 3 Å, b = 4 Å and $\gamma = 115°$. Calculate the distance between the points with fractional coordinates: (i) 0.210, 0.150 and 0.410, 0.050, (ii) 0.210, −0.150 and 0.410. 0.050.

Solution: *Given*: a = 3 Å, b = 4 Å and $\gamma = 115°$. Fractional coordinates: (i) 0.210, 0.150 and 0.410, 0.050, (ii) 0.210, −0.150 and 0.410. 0.050.

Making use of Eq. 3.3, we can obtain the distance between the first set of fractional coordinates as

$$L_1 = \left[\begin{array}{c} (0.410 - 0.210)^2 \times 3^2 + (0.050 - 0.150)^2 \times 4^2 + \\ 2(0.410 - 0.210)(0.050 - 0.150) \times 3 \times 4 \cos 115° \end{array} \right]^{1/2}$$
$$= \left[(0.2)^2 \times 9 + (-0.1)^2 \times 16 + 2(0.2)(-0.1) \times 12 \times (-0.423) \right]^{1/2}$$
$$= [0.04 \times 9 + 0.01 \times 16 + 2(-0.02) \times 12 \times (-0.423)]^{1/2}$$
$$= [0.36 + 0.16 + 24(-0.02)(-0.423)]^{1/2}$$
$$= [0.723]^{1/2} = 0.850 \text{Å}$$

Similarly, for the second set of fractional coordinates, we have

$$L_2 = \left[\begin{array}{c} (0.410 - 0.210)^2 \times 3^2 + (0.050 + 0.150)^2 \times 4^2 + \\ 2(0.410 - 0.210)(0.050 + 0.150) \times 3 \times 4cos115° \end{array} \right]^{1/2}$$

$$= \left[(0.2)^2 \times 9 + (0.2)^2 \times 16 + 2(0.2)(-0.1) \times 12 \times (-0.423) \right]^{1/2}$$

$$= [0.04 \times 9 + 0.04 \times 16 + 24 \times (0.04) \times (-0.423)]^{1/2}$$

$$= [0.36 + 0.64 - 0.406]^{1/2}$$

$$= [0.594]^{1/2} = 0.770 \, Å$$

Example 2 A unit cell has the following dimensions: a = 5 Å, b = 8 Å and γ = 50°. Calculate the length of the vectors with the help of following components: (i) [0.50, 1.0] and (ii) [0.75, −0.25].

Solution: *Given*: a = 5 Å, b = 8 Å and γ = 50°. Vector components: (i) $(x_2 - x_1) = 0.50$, $(y_2 - y_1) = 1.00$. (ii) $(x_2 - x_1) = 0.75$, $(y_2 - y_1) = -0.25$. Making use of Eq. 3.3 in slightly modified form, we can obtain the length of the vectors for the given components as:

$$U = \left[(u_1a)^2 + (u_2b)^2 + 2u_1u_2ab \cos50° \right]^{1/2}$$

$$= \left[(5 \times 0.5)^2 + (8 \times 1)^2 + 2 \times 0.5 \times 1 \times 5 \times 8 \times 0.643 \right]^{1/2}$$

$$= [6.25 + 64 + 25.72]^{1/2} = [95.97]^{1/2} = 9.8 Å$$

Similarly,

$$V = \left[(v_1a)^2 + (v_2b)^2 + 2v_1v_2ab \cos50° \right]^{1/2}$$

$$= \left[(0.75 \times 5)^2 + (-0.25 \times 8)^2 + 2 \times 0.75 \times (-0.25) \times 5 \times 8 \times 0.643 \right]^{1/2}$$

$$= [14.06 + 4 - 9.645]^{1/2} = [8.42]^{1/2} = 2.9 \, Å$$

Example 3 A unit cell has the following dimensions: a = 2 Å, b = 3 Å and γ = 90°. Calculate the distance between the points with fractional coordinates: (i) 0.100, 0.250 and 0.200, 0.050, (ii) 0.100, −0.250 and 0.210. 0.050.

Solution: *Given*: a = 2 Å, b = 3 Å and γ = 90°. Fractional coordinates: (i) 0.100, 0.250 and 0.200, 0.050, (ii) 0.100, −0.250 and 0.210. 0.050.

Making use of Eq. 3.3 for $\gamma = 90°$, we can obtain the distance between the first set of fractional coordinates as

$$L_1 = \left[(0.200 - 0.100)^2 \times 2^2 + (0.050 - 0.250)^2 \times 3^2\right]^{1/2}$$

$$= \left[(0.100)^2 \times 4 + (-0.200)^2 \times 9\right]^{1/2}$$

$$= [(0.010) \times 4 + (0.040) \times 9]^{1/2} = 0.632\ \text{Å}$$

Similarly, for the second set of fractional coordinates, we have

$$L_2 = \left[(0.210 - 0.100)^2 \times 2^2 + (0.050 + 0.250)^2 \times 3^2\right]^{1/2}$$

$$= \left[(0.110)^2 \times 4 + (0.300)^2 \times 9\right]^{1/2}$$

$$= [(0.0121) \times 4 + (0.090) \times 9]^{1/2} = 0.926\text{Å}$$

Example 4 A unit cell has the following dimensions: a = 6 Å, b = 7 Å, c = 8 Å, $\alpha = 90°$, $\beta = 115°$ and $\gamma = 90°$. Calculate the distance between the points with fractional coordinates: (i) 0.200, 0.150, 0.333 and 0.300, 0.050, 0.123, and (ii) 0.200, 0.150, 0.333 and 0.300, 0.050, −0.123.

Solution: *Given*: a = 6 Å, b = 7 Å, c = 8 Å, $\alpha = 90°$, $\beta = 115°$ and $\gamma = 90°$. Fractional coordinates: (i) 0.200, 0.150, 0.333 and 0.300, 0.050, 0.123, and (ii) 0.200, 0.150, 0.333 and 0.300, 0.050, −0.123.

Making use of Eq. 3.8, we can obtain the distance between the first set of fractional coordinates as

$$L_1 = \left[\begin{array}{c}(0.200 - 0.300)^2 \times 6^2 + (0.150 - 0.050)^2 \times 7^2 + (0.333 - 0.123)^2 \times 8^2 + \\ 0 + 0 + 2 \times 8 \times 6(0.333 - 0.123)(0.200 - 0.300)\cos 115°\end{array}\right]^{1/2}$$

$$= \left[(-0.100)^2 \times 36 + (0.100)^2 \times 49 + (0.210)^2 \times 64 + 96(0.210)(-0.100)(-0.423)\right]^{1/2}$$

$$= [0.010 \times 36 + 0.010 \times 49 + 0.044 \times 64 + 96(0.210)(-0.100)(-0.423)]^{1/2}$$

$$= [0.360 + 0.490 + 2.820 + 0.853]^{1/2}$$

$$= [4.523]^{1/2} = 2.13\text{Å}$$

Similarly, for the second set of fractional coordinates, we have

$$L_2 = \left[\begin{array}{c} (0.200 - 0.300)^2 \times 6^2 + (0.150 - 0.050)^2 \times 7^2 + (0.333 + 0.123)^2 \times 8^2 + \\ 0 + 0 + 2 \times 8 \times 6(0.333 + 0.123)(0.200 - 0.300)\cos 115° \end{array} \right]^{1/2}$$

$$= \left[(-0.100)^2 \times 36 + (0.100)^2 \times 49 + (0.456)^2 \times 64 + 96(0.456)(-0.100)(-0.423) \right]^{1/2}$$

$$= [0.100 \times 36 + 0.010 \times 49 + 0.208 \times 64 + 1.852]^{1/2}$$

$$= [0.360 + 0.490 + 13.308 + 1.852]^{1/2}$$

$$= [16.01]^{1/2} = 4.00 \text{Å}$$

Example 5   A unit cell has the following dimensions: a = 5 Å, b = 7 Å, c = 9 Å, $\alpha = 120°$, $\beta = 85°$ and $\gamma = 75°$. Calculate the length of the vectors with the help of following components: (i) [0.2, −0.3, 0.6] and (ii) [0.5, 0.7, −0.1].

Solution: *Given*: a = 5 Å, b = 7 Å, c = 9 Å, $\alpha = 120°$, $\beta = 85°$ and $\gamma = 75°$. Vectors components: (i) $x_2 - x_1 = 0.2$, $y_2 - y_1 = -0.3$, $z_2 - z_1 = 0.6$, (ii) $x_2 - x_1 = 0.5$, $y_2 - y_1 = 0.7$, $z_2 - z_1 = -0.1$.

Making use of Eq. 3.8 in slightly modified form, we can obtain the length of the vectors for the given components as:

$$U = \left[(u_1 a)^2 + (u_2 b)^2 + (u_3 c)^2 + 2u_1 u_2 ab \cos\gamma + 2u_2 u_3 bc \cos\alpha + 2u_3 u_1 ca \cos\beta \right]^{1/2}$$

$$= \left[\begin{array}{c} (5 \times 0.2)^2 + (7 \times -0.3)^2 + (9 \times 0.6)^2 + 2 \times 0.2 \times -0.3 \times 5 \times 7 \cos 75° + \\ 2 \times -0.3 \times 0.6 \times 7 \times 9 \times \cos 120° + 2 \times 0.6 \times 0.2 \times 9 \times 5 \times \cos 85° \end{array} \right]^{1/2}$$

$$= \left[(1.0)^2 + (-2.1)^2 + (5.4)^2 - 4.2 \times 0.259 - 22.68 \times -0.5 + 10.8 \times 0.087 \right]^{1/2}$$

$$= [1 + 4.41 + 29.16 - 1.087 + 11.34 + 0.94]^{1/2}$$

$$= [45.764]^{1/2} = 6.76 \text{Å}$$

Similarly,

$$V = \left[\begin{array}{c} (5 \times 0.5)^2 + (7 \times 0.7)^2 + (9 \times -0.1)^2 + 2 \times 0.5 \times 0.7 \times 5 \times 7 \cos 75° + \\ 2 \times 0.7 \times -0.1 \times 7 \times 9 \times \cos 120° + 2 \times -0.1 \times 0.5 \times 9 \times 5 \times \cos 85° \end{array} \right]^{1/2}$$

$$= \left[(2.5)^2 + (4.9)^2 + (-0.9)^2 + 24.5 \times 0.259 - 8.82 \times -0.5 - 4.5 \times 0.087 \right]^{1/2}$$

$$= [6.25 + 24.01 + 0.81 + 6.34 + 4.41 - 0.39]^{1/2}$$

$$= [41.43]^{1/2} = 6.44 \text{Å}$$

3.3 Unit Cell Volume

In order to calculate the volume of a unit cell, let us make use of the vector algebra to express the basis vectors $[\vec{a}, \vec{b}$ and $\vec{c}]$ in terms of their components as

$$\vec{a} = a_x\widehat{i} + a_y\widehat{j} + a_z\widehat{k}$$
$$\vec{b} = b_x\widehat{i} + b_y\widehat{j} + b_z\widehat{k} \qquad (3.9)$$
$$\text{and } \vec{c} = c_x\widehat{i} + c_y\widehat{j} + c_z\widehat{k}$$

From this, the volume of the unit cell can be written as

$$V = \begin{vmatrix} a_x & a_y & a_z \\ b_x & b_y & b_z \\ c_x & c_y & c_z \end{vmatrix} \qquad (3.10)$$

From the properties of determinant, we know that the value of the determinant remains unchanged when its rows and columns are interchanged. Consequently, it follows that

$$V^2 = \begin{vmatrix} a_x & a_y & a_z \\ b_x & b_y & b_z \\ c_x & c_y & c_z \end{vmatrix} \times \begin{vmatrix} a_x & b_x & c_x \\ a_y & b_y & c_y \\ a_z & b_z & c_z \end{vmatrix} \qquad (3.11)$$

On multiplication, the right hand side of Eq. 3.11 becomes

$$V^2 = \begin{vmatrix} a_xa_x + a_ya_y + a_za_z & a_xb_x + a_yb_y + a_zb_z & a_xc_x + a_yc_y + a_zc_z \\ b_xa_x + b_ya_y + b_za_z & b_xb_x + b_yb_y + b_zb_z & b_xc_x + b_yc_y + b_zc_z \\ c_xa_x + c_ya_y + c_za_z & c_xb_x + c_yb_y + c_zb_z & c_xc_x + c_yc_y + c_zc_z \end{vmatrix} \qquad (3.12)$$

In the reduced form, Eq. 3.12 can be written as

$$V^2 = \begin{vmatrix} a.a & a.b & a.c \\ b.a & b.b & b.c \\ c.a & c.b & c.c \end{vmatrix} \qquad (3.13)$$

where $a.a = a^2$, $b.b = b^2$, $c.c = c^2$

$$a \cdot b = b \cdot a = ab\cos\gamma = ba\cos\gamma$$
$$b \cdot c = c \cdot b = bc\cos\alpha = cb\cos\alpha \qquad (3.14)$$
$$c \cdot a = a \cdot c = ca\cos\beta = ac\cos\beta$$

Table. 3.2 Unit cell volume of different lattice types

Lattice type	Volume
Cubic	a^3
Orthorhombic	abc
Tetragonal	a^2c
Hexagonal	$\frac{\sqrt{3}a^2c}{2}$
Rhombohedral	$a^3\sqrt{1-3\cos^2\alpha+2\cos^3\alpha}$
Monoclinic	abc sinβ
Triclinic	$abc\sqrt{1-\cos^2\alpha-\cos^2\beta-\cos^2\gamma+2\cos\alpha\cos\beta\cos\gamma}$

Substituting these values in Eq. 3.13, we obtain

$$V^2 = \begin{vmatrix} a^2 & ab\cos\gamma & ac\cos\beta \\ ba\cos\gamma & b^2 & bc\cos\alpha \\ ca\cos\beta & cb\cos\alpha & c^2 \end{vmatrix} \tag{3.15}$$

Further, simplifying the determinant, we obtain

$$V^2 = a^2b^2c^2\left(1-\cos^2\alpha-\cos^2\beta-\cos^2\gamma+2\cos\alpha\cos\beta\cos\gamma\right)$$
$$V = abc\left(1-\cos^2\alpha-\cos^2\beta-\cos^2\gamma+2\cos\alpha\cos\beta\cos\gamma\right)^{1/2} \tag{3.16}$$

This is a general equation expressing the volume of a triclinic unit cell. Equations for other unit cells can be obtained by substituting the values of respective axial parameters (axes and angles) in Eq. 3.16. Unit cell volumes of different lattice types are provided in Table 3.2.

3.4   Angle Between Two Crystallographic Directions

The scalar product (also called the dot product) of two vectors $\vec{a}$ and $\vec{b}$ denoted by $\vec{a}.\vec{b}$ is defined as a scalar quantity which is equal to the product of the magnitudes of two vectors and the cosine of angle between their directions (Fig. 3.9), that is,

$$\vec{a}.\vec{b} = |a||b|\cos\theta$$

$$\text{or} \quad \cos\theta = \frac{\vec{a}.\vec{b}}{|a||b|}$$

Here, $\vec{b}.\vec{a} = |b||a|\cos(-\theta) = |a||b|\cos\theta$

Fig. 3.9 Graphical
representation of dot product

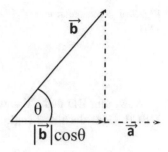

Therefore,

$$\vec{a}.\vec{b} = \vec{b}.\vec{a}$$

Following the principle of dot product, the general formula for angle between two crystallographic directions $[u_1v_1w_1]$ and $[u_2v_2w_2]$ for a given crystal system can be obtained. Formula for some of the common crystal systems are given in Table 3.3.

Solved Example

Example 1 Determine the area of a primitive rectangle whose sides are 4 Å and 3 Å, respectively. Construct another unit cell whose one side is the diagonal of the rectangle. Determine the length of the side and the angle made by this on the other side.

Solution: *Given*: a = 4 Å and b = 3 Å. For other unit cell, a′ = a and b′ = diagonal of rectangle. Also for a rectangle lattice, a ≠ b, and γ = 90°. Construct two unit cells side by side (Fig. 3.10).

Area of the rectangle ABCD = ab sin90° = 4 × 3 = 12 Å².

Table. 3.3 Angle between two crystallographic directions in more common crystal systems

Crystal system	$\cos\theta$
Cubic	$\dfrac{u_1u_2 + v_1v_2 + w_1w_2}{\sqrt{u_1^2 + v_1^2 + w_1^2}\sqrt{u_2^2 + v_2^2 + w_2^2}}$
Tetragonal	$\dfrac{a^2(u_1u_2 + v_1v_2) + c^2w_1w_2}{\sqrt{a^2(u_1^2 + v_1^2) + c^2w_1^2}\sqrt{a^2(u_2^2 + v_2^2) + c^2w_2^2}}$
Orthorhombic	$\dfrac{a^2u_1u_2 + b^2v_1v_2 + c^2w_1w_2}{\sqrt{a^2u_1^2 + b^2v_1^2 + c^2w_1^2}\sqrt{a^2u_2^2 + b^2v_2^2) + c^2w_2^2}}$
Hexagonal	$\dfrac{u_1u_2 + v_1v_2 - \frac{1}{2}(u_1v_2 + v_1u_2) + \frac{c^2}{a^2}w_1w_2}{\sqrt{u_1^2 + v_1^2 - u_1v_1 + \frac{c^2}{a^2}w_1^2}\sqrt{u_2^2 + v_2^2 - u_2v_2 + \frac{c^2}{a^2}w_2^2}}$
Rhombohedral	One can transform the rhombohedral indices into hexagonal indices and then can use the above equation

Fig. 3.10 Rectangular unit
cells

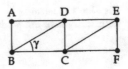

Now, join BD and CE and construct another unit cell. The required unit cell is
DBCE. Consider the triangle BCD, where the $< BCD = 90°$. Therefore,

$$BD = \left(4^2 + 3^2\right)^{1/2} = (25)^{1/2} = 5\text{Å}$$

Further, the angle made by this side on BC is

$$\gamma = \sin^{-1}\left(\frac{3}{5}\right) = 36.87°$$

Example 2 A primitive cell of square lattice has a = 2 Å. A new unit cell is
chosen with edges defined by the vectors from the origin to the points with coordinates 1, 0 and 0, 3. Calculate: (i) area of the original unit cell, (ii) length of the
two edges and angle between them, (iii) area of the new unit cell, and (iv) number
of lattice points in the new unit cell.

Solution: *Given*: a = 2 Å, for a square lattice, $\gamma = 90°$.

(i) Area of the original unit cell, $a^2 = 4\text{Å}^2$
(ii) Length of the edges from the origin:
    (a) For coordinates 0, 0 and 1, 0

$$a = \left[(0-1)^2 \times 2^2 + 0\right]^{1/2} = (4)^{1/2} = 2\text{Å}$$

    (b) For coordinates 0, 0 and 0, 3

$$b = \left[0 + (0-3)^2 \times 2^2\right]^{1/2} = (36)^{1/2} = 6\text{Å}$$

Angle between the lines with end coordinates 1, 0 and 0, 3:

$$\cos\gamma = \frac{0}{\sqrt{1}\sqrt{3}} = 0, \quad \Rightarrow \gamma = 90°$$

(iii) Area of the new unit cell, ab = 2 × 6 = 12 Å^2
(iv) Ratio of the two areas = $\frac{12}{4} = 3$. Therefore, the number of lattice points in
    the new unit cell is 3

Example 3   A primitive orthorhombic unit cell has a = 5 Å, b = 6 Å, and c = 7 Å; α = β = γ = 90°. A new unit cell is chosen with edges defined by the vectors from the origin to the points with coordinates: 3, 1, 0; 1, 2, 0 and 0, 0, 1. Calculate the following:

 (i)   Volume of the original unit cell.
 (ii) The length of the three edges and the three angles of the new unit cell.
 (iii) Volume of the new unit cell.
 (iv) Ratio of the two unit cell volumes and the number of lattice points in the new unit cell.

Solution: *Given*: a = 5 Å, b = 6 Å, and c = 7 Å; α = β = γ = 90°. Coordinates: 0, 0, 0; 3, 1, 0; 1, 2, 0 and 0, 0, 1. From the given data, we can calculate:

 (i)   Volume of the original unit cell, $V_0 = abc = 5 \times 6 \times 7 = 210 \text{Å}^3$
 (ii) Using Eq. 3.8, the length of the vector a between the coordinates 0, 0, 0 and 3, 1, 0, is

$$a = \left[(0-3)^2 5^2 + (0-1)^2 6^2 + 0\right]^{1/2}$$
$$= [9 \times 25 + 36]^{1/2} = (261)^{1/2} = 16.16 \text{ Å}$$

Similarly, the length of the vector b between the coordinates 0, 0, 0 and 1, 2, 0, is

$$b = \left[(0-1)^2 \times 5^2 + (0-2)^2 \times 6^2 + 0\right]^{1/2}$$
$$= [25 + 4 \times 36]^{1/2} = (169)^{1/2} = 13 \text{Å}$$

and the length of the vector c between the coordinates 0, 0, 0 and 0, 0, 1, is

$$c = \left[0 + 0 + (0-1)^2 \times 7^2\right]^{1/2}$$
$$= (49)^{1/2} = 7 \text{Å}$$

Now, the angle between the edges with the end coordinates: 3, 1, 0 and 1, 2, 0

$$\cos \gamma = \frac{3+2}{(9+1+0)^{1/2}(1+4+0)^{1/2}} = \frac{5}{(10)^{1/2}(5)^{1/2}} = \frac{1}{\sqrt{2}}$$

So that, γ = 45°

Similarly, the angle between the edges with the end coordinates: 1, 2, 0 and 0, 0, 1

$$\cos \alpha = \frac{0}{(1+4+0)^{1/2}(0+0+1)^{1/2}} = \frac{0}{(5)^{1/2}(1)^{1/2}} = 0$$

So that, $\alpha = 90°$

and the angle between the edges with the end coordinates: 3, 1, 0 and 0, 0, 1

$$\cos \beta = \frac{0}{(9+1+0)^{1/2}(0+0+1)^{1/2}} = \frac{0}{(10)^{1/2}(1)^{1/2}} = 0$$

So that, $\beta = 90°$

(iii) Volume of the new unit cell with a = 16.16 Å, b = 13 Å, and c = 7 Å; $\alpha = \beta = 90°$, $\gamma = 45°$ is:

$$V_n = 16.16 \times 13 \times 7\left(1 - \cos^2 45\right)^{1/2} = 1470.56 \times \sin 45 = 1039.84 \text{Å}^3$$

(iv) Ratio of the two volumes is

$$\frac{V_n}{V_o} = \frac{1040}{210} = 4.95 \cong 5$$

$\Rightarrow$ There are five lattice points in the new unit cell.

3.5 Atomic Density in Crystals

Sometimes, it is important to know the atomic density along a particular direction or in a particular plane of the given crystal lattice. Similarly, sometimes it is also important to know the atomic density in crystal to determine other related quantities. They are defined as follows:

(i) **Linear Atomic Density**

Linear atomic density ρ_l is defined as the number of atomic diameter intersected by the selected line in the given direction as

$$\rho_l = \frac{\text{Number of atoms in the given direction (line)}}{\text{Length of the given line}} = \frac{n}{L} \qquad (3.17)$$

(ii) **Planar Atomic Density**

Planar atomic density ρ_p is defined as the equivalent number of atoms whose centers are intersected by the selected area in the given plane as

$$\rho_p = \frac{\text{Effective number of atoms in a given plane}}{\text{Area of the given plane}} = \frac{n'}{A} = \frac{nd}{V} \qquad (3.18)$$

where, n = Number of atoms in the unit cell (3-D).
d = Interplanar spacing
V = Volume of the unit cell.

(iii) **Volume Atomic Density**

Volume atomic density ρ_v of a crystal material is defined as

$$\rho_v = \frac{\rho N}{M} = \frac{n}{a^3} \qquad (3.19)$$

where, M = Atomic weight or molecular weight.
n = Number of atoms in the unit cell (or No. of formula unit).
N = 6.023×10^{26} (Avogadro's number in SI system).
a = Side of the (cubic) unit cell.

Mass of the unit cell is given by
M = No. of atoms in the unit cell × mass of each atom = n × m.
Mass of an atom (m), in turn, is given by

$$m = \frac{\text{Atomic mass (or molar mass)}}{\text{Avogadro's number}} = \frac{M}{N} \text{ kg} \qquad (3.20)$$

Solved Example

Example 1 Calculate the linear atomic density (i.e., the number of atoms/m) along [11] of the centered rectangular lattice with a = 3 Å and b = 4 Å.

Solution: *Given*: A centered rectangular lattice with a = 3 Å and b = 4 Å. For a rectangular lattice, $\gamma = 90°$.

$$\text{Length of the diagonal} = \sqrt{3^2 + 4^2} = \sqrt{9 + 16} = 5\text{Å} = 5 \times 10^{-10}\text{m}$$

$$\text{Effective atomic diameters intersected by the diagonal} = \frac{1}{2} + 1 + \frac{1}{2} = 2 \text{ atoms.}$$

Therefore, the linear atomic density is

$$\rho_l = \frac{2 \text{ atoms}}{5\text{Å}} = \frac{2}{5 \times 10^{-10}\text{m}} = \frac{2}{5} \times 10^{10} = 4 \times 10^9 \text{atoms/m}$$

Example 2 Calculate the planar atomic density (i.e., the number of atoms/area) of the centered square lattice with a = $4\sqrt{2}$ Å.

Solution: Given: A centered square lattice with a = $4\sqrt{2}$ Å = $4\sqrt{2} \times 10^{-10}$m. For a square lattice, $\gamma = 90°$.

$$\text{Area of the plane} = \left(4\sqrt{2} \times 10^{-10}\text{m}\right)^2 = 32 \times 10^{-20}\text{m}^2$$

Therefore, the planar atomic density is

$$\rho_p = \frac{2\text{atoms}}{32 \times 10^{-20}\text{m}^2} = 6.25 \times 10^{18}\text{atoms/m}^2$$

Example 3 Calculate the linear atomic density along [100], [110] and [111] directions in fcc copper whose lattice parameter a = 3.61 Å.

Solution: *Given*: Crystal structure is fcc, directions are: [100], [110] and [111], a = 3.61 Å = 3.61 $\times 10^{-10}$m, ρ_l = ?
The number of atoms intersected along [100] or [111] is one, therefore

$$\rho_{[100]} = \frac{1}{a} = \frac{1}{3.61 \times 10^{-10}} = 2.77 \times 10^9 \text{atoms/m}$$

The number of atoms intersected along [110] is 2 (Fig. 3.11a), therefore

$$\rho_{[110]} = \frac{2}{\sqrt{2}a} = \frac{\sqrt{2}}{a} = \frac{\sqrt{2}}{3.61 \times 10^{-10}} = 3.92 \times 10^9 \text{atoms/m}$$

$$\rho_{[111]} = \frac{1}{\sqrt{3}a} = \frac{1}{\sqrt{3} \times 3.61 \times 10^{-10}} = 1.60 \times 10^9 \text{atoms/m}$$

Example 4 Calculate the linear atomic density along [100], [110] and [111] directions in bcc iron whose lattice parameter a = 2.87 Å.

Solution: *Given*: Crystal structure is bcc, directions are: [100], [110] and [111], a = 2.87 Å = 2.87 $\times 10^{-10}$m, ρ_l = ?
The number of atoms intersected along [100] or [110] is one, therefore

$$\rho_{[100]} = \frac{1}{a} = \frac{1}{2.87 \times 10^{-10}} = 3.48 \times 10^9 \text{atoms/m}$$

$$\rho_{[110]} = \frac{1}{\sqrt{2}a} = \frac{1}{\sqrt{2}(2.87 \times 10^{-10})} = 2.46 \times 10^9 \text{atoms/m}$$

The number of atoms intersected along [111] is 2 (Fig. 3.11b), therefore

$$\rho_{[111]} = \frac{2}{\sqrt{3}a} = \frac{2}{\sqrt{3}(2.87 \times 10^{-10})} = 4.02 \times 10^9 \text{atoms/m}$$

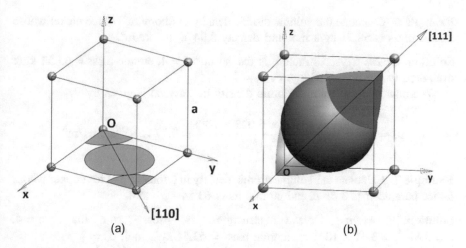

Fig. 3.11 Atomic linear density along **a** [110] direction in fcc **b** [111] direction in bcc

Example 5 Calculate the atomic density in (100), (110) and (111) planes of bcc iron whose lattice parameter is 2.87 Å.

Solution: *Given*: Crystal planes are: (100), (110) and (111). Crystal structure is bcc, so that n = 2, a = 2.87 Å = 2.87×10^{-10}m, ρ_p = ?

We know that the atomic density in a crystal plane is given by

$$\rho_p = \frac{nd}{V}$$

Further, for bcc structure

$$d_{100} = \frac{a}{2}, d_{110} = \frac{a}{\sqrt{2}} \text{ and } d_{111} = \frac{a}{2\sqrt{3}}, \ V = a^3$$

Therefore,

$$\rho_{(100)} = \frac{2 \times d_{100}}{a^3} = \frac{2a}{2a^3} = \frac{1}{a^2} = \frac{1}{(2.87 \times 10^{-10})^2} = 1.21 \times 10^{19} \text{atoms/m}^2$$

Similarly,

$$\rho_{(110)} = \frac{2 \times d_{110}}{a^3} = \frac{2a}{\sqrt{2}a^3} = \frac{\sqrt{2}}{a^2} = \frac{\sqrt{2}}{(2.87 \times 10^{-10})^2} = 1.72 \times 10^{19} \text{atoms/m}^2$$

$$\rho_{(111)} = \frac{2 \times d_{111}}{a^3} = \frac{2a}{2\sqrt{3}a^3} = \frac{1}{\sqrt{3}a^2} = \frac{1}{\sqrt{3}(2.87 \times 10^{-10})^2} = 7.0 \times 10^{18} \text{atoms/m}^2$$

Example 6 Calculate the volume atomic density in (atom/ m^3) of fcc nickel whose atomic mass is 58.71 kg/k mol and density 8.94 $\times$ 10^3 kg/m^3.

Solution: *Given*: Crystal structure is fcc, so that n = 4, atomic mass = 63.54 kg/k mol, density = 8.94 $\times$ 10^3 kg/m^3, ρ_v= ?

We know that the volume atomic density in a crystal is given by

$$\rho_v = \frac{\rho N}{M} = \frac{8.94 \times 10^3 \times 6.023 \times 10^{26}}{58.71} = 9.17 \times 10^{28} \text{atoms/m}^3$$

Example 7 Calculate the volume atomic density in (atom/ m^3) of fcc copper whose lattice parameter is 3.61 Å and atomic mass 63.54 kg/k mol.

Solution: *Given*: Crystal structure is fcc, so that n = 4, a = 3.61 Å = 3.61 $\times$ 10^{-10}m, atomic mass = 63.54 kg/k mol, ρ_v = ?

We know that the volume atomic density in a crystal is given by

$$\rho_v = \frac{\rho N}{M} = \frac{n}{a^3} = \frac{1}{(3.61 \times 10^{-10})^3} = 8.50 \times 10^{28} \text{atoms/m}^3$$

Example 8 Calculate the atomic density in (100), (110) and (111) planes of a simple cubic structure whose lattice parameter is 2.5 Å.

Solution: *Given*: Crystal planes are: (100), (110) and (111). Crystal structure is simple cubic, so that n = 1, a = 2.5 Å = 2.5 $\times$ 10^{-10}m, ρ_p = ?

We know that the atomic density in a crystal plane is given by

$$\rho_p = \frac{nd}{V}$$

Further, for simple cubic structure

$$d_{100} = a, \ d_{110} = \frac{a}{\sqrt{2}} \text{ and } d_{111} = \frac{a}{\sqrt{3}}, V = a^3$$

Therefore,

$$\rho_{(100)} = \frac{1 \times d_{100}}{a^3} = \frac{a}{a^3} = \frac{1}{a^2} = \frac{1}{(2.5 \times 10^{-10})^2} = 1.60 \times 10^{19} \text{atoms/m}^2$$

Similarly,

$$\rho_{(110)} = \frac{1 \times d_{110}}{a^3} = \frac{a}{\sqrt{2}a^3} = \frac{1}{\sqrt{2}a^2} = \frac{1}{\sqrt{2}(2.5 \times 10^{-10})^2} = 1.13 \times 10^{19} \text{atoms/m}^2$$

$$\rho_{(111)} = \frac{1 \times d_{111}}{a^3} = \frac{a}{\sqrt{3}a^3} = \frac{1}{\sqrt{3}a^2} = \frac{1}{\sqrt{3}(2.5 \times 10^{-10})^2} = 9.24 \times 10^{18} \text{atoms/m}^2$$

Example 9 Calculate the atomic density in (100), (110) and (111) planes of fcc aluminum whose lattice parameter is 4.05 Å.

Solution: Given: Crystal planes are: (100), (110) and (111). Crystal structure fcc, so that n = 4, a = 4.05 Å = $4.05 \times 10^{-10}m$, ρ_p = ?

We know that the atomic density in a crystal plane is given by

$$\rho_p = \frac{nd}{V}$$

Further, for fcc structure

$$d_{100} = \frac{a}{2}, \quad d_{110} = \frac{a}{2\sqrt{2}} \text{ and } d_{111} = \frac{a}{\sqrt{3}}, \quad V = a^3$$

Therefore,

$$\rho_{(100)} = \frac{4 \times d_{100}}{a^3} = \frac{4a}{2a^3} = \frac{2}{a^2} = \frac{2}{(4.05 \times 10^{-10})^2} - 1.22 \times 10^{19} \text{atoms/m}^2$$

Similarly,

$$\rho_{(110)} = \frac{4 \times d_{110}}{a^3} = \frac{4a}{2\sqrt{2}a^3} = \frac{\sqrt{2}}{a^2} = \frac{\sqrt{2}}{(4.05 \times 10^{-10})^2} = 8.62 \times 10^{18} \text{atoms/m}^2$$

$$\rho_{(111)} = \frac{4 \times d_{111}}{a^3} = \frac{4a}{\sqrt{3}a^3} = \frac{4}{\sqrt{3}a^2} = \frac{4}{\sqrt{3}(2.5 \times 10^{-10})^2} = 1.41 \times 10^{19} \text{atoms/m}^2$$

Example 10 Calculate the atomic density in (0001) plane of hcp zinc, whose lattice parameter a = b = 2.66 Å and c = 4.95 Å.

Solution: *Given*: Crystal plane is: (0001). Crystal structure hcp, a = b = 2.66 Å = 2.66×10^{-10}m and c = 4.95 Å = 4.95×10^{-10}m, ρ_p = ?

We know that the atomic density in a crystal plane is given by

$$\rho_p = \frac{\text{Effective number of atoms in a given plane}}{\text{Area of the given plane}} = \frac{n'}{A} = \frac{nd}{V}$$

Here, the number of atoms in the basal(0001)plane

$$= \frac{1}{2}(\text{one center atom associated with two unit cells})$$

$$+ 6 \times \frac{1}{6}(\text{each of the six corner atom is surrounded by six unit cells})$$

$$= \frac{1}{2} + 1 = \frac{3}{2}$$

$$\text{Area of the hexagon} = \frac{3\sqrt{3}a^2}{2}$$

Therefore,

$$\rho_{(0001)} = \frac{3/2}{3\sqrt{3}a^2/2} = \frac{1}{\sqrt{3}a^2} = 8.16 \times 10^{18} \text{atoms}/\text{m}^2$$

3.6   Packing Efficiency

In order to ascertain the degree of packing of a given crystal structure, it is needed to know the efficiency with which the available space of its unit cell is filled. In other words, it is to know the relative packing density (also known as packing factor or filling factor) of the given crystal structure. In order to find the packing efficiency of a given crystal structure, the following procedure is adopted.

1. Determine the plane projected area of an atom (in 2-D) or volume of the atom (in 3-D).
2. Determine the number of atoms in the given unit cell.
3. Determine the area (in 2-D) or volume (in 3-D) of the unit cell.

Therefore, the packing efficiency with which the available plane (in 2-D) or space (in 3-D) is filled, is given by

3.6.1   In 2-D

$$\frac{\text{Packing}}{\text{Efficiency}} = \frac{\text{Area occupied by plane projected atoms (in the unit cell)}}{\text{Area of the unit cell}} \qquad (3.21)$$

3.6.2 In 3-D

$$\frac{\text{Packing}}{\text{Efficiency}} = \frac{\text{Volume of the atoms (in the unit cell)}}{\text{Volume of the unit cell}} \tag{3.22}$$

Being a ratio, the packing efficiency is a dimensionless quantity.

Solved Examples

Example 1 Determine the efficiencies of 2-D arrangement of identical spheres on square and hexagonal unit cells.

Solution: Identical spheres in square and hexagonal unit cells (Fig. 3.12). From the figure, we can determine the following:
Area of the circle $= \pi R^2$
Area of the square $= 4R^2$
Therefore,

$$(\text{Efficiency})_s = \frac{\text{Area of the circle}}{\text{Area of the square}} = \frac{\pi R^2}{4R^2} = \frac{\pi}{4} = 78.5\%$$

Similarly, for hexagonal unit cell,
Area of the circle $= \pi R^2$

$$\text{Area of the triangle ABC} = \frac{1}{2} \times \text{base} \times \text{height}$$

$$= \frac{1}{2} \times a \times a \sin 60° \text{ (where } a = 2R \text{ and } R = a \sin 60°$$

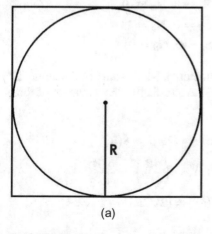

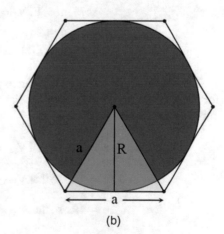

(a)   (b)

Fig. 3.12 Equal spheres arranged in **a** square **b** hexagonal lattice in 2-D

Area of the hexagon $= 6 \times$ area of the triangle ABC

$$= 6 \times \frac{1}{2} \times \left(\frac{R}{\sin 60°}\right) \times R$$

$$= 2\sqrt{3} \times R^2$$

Therefore,

$$(\text{Efficiency})_h = \frac{\text{Area of the circle}}{\text{Area of the hexagon}} = \frac{\pi R^2}{2\sqrt{3}R^2} = \frac{\pi}{2\sqrt{3}} = 90.7\%$$

Example 2 Determine the number of atoms in the unit cells of simple cubic (sc), body-centered cubic (bcc) and face-centered cubic (fcc) crystal structures. Calculate the packing efficiency in each case.

Solution: In a cubic crystal system, the number of atoms per unit cell is given by

$$n = \frac{N_c}{8} + \frac{N_f}{2} + N_b$$

where

$N_c =$ Number of corner atoms

$N_f =$ Number of face $-$ centered atoms

$N_b =$ Number of body $-$ centered atoms $= 1$

With the help of the above equations and the knowledge of corner and face centered atoms, the number of atoms in the given unit cell can be easily obtained. Therefore, for

$$\text{sc,} \quad n = 1 \ (N_c = 8, \ N_f = 0, \ N_b = 0)$$
$$\text{bcc,} \quad n = 2 \ (N_c = 8, \ N_f = 0, \ N_b = 1)$$
$$\text{fcc,} \quad n = 4 (N_c = 8, \ N_f = 6, \ N_b = 0)$$

The relationships between the lattice parameter "a" and radius of the atom "R" for three cubic cases are shown in Fig. 3.13. Accordingly, the volumes of three cubic unit cells are:

$$\text{sc,} \quad a = 2R, \text{ so that } V_{SC} = (a)^3 = (2R)^3 = 8(R)^3$$
$$\text{bcc,} \quad \sqrt{3}a = 4R, \text{ so that } V_{BCC} = (a)^3 = \left(\frac{4R}{\sqrt{3}}\right)^3 = \frac{64R^3}{3\sqrt{3}}$$
$$\text{fcc,} \quad \sqrt{2}a = 4R, \text{ so that } V_{FCC} = (a)^3 = (2\sqrt{2}R)^3 = 16\sqrt{2}R^3$$

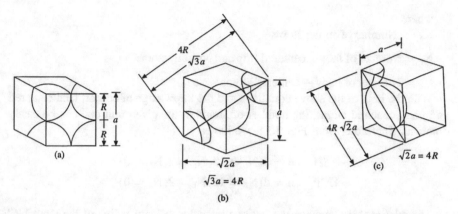

Fig. 3.13 Sectional view of conventional unit cell showing relationships between the lattice constant a and the atomic radius R in **a** sc **b** bcc, and **c** fcc unit cell

Also, volume of the atom in three cases is:

$$V_{atom} = \frac{4}{3}\pi R^3$$

Therefore, efficiencies in three cubic cases are:

$$(\text{Efficiency})_{SC} = \frac{1 \times (4\pi R^3)/3}{8R^3} = \frac{\pi}{6} = 52\%$$

$$(\text{Efficiency})_{BCC} = \frac{2 \times (4\pi R^3)/3}{64R^3/3\sqrt{3}} = \frac{\sqrt{3}\pi}{8} = 68\%$$

$$\text{and} \quad (\text{Efficiency})_{FCC} = \frac{4 \times (4\pi R^3)/3}{16\sqrt{2}R^3} = \frac{\pi}{3\sqrt{2}} = 74\%$$

Example 3 Determine the number of atoms in the unit cells of simple hexagonal (SH) and hexagonal close-packed (HCP) crystal structures. Calculate the packing efficiency in each case.

Solution: In a hexagonal crystal system, the number of atoms per unit cell is given by:

$$n = \frac{N_c}{6} + \frac{N_f}{2} + N_b$$

where

N_c = Number of corner atoms

N_f = Number of face − centered (top and bottom) atoms

N_b = Number of body − centered atoms

With the help of the above equations and the knowledge of corner, face-centered and body-centered atoms, the number of atoms in the given unit cell can be easily obtained. Therefore, from Fig. 3.14, we have:

$$SH, \quad n = 3(N_c = 12, N_f = 2, N_b = 0)$$
$$HCP, \quad n = 3(N_c = 12, N_f = 2, N_b = 0)$$

The relationships between the lattice parameter "a" and radius of the atom "R" for SH and HCP cases are shown in Fig. 3.14. Accordingly, the volumes of the two hexagonal unit cells (where h is the separation between the consecutive layers) are obtained as:

V_{SH} = Area of the hexagonal base × height (h)

$$= 6 \times \frac{1}{2} \times a^2 \sin 60° \times h$$

$$= \frac{3\sqrt{3}a^2h}{2} = \frac{3\sqrt{3} \times 4R^2 \times 2R}{2} = 12\sqrt{3}R^3 \quad \text{(for SH,} \quad a = h = 2R)$$

V_{HCP} = Area of the hexagonal base × height (h)

$$= \frac{3\sqrt{3}a^2h}{2} = \frac{3\sqrt{3} \times 4R^2 \times \sqrt{2} \times 2R}{2\sqrt{3}} = 12\sqrt{2}R^3 \quad \text{(for HCP,} \quad h = \frac{\sqrt{2}}{\sqrt{3}}a, \quad a = 2R)$$

Also, volume of the atom in both cases is:

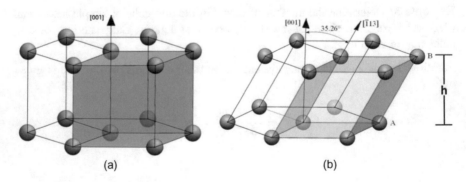

(a)   (b)

Fig. 3.14 Unit cell of **a** SH **b** HCP

$$V_{atom} = \frac{4}{3}\pi R^3$$

Therefore, efficiencies in two hexagonal cases are:

$$(Efficiency)_{SH} = \frac{3 \times (4\pi R^3)/3}{12\sqrt{3}R^3} = \frac{\pi}{3\sqrt{3}} = 60\%$$

$$and \quad (Efficiency)_{HCP} = \frac{3 \times (4\pi R^3)/3}{12\sqrt{2}R^3} = \frac{\pi}{3\sqrt{2}} = 74\%$$

Example 4 The ionic radii of Na^+ and Cl^- are 0.98 Å and 1.81 Å, respectively. Calculate the packing efficiency of NaCl structure.

Solution: *Given*: The ionic radius of Na^+ = 0.98 Å, Cl^- = 1.81 Å. Efficiency of (NaCl) = ?

The NaCl crystal structure is shown in Fig. 3.15. The Na^+ and Cl^- ions are supposed to touch each other, so that the lattice parameter is:

$$a = 2(\text{ radius of } Na^+ + \text{radius of } Cl^-)$$
$$= 2(0.98\text{Å} + 1.81\text{Å}) = 5.58\text{Å}$$

Therefore, the volume of the unit cell is

$$V = a^3 = (5.58)^3 \text{Å}^3$$

Further, from the figure, we observe that there are four NaCl molecules in a unit cell. Therefore, the packing efficiency of NaCl with 4 ions each is

Fig. 3.15 Sodium chloride structure

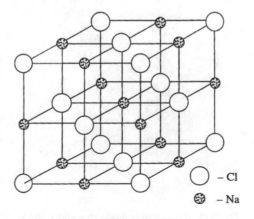

○ – Cl

✸ – Na

$$(\text{Efficiency})_{\text{NaCl}} = \dfrac{\dfrac{4\times\left(4\pi R_{Na}^3\right)}{3} + \dfrac{4\times\left(4\pi R_{Cl}^3\right)}{3}}{a^3}$$

$$= \dfrac{16\pi}{3}\left[\dfrac{(0.98)^3 + (1.81)^3}{(5.58)^3}\right] = 0.663 = 66.3\%$$

Example 5 Calculate the packing efficiency of CsCl structure where the Cs^+ ion is at the body-centered position and Cl^- ions are at the corner positions of the unit cell such that the central Cs^+ ion touches the neighboring 8 Cl^- ions.

Solution: *Given*: CsCl structure where the Cs^+ ion is at the body-centered position and touches the 8 Cl^- ions. Efficiency of CsCl = ?

The CsCl crystal structure is shown in Fig. 3.16. Since the Cs^+ ions touches the 8 Cl^- ions, the lattice parameter a is given by

$$a = 2r_{Cl}, \text{ or } r_{Cl} = a/2$$

Similarly,

$$\sqrt{3}a = 2r_{Cl} + 2r_{Cs}$$
$$\text{or } \sqrt{3} \times 2r_{Cl} = 2r_{Cl} + 2r_{Cs}$$
$$\text{or } \sqrt{3}r_{Cl} = r_{Cl} + r_{Cs}$$
$$\text{or } r_{Cs} = r_{Cl}(\sqrt{3} - 1)$$

Now, the volume of the Cs^+ ion is

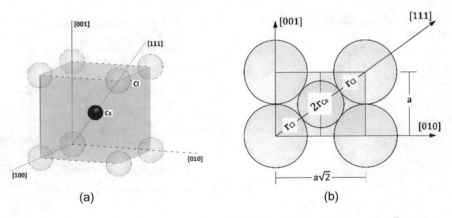

(a)   (b)

Fig. 3.16 **a** Caesium chloride structure with axial directions **b** Cs and Cl atoms in $(\bar{1}\,10)$ plane

$$V(Cl^-) = \frac{4\pi(r_{Cl})^3}{3} = \frac{4\pi}{3}\left(\frac{a}{2}\right)^3 = \frac{\pi a^3}{6}$$

Similarly,

$$V(Cs^+) = \frac{4\pi(r_{Cs})^3}{3} = \frac{4\pi(r_{Cl})^3}{3}\left(\sqrt{3}-1\right)^3 = \frac{\pi a^3}{6}\left(\sqrt{3}-1\right)^3$$

Therefore, the packing efficiency of CsCl with one ion each is

$$(Efficiency)_{CsCl} = \frac{\frac{(1)\pi a^3}{6} + \frac{(1)\pi a^3}{6}\left(\sqrt{3}-1\right)^3}{a^3}$$

$$= 0.52 + 0.21 = 0.76 = 76\%$$

Multiple Choice Questions (MCQ)

1. In a unit cell of plane hexagonal lattice with a = b = 10 Å and γ = 120°, an atom has its fractional coordinates x = 0.5 and y = 0.5. Draw four unit cells on its edges anticlockwise starting from its right, their fractional coordinates in the same order are:
   (a) (1.5, 0.5), (0.5, 1.5), (−0.5, 0.5), (0.5, −0.5)
   (b) (0.5, 0.5), (0.5, 0.5), (0.5, 1.5), (0.5, 1.5)
   (c) (1.5, 1.5), (1.5, 1.5), (0.5, 1.5), (−0.5, −1.5)
   (d) (0.5, 1.5), (0.5, 1.5), (0.5, 1.5), (0.5, 1.5)

2. In a Triclinic unit cell with a = 5 Å, b = 6 Å, c = 8 Å and α, β, γ not equal to 90°, an atom has its fractional coordinates x = 0.6 and y = 0.5 and z = 0.25. Draw six unit cells on its faces, their fractional coordinates counterclockwise are:
   (a) (1.6, 0.5, 0.25), (1.6, 0.5, 1.25), (1.6, 0.5, 0.25), (0.6, 1.5, 0.25), (1.6, 0.5, 1.25), (0.6, −0.5, 0.25)
   (b) (1.6, 0.5, 0.25), (0.6, 1.5, 0.25), (0.6, 0.5, 1.25), (−0.4, 0.5, 0.25), (0.6, −0.5, 0.25), (0.6, 0.5, −0.75)
   (c) (0.6, 0.5, 0.25), (1.6, 0.5, 0.25), (1.6, 1.5, 0.25), (0.6,- 0.5, 0.25), (0.6, 0.5, 1.25), (−1.6, 0.5, 0.25)
   (d) (1.6, 0.5, 0.75), (1.6, 0.5, 0.25), (0.6, 0.5, 0.25), (1.6, 1.5, 0.25), (−0.6, 0.5, 0.25), (1.6, 0.5, 0.25)

3. The set of fractional coordinates of equivalent positions related to a 2-fold rotational symmetry parallel to z-axis in 2-D are:
   (a) (x, y), (x, y)

   (b) (x, y), ($\bar{x}$, y)

   (c) (x, y), ($\bar{x}$, $\bar{y}$)

   (d) (x, y), (x, $\bar{y}$)

4. The set of fractional coordinates of equivalent positions related to a 2-fold rotational symmetry parallel to z-axis in 3-D are:
   (a) (x, y, z), (x, y, $\bar{z}$)
   (b) (x, y, z), ($\bar{x}$, y, z)
   (c) (x, y, z), (x, $\bar{y}$, z)
   (d) (x, y, z), ($\bar{x}$, $\bar{y}$, z)

5. The set of fractional coordinates of equivalent positions related to a 3-fold rotational symmetry parallel to z-axis in 2-D are:
   (a) (x, y), ($\bar{y}$, x–y), (y-x, $\bar{x}$)
   (b) (x, y), ($\bar{x}$, y), (x, $\bar{y}$)
   (c) (x, y), ($\bar{x}$, $\bar{y}$), ($\bar{x}$, y)
   (d) (x, y), ($\bar{x}$, $\bar{y}$), (x, $\bar{y}$)

6. The set of fractional coordinates of equivalent positions related to a 4-fold rotational symmetry parallel to z-axis in 2-D are:
   (a) (x, y), ($\bar{y}$, x), (y, $\bar{x}$), ($\bar{x}$, y)
   (b) (x, y), ($\bar{y}$, x), ($\bar{x}$, $\bar{y}$), (y, $\bar{x}$)
   (c) (x, y), ($\bar{x}$, $\bar{y}$), ($\bar{x}$, y), ($\bar{y}$, x)
   (d) (x, y), ($\bar{x}$, $\bar{y}$), (x, $\bar{y}$) ($\bar{x}$, y)

7. The set of fractional coordinates of equivalent positions related to a 6-fold rotational symmetry parallel to z-axis in 2-D are:
   (a) (x, y), ($\bar{y}$, x–y), (y-x, $\bar{x}$), ($\bar{x}$, $\bar{y}$), (y-x, $\bar{x}$), (y, y-x)
   (b) (x, y), ($\bar{x}$, $\bar{y}$), (y-x, $\bar{x}$), (y, y-x), ($\bar{x}$, y), (x, $\bar{y}$)
   (c) (x, y), (x–y, x), ($\bar{y}$, x–y), ($\bar{x}$, $\bar{y}$), (y-x, $\bar{x}$), (y, y-x)
   (d) (x, y), ($\bar{x}$, $\bar{y}$), ($\bar{x}$, y), (y-x, $\bar{x}$), (y, y-x), (x, $\bar{y}$)

8. The set of fractional coordinates of equivalent positions related to 2 mm symmetry in 2-D are:
   (a) (x, y), ($\bar{y}$, x), (y, $\bar{x}$), ($\bar{x}$, y)
   (b) (x, y), ($\bar{y}$, x), ($\bar{x}$, $\bar{y}$), (y, $\bar{x}$)
   (c) (x, y), ($\bar{x}$, $\bar{y}$), ($\bar{x}$, y), ($\bar{y}$, x)
   (d) (x, y), ($\bar{x}$, y) ($\bar{x}$, $\bar{y}$), (x, $\bar{y}$)

9. The set of fractional coordinates of equivalent positions related to 3 m symmetry in 2-D are:
   (a) (x, y), ($\bar{y}$, x–y), (y-x, $\bar{x}$), ($\bar{y}$, $\bar{x}$), (y-x, y), (x, x–y)
   (b) (x–y, y), ($\bar{x}$, $\bar{y}$), (y-x, $\bar{x}$), (y, y-x), ($\bar{x}$, y), (x, $\bar{y}$)

(c) (x, y), (x–y, x), ($\bar{y}$, x–y), ($\bar{x}$, $\bar{y}$), (y-x, $\bar{x}$), (y, y-x)

(d) (x, y-x), ($\bar{x}$, $\bar{y}$), ($\bar{x}$, y), (y-x, $\bar{x}$), (y, y-x), (y-x, $\bar{y}$)

10. A unit cell has the following dimensions: a = 3 Å, b = 4 Å and γ = 115°. The two distances between the points with fractional coordinates: (i) 0.210, 0.150 and 0.410, 0.050, (ii) 0.210, −0.150 and 0.410. 0.050 are:
    (a) 0.95 Å, 0.87 Å
    (b) 0.85 Å, 0.77 Å
    (c) 0.75 Å, 0.67 Å
    (d) 0.65 Å, 0.57 Å

11. A unit cell has the following dimensions: a = 5 Å, b = 8 Å and γ = 50°. The length of vectors with the help of following components: (i) [0.50, 1.0] and (ii) [0.75, −0.25] can be found to be:
    (a) 6.8 Å, 5.9 Å
    (b) 7.8 Å, 4.9 Å
    (c) 8.8 Å, 3.9 Å
    (d) 9.8 Å, 2.9 Å

12. A unit cell has the following dimensions: a = 2 Å, b = 3 Å and γ = 90°. The two distances between the points with fractional coordinates: (i) 0.100, 0.250 and 0.200, 0.050, (ii) 0.100, −0.250 and 0.210. 0.050 are:
    (a) 0.632 Å, 0.922 Å
    (b) 0.532 Å, 0.822 Å
    (c) 0.432 Å, 0.722 Å
    (d) 0.332 Å, 0.622 Å

13. A unit cell has the following dimensions: a = 5 Å, b = 7 Å, c = 9 Å, α = 120°, β = 85° and γ = 75°. The length of vectors with the help of following components: (i) [0.2, −0.3, 0.6] and (ii) [0.5, 0.7, −0.1] can be found to be:
    (a) 4.76 Å, 4.44 Å
    (b) 5.76 Å, 5.44 Å
    (c) 6.76 Å, 6.44 Å
    (d) 7.76 Å, 7.44 Å

14. The sides of a primitive rectangle are 4 Å and 3 Å. Construct another unit cell whose one side is the diagonal of the rectangle. The angle made by this with the horizontal side is:
    (a) 26.87°
    (b) 36.87°

(c) 46.87°
(d) 56.87°

15. The side of a primitive cell of square lattice has a = 2 Å. A new unit cell is chosen with edges defined by the vectors from the origin to the points with coordinates 1, 0 and 0, 3. Area of the new unit cell is:
    (a) $8Å^2$
    (b) $10Å^2$
    (c) $12Å^2$
    (d) $14Å^2$

16. The side of a primitive cell of square lattice has a = 2 Å. A new unit cell is chosen with edges defined by the vectors from the origin to the points with coordinates 1, 0 and 0, 3. The number of lattice points in the new unit cell is:
    (a) 6
    (b) 5
    (c) 4
    (d) 3

17. A primitive tetragonal unit cell has a = 5 Å, b = 6 Å, and c = 7 Å; α = β = = 90°. A new unit cell is chosen with edges defined by the vectors from the origin to the points with coordinates: 3, 1, 0; 1, 2, 0 and 0, 0, 1.The lengths of the three edges of the new unit cell are:
    (a) 16.16 Å, 13 Å, 7 Å
    (b) 15.16 Å, 12 Å, 7 Å
    (c) 14.16 Å, 11 Å, 7 Å
    (d) 13.16 Å, 10 Å, 7 Å

18. A primitive tetragonal unit cell has a = 5 Å, b = 6 Å, and c = 7 Å; α = β = = 90°. A new unit cell is chosen with edges defined by the vectors from the origin to the points with coordinates: 3, 1, 0; 1, 2, 0 and 0, 0, 1.The three angles of the new unit cell are:
    (a) α = β = γ = 90°
    (b) α = β = 90°, γ = 45°
    (c) α = β = 45°, γ = 90°
    (d) α = β = γ = 45°

19. A primitive tetragonal unit cell has a = 5 Å, b = 6 Å, and c = 7 Å; α = β = = 90°. A new unit cell is chosen with edges defined by the vectors from the origin to the points with coordinates: 3, 1, 0; 1, 2, 0 and 0, 0, 1. Volume of the new unit cell is:
    (a) $839.84Å^3$
    (b) $939.84Å^3$

(c) 1039.84Å^3

(d) 1139.84Å^3

20. A primitive tetragonal unit cell has a = 5 Å, b = 6 Å, and c = 7 Å; $\alpha = \beta = = 90°$. A new unit cell is chosen with edges defined by the vectors from the origin to the points with coordinates: 3, 1, 0; 1, 2, 0 and 0, 0, 1. The number of lattice points in the new unit cell is:

(a) 8

(b) 7

(c) 6

(d) 5

21. The linear atomic density (i.e., the number of atoms/m) along [11] of the centered rectangular lattice with a = 3 Å and b = 4 Å is:

(a) 4×10^9

(b) 5×10^9

(c) 6×10^9

(d) 7×10^9

22. The linear atomic density (i.e., the number of atoms/m) along [110] in fcc copper whose lattice parameter a = 3.61 Å is:

(a) 5.92×10^9

(b) 4.92×10^9

(c) 3.92×10^9

(d) 2.92×10^9

23. The linear atomic density (i.e., the number of atoms/m) along [111] in fcc copper whose lattice parameter a = 3.61 Å is:

(a) 4.60×10^9

(b) 3.60×10^9

(c) 2.60×10^9

(d) 1.60×10^9

24. The linear atomic density (i.e., the number of atoms/m) along [110] in bcc iron whose lattice parameter a = 2.87 Å:

(a) 1.46×10^9

(b) 2.46×10^9

(c) 3.46×10^9

(d) 4.46×10^9

25. The linear atomic density (i.e., the number of atoms/m) along [111] in bcc iron whose lattice parameter a = 2.87 Å:
   (a) 4.02×10^9
   (b) 3.02×10^9
   (c) 2.02×10^9
   (d) 1.02×10^9

26. The planar atomic density (i.e., the number of atoms/area) of the centered square lattice with a = $4\sqrt{2}$ Å is:
   (a) 5.25×10^{18}
   (b) 6.25×10^{18}
   (c) 7.25×10^{18}
   (d) 8.25×10^{18}

27. The atomic density (i.e., the number of atoms/area) in (100) plane of a simple cubic structure with a = 2.5 Å is:
   (a) 4.60×10^{19}
   (b) 3.60×10^{19}
   (c) 2.60×10^{19}
   (d) 1.60×10^{19}

28. The atomic density (i.e., the number of atoms/area) in (110) plane of a simple cubic structure with a = 2.5 Å is:
   (a) 4.13×10^{19}
   (b) 3.13×10^{19}
   (c) 2.13×10^{19}
   (d) 1.13×10^{19}

29. The atomic density (i.e., the number of atoms/area) in (111) plane of a simple cubic structure with a = 2.5 Å is:
   (a) 9.24×10^{18}
   (b) 8.24×10^{18}
   (c) 7.24×10^{18}
   (d) 6.24×10^{18}

30. The atomic density (i.e., the number of atoms/area) in (100) plane of a bcc iron with a = 2.87 Å is:
   (a) 1.21×10^{19}
   (b) 2.21×10^{19}
   (c) 3.21×10^{19}
   (d) 4.21×10^{19}

31. The atomic density (i.e., the number of atoms/area) in (110) plane of a bcc iron
    with a = 2.87 Å is:
    (a) 1.72 × 10^{19}
    (b) 2.72 × 10^{19}
    (c) 3.72 × 10^{19}
    (d) 4.72 × 10^{19}

32. The atomic density (i.e., the number of atoms/area) in (111) plane of a bcc iron
    with a = 2.87 Å is:
    (a) 6.0 × 10^{19}
    (b) 7.0 × 10^{19}
    (c) 8.0 × 10^{19}
    (d) 9.0 × 10^{19}

33. The atomic density (i.e., the number of atoms/area) in (100), plane of fcc
    aluminum with a = 4.05 Å is:
    (a) 4.22 × 10^{19}
    (b) 3. 22 × 10^{19}
    (c) 2. 22 × 10^{19}
    (d) 1. 22 × 10^{19}

34. The atomic density (i.e., the number of atoms/area) in (110) plane of fcc alu-
    minum with a = 4.05 Å is:
    (a) 6.62 × 10^{18}
    (b) 7.62 × 10^{18}
    (c) 8.62 × 10^{18}
    (d) 9.62 × 10^{18}

35. The atomic density (i.e., the number of atoms/area) in (111) plane of fcc alu-
    minum with a = 4.05 Å is:
    (a) 1.41 × 10^{19}
    (b) 2.41 × 10^{19}
    (c) 3.41 × 10^{19}
    (d) 4.41 × 10^{19}

36. The atomic density (i.e., the number of atoms/area) in (0001) plane of an hcp
    zinc with a = b = 2.66 Å and c = 4.95 Å is:
    (a) 9.16 × 10^{18}
    (b) 8.16 × 10^{18}
    (c) 7.16 × 10^{18}
    (d) 6.16 × 10^{18}

37. The volume atomic density (atom/m^3) of fcc nickel whose atomic mass is 58.71 kg/k mol and density 8.94×10^3 kg/m^3 is:
    (a) 9.17×10^{28}
    (b) 8.17×10^{28}
    (c) 7.17×10^{28}
    (d) 6.17×10^{28}

38. The volume atomic density (atom/m^3) of fcc copper whose lattice parameter is 3.61 Å and atomic mass 63.54 kg/k mol is:
    (a) 6.50×10^{28}
    (b) 7.50×10^{28}
    (c) 8.50×10^{28}
    (d) 9.50×10^{28}

39. The efficiency with which a 2-D arrangement of identical spheres (atoms) on a square lattice can pack is:
    (a) 48.5%
    (b) 58.5%
    (c) 68.5%
    (d) 78.5%

40. The efficiency with which a 2-D arrangement of identical spheres (atoms) on a hexagonal lattice can pack is:
    (a) 90.7%
    (b) 80.7%
    (c) 70.7%
    (d) 60.7%

41. The efficiency with which a 3-D arrangement of identical spheres (atoms) on a simple cubic lattice can pack is:
    (a) 42%
    (b) 52%
    (c) 62%
    (d) 72%

42. The efficiency with which a 3-D arrangement of identical spheres (atoms) on a body-centered cubic lattice can pack is:
    (a) 48%
    (b) 58%
    (c) 68%
    (d) 78%

43. The efficiency with which a 3-D arrangement of identical spheres (atoms) on a face-centered cubic lattice can pack is:
    (a) 44%
    (b) 54%
    (c) 64%
    (d) 74%

44. The efficiency with which a 3-D arrangement of identical spheres (atoms) on a simple hexagonal lattice can pack is:
    (a) 60.7%
    (b) 70.7%
    (c) 80.7%
    (d) 90.7%

45. The efficiency with which a 3-D arrangement of identical spheres (atoms) on a hexagonal close-packed (hcp) lattice can pack is:
    (a) 64%
    (b) 74%
    (c) 84%
    (d) 94%

Answers

 1. (a)
 2. (b)
 3. (c)
 4. (d)
 5. (a)
 6. (b)
 7. (c)
 8. (d)
 9. (a)
10. (b)
11. (d)
12. (a)
13. (c)
14. (b)
15. (c)
16. (d)
17. (a)
18. (b)
19. (c)
20. (d)
21. (a)

22. (c)
23. (d)
24. (b)
25. (a)
26. (b)
27. (d)
28. (d)
29. (a)
30. (a)
31. (a)
32. (b)
33. (d)
34. (c)
35. (a)
36. (b)
37. (d)
38. (c)
39. (d)
40. (a)
41. (b)
42. (c)
43. (d)
44. (a)
45. (b)

Chapter 4
Unit Cell Representations of Miller Indices

4.1 Miller Indices of Atomic Sites, Planes and Directions

(a) Miller Indices of Atomic Sites

In order to locate the atomic sites (positions) in cubic or orthogonal unit cells, rectangular axes are used. The positive x-axis is taken as the direction coming out of the paper, positive y-axis along the right side of the paper and positive z-axis towards the top as shown in Fig. 4.1a. Negative directions are just opposite to their counterparts. The atomic positions in the unit cells are located at unit distances along the three axes as indicated in Figs. 4.1b and c. For example, the position coordinates of the eight corners are:

[[000]], [[100]], [[010]], [[001]], [[111]], [[110]], [[101]], [[011]].

(b) Miller Indices of Planes

Miller indices of planes are obtained by adopting the following procedure:

(i) Determine the intercepts of the plane along x, y and z-axes in terms of lattice parameters.
(ii) Divide the intercepts by appropriate unit translations (see Tables 4.1, 4.2 and 4.3).
(iii) Note their reciprocals.
(iv) If fractions result, multiply each of them by the smallest common divisor.
(v) Put the resulting integers in parenthesis, that is, (hkl) to get the required indices of that and all other parallel planes.

Miller indices of six faces of a cube are shown in Fig. 4.2.

(c) Miller Indices of Directions

Miller indices of directions are obtained by adopting the following procedure:

M. A. Wahab, *Numerical Problems in Crystallography*,
https://doi.org/10.1007/978-981-15-9754-1_4

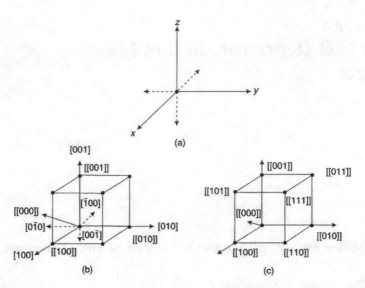

Fig. 4.1 **a** Orthogonal axes, **b** site indices and principal directions, **c** site indices in a cube

Table 4.1 Unit translation for low index directions of cubic system

Family of directions	Unit translation		
	p	I	F
<100>	a	a	a
<110>	$\sqrt{2}a$	$\sqrt{2}a$	$\frac{a}{\sqrt{2}}$
<111>	$\sqrt{3}a$	$\frac{\sqrt{3}a}{2}$	$\sqrt{3}a$

Table 4.2 Unit translation for low index directions of a tetragonal system

Family of directions	Unit translation	
	P	I
<100>	a	a
<001>	c	c
<110>	$\sqrt{2}a$	$\sqrt{2}a$
<101>	$\sqrt{(a^2+c^2)}$	$\sqrt{(a^2+c^2)}$
<111>	$\sqrt{(2a^2+c^2)}$	$\frac{1}{2}\sqrt{(2a^2+c^2)}$

(i)   Note down the coordinates of the lattice site nearest to the origin corresponding to a given direction.

(ii) Divide the coordinates by appropriate unit translations.

(iii) If fractions result, multiply each of them by the smallest common divisor.

(iv) Put the resulting integers in a set of square brackets, that is, [hkl] to get the required indices of that and all other parallel directions.

Miller indices of some principal directions are shown in Fig. 4.3.

Table 4.3 Unit translation for low index directions of an orthorhombic system

Family of direction	Unit translation			
	P	C	I	F
<100>	a	a	a	a
<010>	b	b	b	b
<001>	c	c	c	c
<110>	$\sqrt{(a^2+b^2)}$	$\sqrt{(a^2+b^2)}$	$\sqrt{(a^2+b^2)}$	$\frac{1}{2}\sqrt{(a^2+b^2)}$
<101>	$\sqrt{(a^2+c^2)}$	$\sqrt{(a^2+c^2)}$	$\sqrt{(a^2+c^2)}$	$\frac{1}{2}\sqrt{(a^2+c^2)}$
<011>	$\sqrt{(b^2+c^2)}$	$\sqrt{(b^2+c^2)}$	$\sqrt{(b^2+c^2)}$	$\frac{1}{2}\sqrt{(b^2+c^2)}$
<111>	$\sqrt{(a^2+b^2+c^2)}$	$\sqrt{(a^2+b^2+c^2)}$	$\frac{1}{2}\sqrt{(a^2+b^2+c^2)}$	$\sqrt{(a^2+b^2+c^2)}$

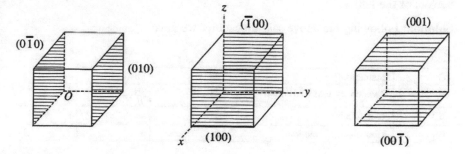

Fig. 4.2 Indices of six faces of a cube

Solved Examples

Example 1 Show how the atomic positions are located in three cubic unit cells? Indicate the position coordinates for the atoms in bcc and fcc unit cells.

Solution: The corner atomic positions in the cubic unit cells are located at unit distances from the origin along the three crystallographic axes as indicated above in

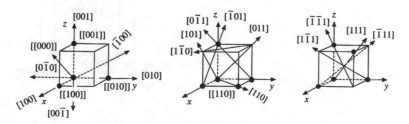

Fig. 4.3 Indices of principal directions in a cubic crystal

Fig. 4.1. The position coordinates of the eight corner atoms will remain the same for the three cubic cases.

The position coordinates of the central atom in a bcc unit cell are $\left[\left[\frac{1}{2}\frac{1}{2}\frac{1}{2}\right]\right]$ as shown in Fig. 4.4. For simplicity, sometimes only two position coordinates, for example, [[000]] and $\left[\left[\frac{1}{2}\frac{1}{2}\frac{1}{2}\right]\right]$ are specified to represent a bcc unit cell while another position coordinates (for corner atoms) are assumed to be understood.

Similarly, the position coordinates of six face-centered atoms in an fcc unit cell are: $\left[\left[\frac{1}{2}\frac{1}{2}0\right]\right]$, $\left[\left[0\frac{1}{2}\frac{1}{2}\right]\right]$, $\left[\left[\frac{1}{2}0\frac{1}{2}\right]\right]$, $\left[\left[\frac{1}{2}\frac{1}{2}1\right]\right]$, $\left[\left[1\frac{1}{2}\frac{1}{2}\right]\right]$ and $\left[\left[\frac{1}{2}1\frac{1}{2}\right]\right]$.

They are shown in Fig. 4.5. Like bcc, only four position coordinates, for example, [[000]]. $\left[\left[\frac{1}{2}\frac{1}{2}0\right]\right]$, $\left[\left[0\frac{1}{2}\frac{1}{2}\right]\right]$ and $\left[\left[\frac{1}{2}0\frac{1}{2}\right]\right]$ are specified to represent an fcc unit cell while other position coordinates are assumed to be understood.

Example 2 Find the Miller indices of a plane that makes intercepts of 2a, 3b and 4c along the three crystallographic axes, where a, b, c are primitive translation vectors of the lattice.

Solution: Following the above said procedure, we have

(i)	Intercepts	2a	3b	4c
(ii)	Division by unit translation	$\frac{2a}{a}=2$	$\frac{3b}{b}=3$	$\frac{4c}{c}=4$
(iii)	Reciprocals	$\frac{1}{2}$	$\frac{1}{3}$	$\frac{1}{4}$
(iv)	After clearing fraction	6	4	3

$\Longrightarrow$ The required Miller indices of the plane are (643).

Example 3 Find the Miller indices of a plane that makes intercepts as 3a: 4b on the x and y axes, and is parallel to z-axis.

Fig. 4.4 The position coordinates of the central atom and corner atoms in a bcc unit cell

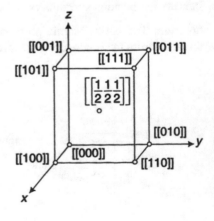

Fig. 4.5 The position coordinates of face-centered atoms in a fcc unit cell

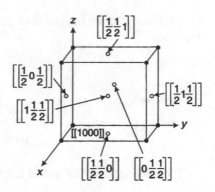

Solution: Following the above said procedure, we have

(i)	Intercepts	3a	4b	∞
(ii)	Division by unit translation	$\frac{3a}{a} = 3$	$\frac{4b}{b} = 4$	∞
(iii)	Reciprocals	$\frac{1}{3}$	$\frac{1}{4}$	$\frac{1}{\infty}$
(iv)	After clearing fraction	4	3	0

$\implies$ The required Miller indices of the plane are (430).

Example 4 In an orthorhombic crystal with a: b: c = 1: 2: 5, a plane cuts intercepts of 3Å, 4Å, and 5Å on its coordinate axes. Find the Miller indices of the plane.

Solution: Following the above said procedure, we have

(i)	Intercepts	3	4	5
(ii)	Division by unit translation	$\frac{3}{1} = 3$	$\frac{4}{2} = 2$	$\frac{5}{5} = 1$
(iii)	Reciprocals	$\frac{1}{3}$	$\frac{1}{2}$	1
(iv)	After clearing fraction	2	3	6

$\implies$ The required Miller indices of the plane are (236).

Example 5 In a crystal whose primitive translations are 1.2Å, 1.8Å and 2Å, a plane with Miller indices (231) cuts an intercept of 1.2Å along the x-axis. Determine the lengths of intercepts along y and z-axes.

Solution: *Given*: A plane with Miller indices (231), intercept along x-axis 1.2Å, intercepts along y and z-axes = ?

We know that the reciprocal of the coefficients of unit translation vectors are in the ratio:

$$p^{-1} : q^{-1} : r^{-1} = h : k : l = 2 : 3 : 1$$

Therefore,

$$p : q : r = \frac{1}{2} : \frac{1}{3} : 1$$

Then, the ratios of the actual lengths of the intercepts are:

$$l_1 : l_2 : l_3 = pa : qb : rc$$
$$= 1.2 \times \frac{1}{2} : 1.8 \times \frac{1}{3} : 2 \times 1$$

Since l_1 is given as 1.2Å, we therefore multiply the RHS by 2, so that

$$l_1 : l_2 : l_3 = 1.2 : 1.2 : 4$$
$$\Rightarrow l_2 = 1.2\text{Å and } l_3 = 4\text{Å}$$

Example 6 In a cubic crystal, find the lengths of the intercepts made on three axes by a plane with Miller indices $(1\bar{3}2)$.

Solution: *Given*: A cubic crystal, so that a = b = c, plane with Miller indices $(1\bar{3}2)$, intercepts along three axes = ?

We know that the reciprocal of the coefficients of unit translation vectors are in the ratio:

$$p^{-1} : q^{-1} : r^{-1} = h : k : l = 1 : -3 : 2$$

Therefore,

$$p : q : r = 1 : -\frac{1}{3} : \frac{1}{2}$$

Then, the ratios of the actual lengths of the intercepts are:

$$l_1 : l_2 : l_3 = pa : qb : rc$$
$$= 6a : -2a : 3a = 6 : -2 : 3$$

Example 7 $\left[\left[\frac{a}{2} b 0\right]\right]$ are the coordinates of the lattice site nearest to the origin. Find the indices of its direction.

Solution: Following the above said procedure, we have

(i)	Coordinates	$\frac{a}{2}$	b	0
(ii)	Division by unit translation	$\frac{a}{2a} = \frac{1}{2}$	$\frac{b}{b} = 1$	$\frac{0}{c} = 0$
(iii)	After clearing fraction	1	2	0

$\implies$ The required Miller indices of the given direction are [120].

Example 8 The coordinates of a lattice site nearest to the origin are: $x = -2a$, $y = 1b$ and $z = \frac{3c}{2}$. Determine the indices of direction.

Solution: Following the above said procedure, we have

(i)	Coordinates	$-2a$	$1b$	$\frac{3c}{2}$
(ii)	Division by unit translation	$\frac{-2a}{a} = -2$	$\frac{1b}{b} = 1$	$\frac{3c}{2c} = \frac{3}{2}$
(iii)	After clearing fraction	-4	2	3

$\implies$ The required Miller indices of direction are $[\bar{4}23]$.

Example 9 In a cubic system, different family of directions are symbolically represented as <100>, <110> and <111>. Write all their members.

Solution: The family <100> comprises of 6 crystallographically equivalent directions. They are:

$$[100], [\bar{1}00], [010], [0\bar{1}0], [001] \text{ and } [00\bar{1}].$$

Similarly, the family <110> comprises of 12 crystallographically equivalent directions. They are:

$$[110], [\bar{1}10], [1\bar{1}0], [\bar{1}\bar{1}0], [101], [\bar{1}01], [10\bar{1}], [\bar{1}0\bar{1}], [011], [0\bar{1}1], [01\bar{1}] \text{ and } [0\bar{1}\bar{1}].$$

Further, the family <111> comprises of 8 crystallographically equivalent directions. They are:

$$[111], [\bar{1}11], [1\bar{1}1], [11\bar{1}], [\bar{1}\bar{1}1], [\bar{1}1\bar{1}], [1\bar{1}\bar{1}] \text{ and } [\bar{1}\bar{1}\bar{1}].$$

Example 10 Classify the given members of the family of directions <100>, <110> and <111> according to a tetragonal crystal system.

Solution: For a tetragonal crystal system, we know that a = b ≠ c. Thus, "a" is the unit translation for directions $[100], [\bar{1}00], [010]$ and $[0\bar{1}0]$, whereas "c" is the unit translation for directions $[001]$ and $[00\bar{1}]$. Accordingly, the first four directions are the members of <100> family and the last two are the members of <001> family.

In a cube, the 12 directions (face diagonals) lie in the three different planes (viz. x–y, y–z and z–x) but they are crystallographically equivalent because a = b = c. Unlike this, in tetragonal (a = b ≠ c) all 12 directions are not equivalent. For x–y plane, a = b and the corresponding unit translation is $\sqrt{2}a$ (Table 4.2). Accordingly, <110> family has only four members: $[110], [\bar{1}10], [1\bar{1}0], [\bar{1}\bar{1}0]$. On the other hand, the y–z and z–x planes have the unit translation as $\sqrt{a^2 + c^2}$. Therefore, other eight members: $[101], [\bar{1}01], [10\bar{1}], [\bar{1}0\bar{1}], [011], [0\bar{1}1], [01\bar{1}]$ and $[0\bar{1}\bar{1}]$ belong to <101> family or equivalently <011> family.

Since all <111> directions have the same unit translation $\sqrt{2a^2 + c^2}$ in tetragonal (like $\sqrt{3}a$ in cubic) systems, therefore, this family contains all eight members

$$[111], [\bar{1}11], [1\bar{1}1], [11\bar{1}], [\bar{1}\bar{1}1], [\bar{1}1\bar{1}], [1\bar{1}\bar{1}] \text{ and } [\bar{1}\bar{1}\bar{1}].$$

Example 11 Classify the members of the family of directions <100>, <110> and <111> according to an orthorhombic crystal system.

Solution: For an orthorhombic crystal system, we know that a ≠ b ≠ c. Thus, the unit translations for the directions $[100]$ and $[\bar{1}00]$ is a (Table 4.3), for $[010]$ and $[0\bar{1}0]$ is b, whereas for $[001]$ and $[00\bar{1}]$ is c, accordingly the first two directions are the members of <100> family, the next two are the members of <010> family and the last two are the members of <001> family.

Unlike cubic or tetragonal cases, in an orthorhombic system, a ≠ b ≠ c. This implies that x–y, y–z and z–x planes are crystallographically not equivalent. In other words, the unit translations for different set of equivalent planes are different. Accordingly, <110> family has four members: $[110], [\bar{1}10], [1\bar{1}0], [\bar{1}\bar{1}0]$. The next four members, $[101], [\bar{1}01], [10\bar{1}], [\bar{1}0\bar{1}]$ belong to <101> family and the last four, $[011], [0\bar{1}1], [01\bar{1}]$ and $[0\bar{1}\bar{1}]$ belong to <011> family, respectively.

Since all <111> directions have the same unit translation $\sqrt{a^2 + b^2 + c^2}$ in orthorhombic crystal system, therefore this family contains all eight members. They are:

$$[111], [\bar{1}11], [1\bar{1}1], [11\bar{1}], [\bar{1}\bar{1}1], [\bar{1}1\bar{1}], [1\bar{1}\bar{1}] \text{ and } [\bar{1}\bar{1}\bar{1}].$$

4.2 Representation of Planes and Directions of Known Miller Indices in Cubic Unit Cell

(a) Representation of Planes

Planes of known Miller indices can be represented in a unit cell by using the following procedure:

(i) Take the reciprocal of given Miller indices. They will represent the intercepts in terms of axial units.

(ii) If a Miller index is negative, shift the origin by moving it in the positive direction along that axis.

(iii) Mark the length of the intercepts on the respective coordinate axes, each one starting from the origin. Join their end points; the resulting sketch will represent the required (hkl) plane.

(b) Representation of Directions

Directions of known Miller indices can be represented in a unit cell by using the following procedure:

(i) Divide the given Miller indices by a number such that the resulting indices become ≤ 1, they represent the coordinates of the lattice site nearest to the origin in the given direction and lie within the unit cell.

(ii) If a lattice site nearest to the origin contains fractional coordinates, remove the fraction by multiplying them with suitable number.

(iii) Mark the length of the position vector along the respective coordinate axes without disconnection. Join the origin with the end point to get the required [hkl] direction.

Solved Examples

Example 1 Show $(\bar{1}11)$, $(1\bar{1}2)$ and (210) planes in a cubic unit cell.

Solution: Construct a cubic unit cell. Select the origin and the three crystallographic axes. Follow the above said procedure and draw the given planes within the cubic unit cells as shown in Fig. 4.6.

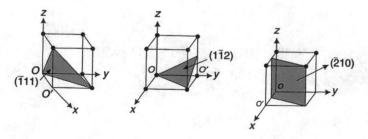

Fig. 4.6 $(\bar{1}11)$, $(1\bar{1}2)$ and $(\bar{2}10)$ planes

Fig. 4.7 $(1\bar{1}\bar{1})$, $(\bar{1}\bar{1}1)$ planes

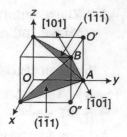

Example 2 Show the planes $(1\bar{1}\bar{1})$ and $(\bar{1}\bar{1}1)$ in a cubic unit cell. Draw the line of along their intersection. Assign the Miller indices of direction to the line.

Solution: Construct a cubic unit cell. Select its origin and the three crystallographic axes. Now, imagine the origin to be at O′ and draw $(1\bar{1}\bar{1})$ plane. Next, imagine the origin to be at O″ and draw $(\bar{1}\bar{1}1)$ plane. Common direction of the two planes is shown as AB or BA (Fig. 4.7) whose Miller indices are: [101] or $[\bar{1}0\bar{1}]$.

Example 3 Show (100), (110) and (111) planes in different bcc unit cells and list the position coordinates of the atoms whose centers are intersected by each of the three planes.

Solution: Construct three bcc unit cells, select their origin and the three crystallographic axes. Follow the above said procedure and draw the given planes within the unit cells as shown in Fig. 4.8.

(a) The position coordinates of the atoms whose centers are intersected by the (100) plane are: [[100]], [[110]], [[101]] and [[111]]. They are shown in Fig. 4.8a.

(b) The position coordinates of the atoms whose centers are intersected by the (110) plane are: [[100]], [[010]], [[011]], [101]] and $\left[\left[\frac{1}{2}\frac{1}{2}\frac{1}{2}\right]\right]$. They are shown in Fig. 4.8b.

(c) The position coordinates of the atoms whose centers are intersected by the (111) plane are: [[100]], [[010]], and [[001]]. They are shown in Fig. 4.8c.

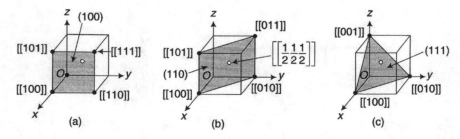

Fig. 4.8 Position coordinates intersected by **a** (100) plane, **b** (110) plane, and **c** (111) plane

Fig. 4.9 [100], [110] and [111] directions in an fcc unit cell

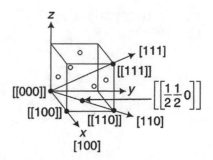

Example 4 Show [100], [110] and [111] directions in a fcc unit cell and list the position coordinates of the atoms whose centers are intersected by each of the three directions.

Solution: Construct an fcc unit cell, select its origin and the three crystallographic axes. Follow the above said procedure and draw the given directions as shown in Fig. 4.9.

(a) The position coordinates of the atoms whose centers are intersected by the [100] direction are: [[000]] and [[100]].
(b) The position coordinates of the atoms whose centers are intersected by the [110] direction are: [[000]], $\left[\left[\frac{1}{2}\frac{1}{2}0\right]\right]$ and [[110]].
(c) The position coordinates of the atoms whose centers are intersected by the [111] direction are: [[000]], and [[111]]. They are shown in Fig. 4.9.

Example 5 Show [112], [121] and [220] directions in a cubic unit cell.

Solution: We know that the Miller indices of a direction and the position coordinates are identical if the digits are zero and 1 only. For other digits, the position coordinates can be obtained by dividing the indices of direction by the largest number so that they lie within the unit cube.

(a) The position coordinates for [112], therefore can be obtained by dividing each index by 2. Thus the position coordinates for the atom nearest to the origin is $\left[\left[\frac{1}{2}\frac{1}{2}1\right]\right]$. The location of the position coordinates can be found by moving half the unit distance along x and y directions and a unit distance along z direction, respectively. The corresponding direction is shown in Fig. 4.10.
(b) Similar to the above, the position coordinates for [121] are found to be $\left[\left[\frac{1}{2}1\frac{1}{2}\right]\right]$ when the indices of direction are divided by 2. The position coordinates and the corresponding direction are shown in Fig. 4.10.
(c) Similarly, the position coordinates for [220] are found to be [[110]] when the indices of direction are divided by 2. The position coordinates and the corresponding direction are shown in Fig. 4.10.

Example 6 Draw a $(1\bar{1}0)$ plane in a cubic unit cell. Show the directions that lie on this plane and find their Miller indices.

Fig. 4.10 [112], [121] and
[220] directions

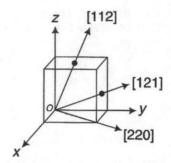

Solution: Construct a cubic unit cell; select its origin and the three crystallographic axes. Follow the above said procedure and draw the given plane by shifting the origin from O to O' as shown in Fig. 4.11. The Miller indices of directions are also shown. They are: $[11\bar{1}], [111]$ and their opposite $[\bar{1}\bar{1}1]$ and $[\bar{1}\bar{1}\bar{1}]$.

Example 7 Draw the planes (110) and (111) in a cubic unit cell. Determine the Miller indices of the line of their intersection.

Solution: Consider the conventional orthogonal axes and construct a unit cell with O as the origin. Follow the above said procedure and draw the given planes as shown in Fig. 4.12. The Miller indices of the common direction of the two planes are shown as $[1\bar{1}0]$ or $[\bar{1}10]$.

Example 8 Determine the directions connecting the origin with the nearest lattice sites whose coordinates are $\left[\left[\frac{1}{3}\frac{1}{3}\frac{2}{3}\right]\right]$, $\left[\left[\frac{1}{3}\frac{2}{3}\frac{1}{3}\right]\right]$ and $\left[\left[\frac{1}{2}\frac{1}{2}0\right]\right]$. Show them in a cubic unit cell.

Solution: *Given:* The nearest lattice sites whose fractional coordinates are: $\left[\left[\frac{1}{3}\frac{1}{3}\frac{2}{3}\right]\right]$, $\left[\left[\frac{1}{3}\frac{2}{3}\frac{1}{3}\right]\right]$ and $\left[\left[\frac{1}{2}\frac{1}{2}0\right]\right]$. Removing fractions by multiplying with 3, 3 and 2 respectively, their position coordinates and the directions become:

$$\left[\left[\frac{1}{3}\frac{1}{3}\frac{2}{3}\right]\right] \rightarrow [[112]] \rightarrow [112]$$

$$\left[\left[\frac{1}{3}\frac{2}{3}\frac{1}{3}\right]\right] \rightarrow [[121]] \rightarrow [121]$$

Fig. 4.11 $(1\bar{1}0)$ plane and
the directions that lie on this
plane

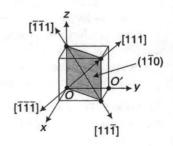

Fig. 4.12 (110) and
(111) planes and their
intersection

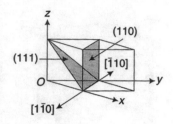

$$\left[\left[\frac{1}{2}\frac{1}{2}0\right]\right] \rightarrow [[110]] \rightarrow [110]$$

They are shown in Fig. 4.10.

Example 9 Draw (100), (110), (111) and their next order planes in sc, bcc and fcc unit cells.

Solution: Let us select the suitable axes and draw the required planes as per the set procedure so that we have a clear view of them as shown in Fig. 4.13.

4.3 Miller–Bravais Indices

We know that non-hexagonal crystal systems require a three index system (Miller indices) to represent a plane or a direction. However for a hexagonal crystal system, a four index system (Miller–Bravais indices) is required: three coplanar axes in the basal plane of the hexagon and a perpendicular axis as shown in Fig. 4.14.

Indices of Planes.

Miller–Bravais indices of a plane can be obtained as Miller indices as before. However, because the three coplanar axes a_1, a_2, a_3 are related to each other by a rotation of 120° (about c-axis), the indices h, k and l are also related to one another according to the equation

$$i = -(h+k) \quad \text{or} \quad h+k+i = 0$$

Therefore, the Miller–Bravais indices are symbolized as (hkil) or (hk.l).

Indices of Directions.

The conversion from three index system Miller indices [HKL] to four index system Miller–Bravais indices [hkil] for a crystallographic direction is not as simple as it is for a crystal plane. We need to derive the relationships between [HKL] and [hkil] using the fact that for a vector specified in both systems must be identical. Therefore, we can write

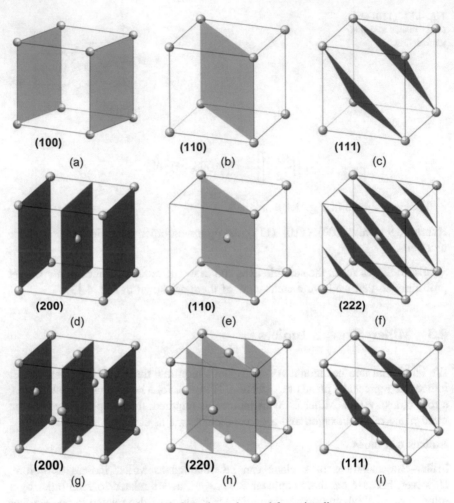

Fig. 4.13 Representation of crystal planes in sc, bcc and fcc unit cells

Fig. 4.14 Miller–Bravais
axes

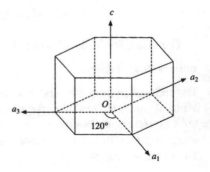

$$Ha_1 + Ka_2 + Lc = ha_1 + ka_2 + ia_3 + lc \tag{4.1}$$

Since, the axes a_1, a_2 and a_3 are related to each other by a rotation of 120°, the vector sum

$$a_1 + a_2 + a_3 = 0 \quad \text{or} \quad a_3 = -(a_1 + a_2)$$

Substituting the value of a_3, Eq. 4.1 becomes

$$Ha_1 + Ka_2 + Lc = ha_1 + ka_2 - i(a_1 + a_2) + lc \tag{4.2}$$

Comparing the coefficients of a_1, a_2 and c in Eq. 4.2, we obtain

$$H = h - i, \quad K = k - i \quad \text{and} \quad L = 1$$

An inverse transformation gives us

$$h = \frac{1}{3}(2H - K), k = \frac{1}{3}(2K - H), i = -(h + k) = -\frac{1}{3}(H + K), l = L$$

Figure 4.15 shows the Miller–Bravais indices of principal directions in the basal plane of hexagonal crystal system.

Solved Examples

Example 1 Change the three index system of Miller indices (310), ($\bar{1}$23), (011), (346), and ($4\bar{2}3$) into four index system of Miller–Bravais indices.

Solution: We can determine the value of the index i and hence the Miller–Bravais indices using the formula $i = -(h + k)$. The Miller–Bravais indices are shown in the right column.

Fig. 4.15 Miller–Bravais indices of principal directions in the basal plane of hexagonal crystal system

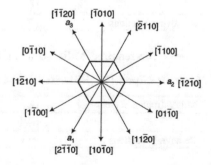

Miller indices	Miller–Bravais indices
(310)	$(31\bar{4}0)$ or (31.0)
$(\bar{1}23)$	$(\bar{1}2\bar{1}3)$ or $(\bar{1}2.3)$
(011)	$(01\bar{1}1)$ or (01.1)
$(3\bar{4}6)$	$(3\bar{4}\bar{7}6)$ or (34.6)
$(4\bar{2}3)$	$(4\bar{2}\bar{2}3)$ or $(4\bar{2}.3)$

Example 2 Replace the dots by numerals from the following shortened Miller–Bravais notations: (11.2), (10.3), $(1\bar{1}.4)$, (12.6), $(2\bar{4}.5)$, (21.3), (01.2), $(\bar{1}3.2)$, $(\bar{1}\bar{1}.2)$, $(2\bar{2}.3)$ and $(1\bar{4}.4)$.

Solution: We can determine the value of i and hence the value of dot by using the formula i = − (h + k). The full Miller–Bravais notations of the above shortened notations are: $(11\bar{2}2)$, $(10\bar{1}3)$, $(1\bar{1}04)$ $(12\bar{3}6)$ $(2\bar{4}25)$, $(21\bar{3}3)$, $(01\bar{1}2)$, $(\bar{1}3\bar{2}2)$, $(\bar{1}\bar{1}22)$, $(2\bar{2}03)$ and $(1\bar{4}34)$.

Example 3 Determine the Miller–Bravais indices for the following directions: [100], [010] and [110]. Show that they belong to the same family.

Solution: *Given*: Three directions: [100], [010] and [110] in three index system. The given directions can be changed into four index system Miller–Bravais indices by using the following conversion formula:

$$h = \frac{1}{3}(2H - K), k = \frac{1}{3}(2K - H), i = -(h + k) = -\frac{1}{3}(H + K), l = L$$

Let us find them one by one.

Case I: Given: [HKL] ≡ [100]. Using the conversion formula, we can write

Miller–Bravais Indices	h	k	i	l
	$\frac{2}{3}$	$-\frac{1}{3}$	$-\frac{1}{3}$	0
Removing fractions	2	−1	−1	0

⟹ The required Miller–Bravais indices are: $[2\bar{1}\bar{1}0]$.

Case II: *Given*: [HKL] ≡ [010]. Using the conversion formula, we can write.

Miller–Bravais indices	h	k	i	l
	$-\frac{1}{3}$	$\frac{2}{3}$	$-\frac{1}{3}$	0
Removing fractions	−1	2	−1	0

⟹ The required Miller–Bravais indices are: $[\bar{1}2\bar{1}0]$.

Case III: *Given*: [HKL] ≡ [110]. Using the conversion formula, we can write.

Miller–Bravais indices	h	k	i	l
	$\frac{1}{3}$	$\frac{1}{3}$	$-\frac{2}{3}$	0
Removing fractions	1	1	−2	0

⟹ The required Miller–Bravais indices are: $[11\bar{2}0]$.

The above calculations show that: $[2\bar{1}\bar{1}0]$, $[\bar{1}2\bar{1}0]$ and $[11\bar{2}0]$ belong to the same family.

Example 4 Show that in a hexagonal crystal system, the Miller–Bravais indices are related as

$$i = -(h+k) \quad \text{or} \quad h+k+i = 0.$$

Proof We know that in a hexagonal unit cell a = b ≠ c and α = β = 90° ≠ γ = 120°. The basal plane of the same is shown in Fig. 4.16.

Let an arbitrary plane (hkil) makes intercepts p on a_1, q on a_2, − r on a_3 (and s on c) axes, respectively. Since the axes a_1, a_2, and a_3 are related to one another by a rotation of 120° (about c-axis), they represent equivalent directions. The unit translation on each of these axes is the same, that is, a. Therefore, the intercepts along the four axes are:

$$p = \frac{a}{h}, q = \frac{a}{k}, r = \frac{-a}{i} \text{ and } s = \frac{c}{l}$$

Fig. 4.16 Intercepts of (hkil) plane on different axes

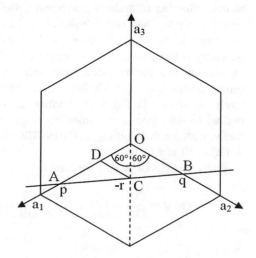

Draw a line DC parallel to OB, so that the Δ OCD is an equilateral triangle and the length of its sides, OD = DC = r (Fig. 4.16). Further, the Δ AOB and Δ ADC are similar, so that

$$\frac{DC}{AD} = \frac{OB}{OA}$$

$$\text{or} \quad \frac{r}{p - r} = \frac{q}{p}$$

$$\text{or} \quad rp = q(p - r) = qp - qr$$

Substituting the values of p, q, and r, we have

$$\left(\frac{-a}{i} \times \frac{a}{h}\right) = \left(\frac{a}{k} \times \frac{a}{h}\right) - \left(\frac{a}{k} \times \frac{-a}{i}\right) \text{or} \quad \left(\frac{-1}{ih}\right) = \left(\frac{1}{hk}\right) + \left(\frac{1}{ki}\right)$$

Now, multiplying both sides by hki, we obtain

$$i = -(h + k) \quad \text{or} \quad h + k + i = 0.$$

4.4   Interplanar Spacing

Two methods are used to determine the interplanar spacing between two consecutive parallel planes. They are briefly described below.

1. Using Cartesian geometry

With the knowledge of indexing of crystal planes and directions, it is now possible to determine the formula of interplanar spacing between two consecutive parallel planes in a given unit cell. We shall limit our discussion to the unit cells which are expressed in terms of the orthogonal coordinate axes, so that simple Cartesian geometry is applicable. Thus let us consider three mutually perpendicular axes Ox, Oy and Oz, and assume that a plane (hkl) parallel to the plane passing through the origin, makes intercepts a/h, b/k and c/l on the three axes at A, B and C, respectively, as shown in Fig. 4.17. Further, let OP (= d, the interplanar spacing) be normal to the plane drawn from the origin and makes angles α, β and γ, respectively, with the three orthogonal axes. Therefore, we can write: OA = a/h, OB = b/k, OC = c/l and OP = d.

From triangle OPA etc., we have

$$\cos \alpha = \frac{OP}{OA} = \frac{d}{a/h}, \cos \beta = \frac{OP}{OB} = \frac{d}{b/k} \text{ and } \cos \gamma = \frac{OP}{OC} = \frac{d}{c/l}$$

Fig. 4.17 (hkl) plane intercepting x, y and z axes at A, B and C, respectively

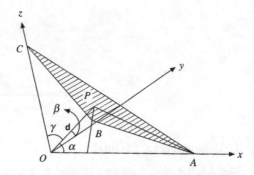

Now, making use of the direction cosine, which states that

$$\cos^2 \alpha + \cos^2 \beta + \cos^2 \gamma = 1 \tag{4.3}$$

and substituting the values of $\cos \alpha$, $\cos \beta$ and $\cos \gamma$ in Eq. 4.3, we obtain

$$\frac{d^2}{(a/h)^2} + \frac{d^2}{(b/k)^2} + \frac{d^2}{(c/l)^2} = 1$$

or

$$d^2 \left[\frac{h^2}{a^2} + \frac{k^2}{b^2} + \frac{l^2}{c^2} \right] = 1$$

So that

$$d = \left[\frac{h^2}{a^2} + \frac{k^2}{b^2} + \frac{l^2}{c^2} \right]^{-1/2} \tag{4.4}$$

This is a general formula and applicable to the primitive lattice of orthorhombic, tetragonal and cubic systems.

(i) Tetragonal system: $a = b \neq c$, the above Eq. 4.4, reduces to

$$d = \left[\frac{h^2 + k^2}{a^2} + \frac{l^2}{c^2} \right]^{-1/2}$$

(ii) Cubic system: $a = b = c$, the above Eq. 4.4, reduces to

$$d = a\left(h^2 + k^2 + l^2\right)^{-1/2}$$

or

$$d = \frac{a}{\left(h^2 + k^2 + l^2\right)^{1/2}}$$

2. Using General Method

Simple Cartesian geometry is used to determine the interplanar spacing in the unit cells involving orthogonal coordinate system. However, for the unit cells involving oblique system, simple Cartesian geometry is not very helpful. Therefore, let us use the reciprocal lattice concept to get the general expression for interlayer spacing in complex cases. Accordingly, let us write

$$
\begin{aligned}
\frac{1}{d_{hkl}^2} = \sigma_{hkl}.\sigma_{hkl} &= (ha^* + kb^* + lc^*).(ha^* + kb^* + lc^*) \\
&= h^2 a^*.a^* + hk\, a^*.b^* + hl\, a^*.c^* + kh\, b^*.a^* + k^2 b^*.b^* \\
&\quad + kl\, b^*.c^* + lh\, c^*.a^* + lk\, c^*.b^* + l^2 c^*.c^*
\end{aligned}
\tag{4.5}
$$

Making use of the identities $a^*.a^* = a^{*\,2}$, etc. and $a^*.b^* = b^*.a^* = a^*\,b^* \cos\gamma^*$, etc. and collecting the like terms, Eq. 4.5 becomes

$$
\begin{aligned}
\frac{1}{d_{hkl}^2} &= h^2 a^{*2} + k^2 b^{*2} + l^2 c^{*2} \\
&\quad + 2hk\, a^* b^* \cos\gamma^* + 2kl\, b^* c^* \cos\alpha^* + 2lh\, c^* a^* \cos\beta^*
\end{aligned}
\tag{4.6}
$$

Substituting the values of $a^* = \frac{b \times c}{V}$, b^*, c^*, etc. and simplifying Eq. 4.6 will become

$$
\frac{1}{d_{hkl}^2} = \frac{1}{V^2}
\left[
\begin{array}{l}
h^2 b^2 c^2 \sin^2\alpha + k^2 c^2 a c^2 \sin^2\beta + l^2 a^2 b^2 \sin^2\gamma\; + \\
2hkabc^2(\cos\alpha\,\cos\beta - \cos\gamma)\; + \\
2kla^2 bc(\cos\beta\,\cos\gamma - \cos\alpha)\; + \\
2lhkab^2 c(\cos\gamma\,\cos\alpha - \cos\beta)
\end{array}
\right]
\tag{4.7}
$$

where $V^2 = a^2 b^2 c^2 (1 - \cos^2\alpha - \cos^2\beta - \cos^2\gamma + 2\cos\alpha\cos\beta\cos\gamma)$.

The general expression for interplanar spacing given by Eq. 4.6 is valid for triclinic crystal system and the equations for other crystal systems can be obtained by substituting respective axial parameters (i.e., axes and angles). They are provided in Table 4.4.

Table 4.4 Interplanar spacing in various crystal systems

Crystal system	d_{hkl}
Cubic	$a\left(h^2 + k^2 + l^2\right)^{-1/2}$
Tetragonal	$\left[\frac{h^2+k^2}{a^2} + \frac{l^2}{c^2}\right]^{-1/2}$
Orthorhombic	$\left[\frac{h^2}{a^2} + \frac{k^2}{b^2} + \frac{l^2}{c^2}\right]^{-1/2}$
Hexagonal	$\left[\frac{4/3(h^2+hk+k^2)}{a^2} + \frac{l^2}{c^2}\right]^{-1/2}$
Rhombohedral	$\dfrac{a\left(1-3\cos^2\alpha+2\cos^3\alpha\right)^{1/2}}{\left[\left(h^2+k^2+l^2\right)\sin^2\alpha+2\left(hk+kl+lh\right)(\cos^2\alpha-\cos\alpha)\right]^{1/2}}$
Monoclinic	$\left[\dfrac{\frac{h^2}{k^2}+\frac{l^2}{c^2}-\frac{(2hl\cos\beta)}{ac}}{\sin^2\beta} + \frac{k^2}{b^2}\right]^{-1/2}$
Triclinic	$\left[\begin{vmatrix} h/a & \cos\gamma & \cos\beta \\ \frac{h}{a}\ k/b & 1 & \cos\alpha \\ l/c & \cos\alpha & 1 \end{vmatrix} + \begin{vmatrix} 1 & h/a & \cos\beta \\ \frac{k}{b}\cos\gamma & k/b & \cos\alpha \\ \cos\beta & l/c & 1 \end{vmatrix} + \begin{vmatrix} 1 & \cos\gamma & h/a \\ \frac{l}{c}\cos\gamma & 1 & k/b \\ \cos\beta & \cos\alpha & l/c \end{vmatrix}\right]^{-1/2}$
	$\begin{vmatrix} 1 & \cos\gamma & \cos\beta \\ \cos\gamma & 1 & \cos\alpha \\ \cos\beta & \cos\alpha & 1 \end{vmatrix}^{-1/2}$

Solved Examples

Example 1 In a cubic crystal system, the distance between the consecutive (111) plane is 2Å. Determine its lattice parameter.

Solution: *Given*: Crystal plane (111), d = 2Å, a = ?
For a cubic crystal (Table 4.3), we have

$$d = \frac{a}{\left(h^2 + k^2 + l^2\right)^{1/2}}$$

Substituting different values, we get

$$d = \frac{a}{\left(1^2 + 1^2 + 1^2\right)^{1/2}} = \frac{a}{\sqrt{3}}$$

$$\text{or} \quad a = d\sqrt{3} = 2\sqrt{3}\text{Å} = 3.46\text{Å}$$

Example 2 In a tetragonal crystal, the lattice parameters a = b = 2.42Å, and c = 1.74Å. Deduce the interplanar spacing between the consecutive (101) planes.

Solution: *Given*: Tetragonal crystal whose lattice parameters are: a = b = 2.42Å, and c = 1.74Å. $d_{101} = ?$

From Table 4.3, we have

$$d_{hkl} = \left[\frac{h^2 + k^2}{a^2} + \frac{l^2}{c^2}\right]^{-1/2}$$

$$\text{and} \quad d_{101} = \left[\frac{1^2 + 0^2}{(2.42)^2} + \frac{1^2}{(1.74)^2}\right]^{-1/2} = 1.41\text{Å}$$

Example 3 Determine the interplanar spacing in sc, bcc and fcc unit cells. Analyze the results obtained from them.

Solution: Let us take the three cases one by one.

Case I: Simple Cubic System.

In a simple cubic system, the lattice parameter a = b = c but the lattice points are situated only at the corners of the unit cell. Thus the interplanar spacing can be determined by simply using the equation

$$d = \frac{a}{\left(h^2 + k^2 + l^2\right)^{1/2}}$$

The interplanar spacing corresponding to three low index planes (100), (110) and (111) shown in Fig. 4.18 is:

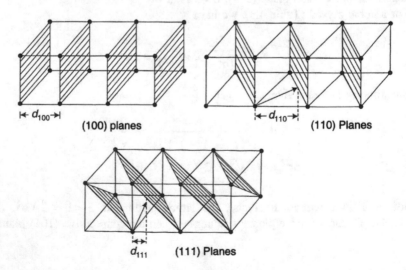

Fig. 4.18 Low index planes in simple cubic crystal: **a** (100) planes **b** (110) planes **c** (111) planes

$$d_{100} = a, d_{110} = \frac{a}{\sqrt{2}} \text{ and } d_{111} = \frac{a}{\sqrt{3}}$$

Hence, their ratio is:

$$d_{100} : d_{110} : d_{111} = 1 : \frac{1}{\sqrt{2}} : \frac{1}{\sqrt{3}}$$

Case II: Body-Centered Cubic System.

In a body-centered cubic system, the lattice points are situated at the eight corners and at the body center of the unit cell. Because of the presence of an additional point at the body center, the interplanar spacing for three low index planes (100), (110) and (111) give slightly different results. There appears an additional plane halfway between (100) and (111) planes as shown in Fig. 4.19, while no new planes appear between (110) planes when compared to simple cubic system. Therefore, the interplanar spacing for the low index planes in the body-centered cubic system is:

$$d_{100} = \frac{1}{2} (d_{100}) \text{simple cubic lattice} = \frac{a}{2}$$

$$d_{110} = (d_{110}) \text{simple cubic lattice} = \frac{a}{\sqrt{2}}$$

$$d_{111} = \frac{1}{2} (d_{111}) \text{simple cubic lattice} = \frac{a}{2\sqrt{3}}$$

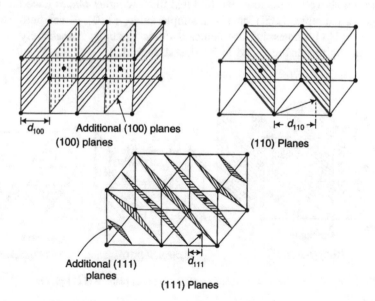

Fig. 4.19 Low index planes in bcc crystal: **a** (100) planes **b** (110) planes **c** (111) planes

Hence, their ratio is:

$$d_{100} : d_{110} : d_{111} = 1 : \sqrt{2} : \frac{1}{\sqrt{3}}$$

Case III: Face-Centered Cubic System.

In a face-centered cubic system, the lattice points are situated at the eight corners and at six face centers of the unit cell. Figure 4.20 shows the appearance of additional planes halfway between (100) and (110) planes, while no new planes appear between (111) planes when compared to simple cubic system. Therefore, the interplanar spacing for the low index planes in the face-centered cubic system is:

$$d_{100} = \frac{1}{2} (d_{100}) \text{simple cubic lattice} = \frac{a}{2}$$

$$d_{110} = \frac{1}{2} (d_{110}) \text{simple cubic lattice} = \frac{a}{2\sqrt{2}}$$

$$d_{111} = (d_{111}) \text{simple cubic lattice} = \frac{a}{\sqrt{3}}$$

Hence, their ratio is:

$$d_{100} : d_{110} : d_{111} = 1 : \frac{1}{\sqrt{2}} : \frac{2}{\sqrt{3}}$$

From the above calculations, we find that the low index planes have the widest spacing, for example, {100} planes in simple cubic, {110} planes body-centered cubic and {111} planes in face-centered cubic systems, respectively. Further, widest spacing planes are found to be closest packed.

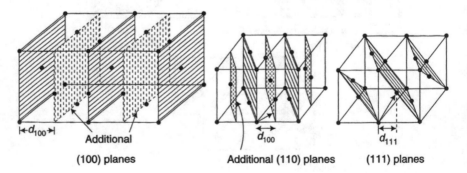

Fig. 4.20 Low index planes in fcc crystal: **a** (100), **b** (110) plane **c** (111) planes

Example 4 A tetragonal crystal has the c/a ratio as 1.5. Determine the interplanar spacing in (100), (110) and (111) planes.

Solution: *Given*: A tetragonal crystal system with c/a = 1.5. d_{100} = ?$0.1cmd_{110}$ = ?$0.1cmd_{111}$ = ?

For a tetragonal crystal system, we know that a = b ≠ c, therefore, c = 1.5a or c = 1.5 (if a = 1). Further, the interplanar spacing for tetragonal crystal system is given by

$$d_{(hkl)} = \left[\frac{h^2 + k^2}{a^2} + \frac{l^2}{c^2}\right]^{-1/2}$$

Therefore,

$$d_{(100)} = [1]^{-1/2} = 1\text{Å}$$

$$d_{(110)} = [2]^{-1/2} = 0.707\text{Å}$$

$$d_{(111)} = \left[2 + \frac{1}{1.5}\right]^{-1/2} = [2 + 0.66]^{-1/2} = 0.61\text{Å}$$

Example 5 Lattice parameters of an orthorhombic crystal are 2Å, 3Å and 4Å, respectively. Determine the interplanar spacing in (100), (110) and (111) planes.

Solution: Given: An orthorhombic crystal system with a = 2Å, b = 3Å, c = 4Å. d_{100} = ?$0.1cmd_{110}$ = ?$0.1cmd_{111}$ = ?

For an orthorhombic crystal system, we know that a = b = c and the interplanar spacing is given by

$$d_{(hkl)} = \left[\frac{h^2}{a^2} + \frac{k^2}{b^2} + \frac{l^2}{c^2}\right]^{-1/2}$$

Therefore,

$$d_{(100)} = \left[\frac{1}{4}\right]^{-1/2} = 2\text{Å}$$

$$d_{(110)} = \left[\frac{1}{4} + \frac{1}{9}\right]^{-1/2} = \left[\frac{13}{36}\right]^{-1/2} = 1.66\text{Å}$$

$$d_{(111)} = \left[\frac{1}{4} + \frac{1}{9} + \frac{1}{16}\right]^{-1/2} = \left[\frac{61}{144}\right]^{-1/2} = 1.54\text{Å}$$

Example 6 Lattice parameters of a hexagonal crystal are a = b = 4Å, and c = 6Å, respectively. Determine the interplanar spacing in $(10\bar{1}0)$, $(01\bar{1}0)$, $(\bar{1}100)$ and $(11\bar{2}0)$ planes.

Solution: *Given*: A hexagonal crystal system with a = b = 4Å, c = 6Å. Determine the interplanar spacing in $(10\bar{1}0)$, $(01\bar{1}0)$, $(\bar{1}100)$ and $(11\bar{2}0)$ planes.

For a hexagonal crystal system, we know that $a = b \neq c$ and the interplanar spacing is given by

$$d_{(hkil)} = \left[\frac{4}{3} \cdot \frac{(h^2 + hk + k^2)}{a^2} + \frac{l^2}{c^2} \right]^{-1/2}$$

Therefore,

$$d_{(10\bar{1}0)} = \left[\frac{4}{3} \times \frac{1}{16} \right]^{-1/2} = \left[\frac{1}{12} \right]^{-1/2} = 3.46\text{Å}$$

$$d_{(01\bar{1}0)} = \left[\frac{4}{3} \times \frac{1}{16} \right]^{-1/2} = \left[\frac{1}{12} \right]^{-1/2} = 3.46\text{Å}$$

$$d_{(\bar{1}100)} = \left[\frac{4}{3} \times \frac{(1-1+1)}{16} \right]^{-1/2} = \left[\frac{1}{12} \right]^{-1/2} = 3.46\text{Å}$$

$$d_{(11\bar{2}0)} = \left[\frac{4}{3} \times \frac{(1+1+1)}{16} \right]^{-1/2} = \left[\frac{1}{4} \right]^{-1/2} = 2\text{Å}$$

Example 7 The lattice parameter of an ideal rhombohedrdron is 3Å. Determine the interplanar spacing in (100), (110) and (111) planes.

Solution: *Given*: An ideal rhombohedron with $a = b = c = 3\text{Å}$. Also, we know that for an ideal rhombohedron, $\alpha = 60°$. $d_{100} = ?0.1cmd_{110} = ?0.1cmd_{111} = ?$

The interplanar spacing for rhombohedral crystal system is given by

$$d_{(hkl)} = \frac{a(1 + \cos^3 \alpha - 3\cos^2 \alpha)^{1/2}}{\left[(h^2 + k^2 + l^2) \sin^2 \alpha + 2(hk + kl + lh)(\cos^2 \alpha - \cos \alpha) \right]^{1/2}}$$

Therefore,

$$d_{(100)} = \frac{3(1 + \cos^3 60 - 3\cos^2 60)^{1/2}}{\sin 60}$$

$$= \frac{3(1 + 2 \times 0.125 - 3 \times 0.25)^{1/2}}{0.886} = \frac{2.12}{0.866} = 2.45\text{Å}$$

$$d_{(110)} = \frac{2.12}{\left[2\sin^2 60 + 2(\cos^2 60 - \cos 60) \right]^{1/2}}$$

$$= \frac{2.12}{\left[2 \times 0.75 + 2(0.25 - 0.5) \right]^{1/2}}$$

$$= \frac{2.12}{[1.5 - 0.5]^{1/2}} = 2.12\text{Å}$$

$$d_{(111)} = \frac{2.12}{\left[3\sin^2 60 + 6(\cos^2 60 - \cos 60)\right]^{1/2}}$$

$$= \frac{2.12}{\left[2.25 + 6(0.25 - 0.5)\right]^{\frac{1}{2}}}$$

$$= \frac{2.12}{[2.25 - 1.5]^{1/2}} = \frac{2.12}{[0.75]^{1/2}} = 2.45\text{Å}$$

4.5 Zone and Zone Axis

(a) Zone

A zone in crystal may be defined as a set of crystal planes (or faces) which when intersect each other, produce parallel edges. For example faces of a pencil (hexagon) intersect each other along one direction (the direction of the lead) constitutes a zone (Fig. 4.21). Further, in a zone, there is neither any limitation in the number of planes (or faces) nor they need to be crytallographically equivalent. For example, the edges of the pencil may be shaved flat to get a 12-sided pencil, that is, an additional six faces in the zone. On the other hand, a plane in a crystal belongs to a limited number of zones, that is, 2–4.

(b) Zone Axis

A zone axismay be defined as the common direction of the intersection of the faces in the zone. For example, in a pencil, the direction of the lead is the zone axis corresponding to the six faces of the zone. In fact, all crystal directions are zone axes, so the term "crystal direction" and "zone axis" are synonymous. Thus a zone axis like crystal direction is represented as [uvw]. It is defined as a vector R (measured from the origin) according to the equation:

$$R = ua + vb + wc \tag{4.8}$$

where a, b, c are basis vectors as before.

Fig. 4.21 Parallel edges in a pencil

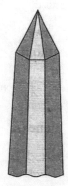

There exist some geometrical relationships among the zone axis and the plane in the zone. Let us derive them with the help of reciprocal lattice concept using vector algebra.

1. *The Weiss Zone Law*

This states that the indices of a zone axis [uvw] and a crystal plane (hkl) in the zone must obey the algebraic relation:

$$hu + kv + lw = 0 \tag{4.9}$$

Proof If the plane (hkl) and the zone axis [uvw] are in the same zone, then the reciprocal lattice vector d^*_{hkl} must be perpendicular to the vector R_{uvw}. Therefore, we can write.

$$d^*hkl.R_{uvw} = 0$$
$$\text{or } (ha^* + kb^* + lc^*).(ua + vb + wc) = 0$$
$$\Rightarrow hu + kv + lw = 0$$

where $a^*.a = b^*.b = c^*.c = 1$ and $a^*.b = a^*.c = 0$, etc.

2. *Zone Axis Lying at the Intersection of Two Planes*

This states that the indices of the zone axis   [uvw] of two intersection planes $(h_1k_1l_1)$ and $(h_2k_2l_2)$ in a zone   can be determined according to the following relations:

$$u = \begin{vmatrix} k_1 & l_1 \\ k_2 & l_2 \end{vmatrix}, v = \begin{vmatrix} l_1 & h_1 \\ l_2 & h_2 \end{vmatrix}, w = \begin{vmatrix} h_1 & k_1 \\ h_2 & k_2 \end{vmatrix} \tag{4.10}$$

where the determinant $\begin{vmatrix} k_1 & l_1 \\ k_2 & l_2 \end{vmatrix} = (k_1l_2 - l_1k_2)$, etc.

Proof Since the zone axis R_{uvw} is lying at the intersection of the planes $(h_1k_1l_1)$ and $(h_2k_2l_2)$, therefore, R_{uvw} will be perpendicular to their normal, that is, $d^*_{h_1k_1l_1}$ and $d^*_{h_2k_2l_2}$. From simple geometry, the zone axis is expressed as.

$$R_{uvw} = \frac{d^*_{h_1k_1l_1} \times d^*_{h_2k_2l_2}}{V^*} \tag{4.11}$$

where V^* is the volume of the reciprocal unit cell. Equation 4.11 can further be written as

$$ua + vb + wc = \frac{(h_1a^* + k_1b^* + l_1c^*) \times (h_2a^* + k_2b^* + l_2c^*)}{V^*}$$

$$= \frac{1}{V^*}\left[\begin{array}{l} h_1k_2(a^* \times b^*) + h_1l_2(a^* \times c^*) + k_1h_2(b^* \times a^*) + \\ k_1l_2(b^* \times c^*) + l_1h_2(c^* \times a^*) + l_1k_2(c^* \times b^*) \end{array}\right]$$

$$= \frac{1}{V^*}[(a^* \times b^*)(h_1k_2 - k_1h_2) + (b^* \times c^*)(k_1l_2 - l_1k_2) + (c^* \times a^*)(l_1h_2 - h_1l_2)]$$

Using the identities such as $(b^* \times a^*) = -(a^* \times b^*)$, etc. and $a = \frac{b^* \times c^*}{V^*}$, etc. the above equation reduces to

$$ua + vb + wc = c(h_1k_2 - k_1h_2) + a(k_1l_2 - l_1k_2) + b(l_1h_2 - h_1l_2)$$

Comparing the coefficients of a, b and c, we obtain

$$u = (k_1l_2 - l_1k_2), \ v = (l_1h_2 - h_1l_2), \ w = (h_1k_2 - k_1h_2)$$

3. *A Plane (hkl) Containing Two Directions* $[u_1v_1w_1]$ *and* $[u_2v_2w_2]$

This states that if a plane (hkl) contains two directions $[u_1v_1w_1]$ and $[u_2v_2w_2]$, then the indices of the plane can be obtained according to the following relations:

$$h = \begin{vmatrix} v_1 & w_1 \\ v_2 & w_2 \end{vmatrix}, \ k = \begin{vmatrix} w_1 & u_1 \\ w_2 & u_2 \end{vmatrix}, \ l = \begin{vmatrix} u_1 & v_1 \\ u_2 & v_2 \end{vmatrix}$$

Proof Since the plane (hkl) defined by its normal d^*_{hkl} is perpendicular to both directions $R_{u_1v_1w_1}$ and $R_{u_2v_2w_2}$, therefore, d^*_{hkl} can be expressed as

$$d^*_{hkl} = \frac{R_{u_1v_1w_1} \times R_{u_2v_2w_2}}{V} \tag{4.12}$$

Here, V is the volume of direct unit cell. Equation 4.12 can further be written as

$$ua^* + vb^* + wc^* = \frac{(u_1a + v_1b + w_1c) \times (u_2a + v_2b + w_2c)}{V}$$

$$= \frac{1}{V}[u_1v_2(a \times b) + u_1w_2(a \times c) + v_1u_2(b \times a)$$

$$+ v_1w_2(b \times c) + w_1u_2(c \times a) + w_1v_2(c \times b)]$$

$$= \frac{1}{V}[(a \times b)(u_1v_2 - v_1u_2) + (b \times c)(v_1w_2 - w_1v_2)$$

$$+ (c \times a)(w_1u_2 - u_1w_2)]$$

Using the identities such as $a^* = \frac{b \times c}{V}$, etc. above equation reduces to

$$ha^* + kb^* + lc^* = c^*(u_1v_2 - v_1u_2) + a^*(v_1w_2 - w_1v_2) + b^*(w_1u_2 - u_1w_2)$$

Comparing the coefficients of a*, b* and c*, we obtain

$$h = (v_1w_2 - w_1v_2), \; k = (w_1u_2 - u_1w_2), \; l = (u_1v_2 - v_1u_2)$$

4. *The Addition Rule*

The Miller indices of a plane (hkl) sandwiched between two planes of indices $(h_1k_1l_1)$ and $(h_2k_2l_2)$ in a zone can be determined according to the following relations:

$$h = mh_1 \pm nh_2$$

$$k = mk_1 \pm nk_2$$

$$l = ml_1 \pm nl_2$$

where m and n can be any positive integer such as 1, 2, etc.

Solved Examples

Example 1 Identify the planes from (112), (321), (123), (212) and $(23\bar{1})$ belonging to the zone $[1\bar{1}1]$.

Solution: *Given*: A set of planes: (112), (321), (123), (212) and $(23\bar{1})$, the zone axis is $[1\bar{1}1]$.

We know that a plane and a direction can belong to the same zone if they satisfy the condition:

$$hu + kv + lc = 0$$

Now, checking with all the given planes one by one, we have

$$1 \times 1 + 1 \times (-1) + 2 \times (-1) = 1 - 1 - 2 = -2$$
$$3 \times 1 + 2 \times (-1) + 1 \times (-1) = 3 - 2 - 1 = 0$$
$$1 \times 1 + 2 \times (-1) + 3 \times (-1) = 1 - 2 - 3 = -4$$
$$2 \times 1 + 1 \times (-1) + 2 \times (-1) = 2 - 1 - 2 = -1$$
$$2 \times 1 + 3 \times (-1) + (-1) \times (-1) = 2 - 3 + 1 = 0$$

$\Longrightarrow$ The planes (321) and $(23\bar{1})$ belong to the zone $[1\bar{1}1]$.

Example 2 Identify the zone axes from $[\bar{1}11]$, $[1\bar{1}1]$, $[11\bar{1}]$, $[1\bar{2}1]$, $[\bar{1}21]$ and $[112]$ which are parallel to the plane (123).

Solution: *Given*: A set of z4one axes: $[\bar{1}11]$, $[1\bar{1}1]$, $[11\bar{1}]$, $[1\bar{2}1]$, $[\bar{1}21]$ and $[112]$, a plane (123).

We know that a zone axis is parallel to a plane if they satisfy the condition:

$$hu + kv + lc = 0$$

Now, checking with all the given planes one by one, we have

$$1 \times (-1) + 2 \times 1 + 3 \times 1 = -1 + 2 + 3 = 4$$
$$1 \times 1 + 2 \times (-1) + 3 \times 1 = 1 - 2 + 3 = 2$$
$$1 \times 1 + 2 \times 1 + 3 \times (-1) = 1 + 2 - 3 = 0$$
$$1 \times 1 + 2 \times (-2) + 3 \times 1 = 1 - 4 + 3 = 0$$
$$1 \times (-1) + 2 \times 2 + 3 \times 1 = -1 + 4 + 3 = 6$$
$$1 \times 1 + 2 \times 1 + 3 \times 2 = 1 + 2 + 6 = 9$$

$\Longrightarrow$ The zone axes $[11\bar{1}]$ and $[1\bar{2}1]$ lie parallel to the plane (123).

Example 3 Find the zone axes lying at the intersection of the two sets of planes: (i) (342) and $(10\bar{3})$, (ii) $(21\bar{3})$ and (110).

Solution: *Given*: Two sets of planes: (i) (342) and $(10\bar{3})$, (ii) $(21\bar{3})$ and (110), [uvw] = ?

We can determine the indices of the zone axis [uvw] by using the following relations:

$$u = \begin{vmatrix} k_1 & l_1 \\ k_2 & l_2 \end{vmatrix}, \quad v = \begin{vmatrix} l_1 & h_1 \\ l_2 & h_2 \end{vmatrix}, \quad w = \begin{vmatrix} h_1 & k_1 \\ h_2 & k_2 \end{vmatrix}$$

For the first set of planes: (342) and $(10\bar{3})$, we have

$$u = \begin{vmatrix} 4 & 2 \\ 0 & -3 \end{vmatrix} = -12, \quad v = \begin{vmatrix} 2 & 3 \\ -3 & 1 \end{vmatrix} = 11, \quad w = \begin{vmatrix} 3 & 4 \\ 1 & 0 \end{vmatrix} = -4$$

$\Longrightarrow$ [uvw] $\equiv [\bar{1}211\bar{4}]$.

For the second set of planes: $(21\bar{3})$ and (110), we have

$$u = \begin{vmatrix} 1 & -3 \\ 1 & 0 \end{vmatrix} = 3, \quad v = \begin{vmatrix} -3 & 2 \\ 0 & 1 \end{vmatrix} = -3, \quad w = \begin{vmatrix} 2 & 1 \\ 1 & 1 \end{vmatrix} = 1$$

$\Longrightarrow$ [uvw] $\equiv [3\bar{3}1]$.

Example 4 Find the (hkl) plane containing two sets of directions: (i) [131] and $[0\bar{1}1]$, (ii) [102] and $[\bar{1}11]$.

Solution: *Given*: Two sets of directions: (i) [131] and $[0\bar{1}1]$, (ii) [102] and $[\bar{1}11]$, (hkl) = ?

We can determine the indices of the plane (hkl) by using the following relations:

$$h = \begin{vmatrix} v_1 & w_1 \\ v_2 & w_2 \end{vmatrix}, k = \begin{vmatrix} w_1 & u_1 \\ w_2 & u_2 \end{vmatrix}, l = \begin{vmatrix} u_1 & v_1 \\ u_2 & v_2 \end{vmatrix}$$

For the first set of directions: [131] and $[0\bar{1}1]$, we have

$$h = \begin{vmatrix} 3 & 1 \\ -1 & 1 \end{vmatrix} = 4, k = \begin{vmatrix} 1 & 1 \\ 1 & 0 \end{vmatrix} = -1, l = \begin{vmatrix} 1 & 3 \\ 0 & -1 \end{vmatrix} = -1$$

$\Longrightarrow$ (hkl) $\equiv (4\bar{1}\bar{1})$.

For the second set of directions: [102] and $[\bar{1}11]$, we have

$$h = \begin{vmatrix} 0 & 2 \\ -1 & 1 \end{vmatrix} = -2, k = \begin{vmatrix} 2 & 1 \\ 1 & -1 \end{vmatrix} = -3, l = \begin{vmatrix} 1 & 0 \\ -1 & -1 \end{vmatrix} = 1$$

$\Longrightarrow$ (hkl) $\equiv (\bar{2}\bar{3}1)$.

Example 5 Determine the Miller indices (hkl) of a plane marked P in Fig. 4.22, which is sandwiched between two sets of planes (i) $(1\bar{1}\bar{1})$ and (111), (ii) $(1\bar{1}1)$ and $(11\bar{1})$.

Solution: *Given*: Two sets of planes: (i) $(1\bar{1}\bar{1})$ and (111), (ii) $(1\bar{1}1)$ and $(11\bar{1})$.
(hkl) = ?
According to the addition rule, we can obtain:
For the first set of planes: $(1\bar{1}\bar{1})$ and (111), we have

$$(hkl) = (1\bar{1}\bar{1}) + (111) = (200) \text{ with } m = 1 \text{ and } n = 1$$

Similarly, for the second set of planes: $(1\bar{1}1)$ and $(11\bar{1})$, we have

$$(hkl) = (1\bar{1}1) + (11\bar{1}) = (200) \text{ with } m = 1 \text{ and } n = 1$$

Therefore, the required (hkl) plane is (200) $\equiv$ 2(100).

Fig. 4.22 Determination of required plane in the polyhedron

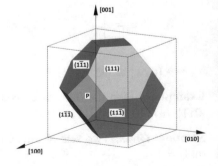

Multiple Choice Questions (MCQ)

1. The Miller indices of a plane that makes intercepts of 2a, 3b and 4c along the three crystallographic axes (where a, b, c are primitive translation vectors of the lattice) are:
   (a) (643)
   (b) (346)
   (c) (234)
   (d) (432)

2. The Miller indices of a plane that makes intercepts as 3a: 4b on the x and y axes, while parallel to z-axis is:
   (a) (340)
   (b) (430)
   (c) (304)
   (d) (403)

3. In an orthorhombic crystal with a: b: c = 1: 2: 5, a plane cuts intercepts of 3Å, 4Å, and 5Å on its coordinate axes. The Miller indices of the plane is:
   (a) (125)
   (b) (521)
   (c) (236)
   (d) (345)

4. In a crystal whose primitive translations are 1.2Å, 1.8Å and 2Å, a plane with Miller indices (231) cuts an intercept of 1.2Å along the x-axis. The lengths of intercepts along y and z-axes are:
   (a) 1.2Å and 1.2Å
   (b) 1.2Å and 1.8Å
   (c) 1.2Å and 2Å
   (d) 1.2Å and 4Å

5. In a cubic crystal, the lengths of the intercepts made by a plane with Miller indices ($1\bar{3}2$) on three axes are:
   (a) 6:2:3
   (b) 6:3:2
   (c) 3:2:1
   (d) 1:3:2

6. In a cubic crystal, a plane makes intercepts 1, -3, 1 on x, y and z axes, respectively. The Miller indices corresponding to this plane is:
   (a) $(1\bar{3}1)$
   (b) $(3\bar{1}3)$
   (c) $(1\bar{3}\bar{1})$
   (d) $(\bar{3}1\bar{3})$

7. The intercepts made by a plane with the Miller indices (123) on the three crystallographic axes x, y and z are:
   (a) a, 2b, 3c
   (b) 2a, 4b, 6c
   (c) 6a, 3b, 2c
   (d) 3a, 2b, c

8. The number of crystallogaphically equivalent planes in the {100} family of a cubic crystal system is:
   (a) 4
   (b) 6
   (c) 8
   (d) 12

9. The number of crystallogaphically equivalent planes in the {110} family of a cubic crystal system is:
   (a) 4
   (b) 6
   (c) 8
   (d) 12

10. The number of crystallogaphically equivalent planes in the {111} family of a cubic crystal system is:
    (a) 4
    (b) 6
    (c) 8
    (d) 12

11. $\left[\left[\frac{a}{2}b0\right]\right]$ are the coordinates of the lattice site nearest to the origin. Miller indices of the corresponding direction are:
    (a) [120]
    (b) [210]
    (c) $\left[\frac{1}{2}10\right]$
    (d) $\left[1\frac{1}{2}0\right]$

12. The coordinates of a lattice site nearest to the origin is given by $x = -2a$, $y = 1b$ and $z = \frac{3c}{2}$. the Miller indices of its direction are:
    (a) $[4\bar{2}3]$
    (b) $[\bar{4}23]$
    (c) $[\bar{2}13]$
    (d) $[1\bar{2}3]$

13. The direction that connects the origin and the point $\left[\left[\frac{1}{3}\frac{1}{3}\frac{2}{3}\right]\right]$ is:
    (a) $[211]$
    (b) $[121]$
    (c) $[112]$
    (d) $[123]$

14. $[211]$ is the direction that connects the origin and the point:
    (a) $[112]$
    (b) $[121]$
    (c) $\left[11\frac{1}{2}\right]$
    (d) $\left[1\frac{1}{2}\frac{1}{2}\right]$

15. The number of crystallogaphically equivalent directions in the <100> family of a cubic crystal system is:
    (a) 4
    (b) 6
    (c) 8
    (d) 12

16. The number of crystallogaphically equivalent directions in the <110> family of a cubic crystal system is:
    (a) 4
    (b) 6
    (c) 8
    (d) 12

17. The number of crystallogaphically equivalent directions in the <111> family of a cubic crystal system is:
    (a) 4
    (b) 6
    (c) 8
    (d) 12

18. The number of crystallogaphically equivalent directions in the <100> family of
    a tetragonal crystal system is:
    (a) 4
    (b) 6
    (c) 8
    (d) 12

19. The number of crystallogaphically equivalent directions in the <110> family of
    a tetragonal crystal system is:
    (a) 4
    (b) 6
    (c) 8
    (d) 12

20. The number of crystallogaphically equivalent directions in the <101> or
    <011> family of a tetragonal crystal system is:
    (a) 4
    (b) 6
    (c) 8
    (d) 12

21. The number of crystallogaphically equivalent directions in the <111> family of
    a tetragonal crystal system is:
    (a) 4
    (b) 6
    (c) 8
    (d) 12

22. The Miller indices of the line of intersection of the planes $(1\bar{1}\bar{1})$ and $(\bar{1}\bar{1}1)$ are:
    (a) $[011]$ or $[0\bar{1}\bar{1}]$
    (b) $[101]$ or $[\bar{1}0\bar{1}]$
    (c) $[110]$ or $[\bar{1}\bar{1}0]$
    (d) $[10\bar{1}]$ or $[\bar{1}01]$

23. The Miller indices of the line of intersection of the planes (110) and (111) are:
    (a) $[011]$ or $[0\bar{1}\bar{1}]$
    (b) $[101]$ or $[\bar{1}0\bar{1}]$
    (c) $[110]$ or $[\bar{1}\bar{1}0]$
    (d) $[1\bar{1}0]$ or $[\bar{1}01]$

24. The four index Miller–Bravais indices of the Miller plane (210) are:
    (a) $(21\bar{3}0)$ or (21.0)
    (b) $(3\bar{1}20)$ or $(3\bar{1}.0)$
    (c) $(\bar{1}320)$ or $(\bar{1}3.0)$
    (d) $(\bar{3}1\bar{2}0)$ or $(\bar{3}1.0)$

25. The four index Miller–Bravais indices of the Miller plane $(\bar{1}13)$ are:
    (a) $(11\bar{2}3)$ or (11.3)
    (b) $(\bar{1}103)$ or $(\bar{1}1.3)$
    (c) $(\bar{1}323)$ or $(\bar{1}3.3)$
    (d) $(\bar{3}1\bar{2}3)$ or $(\bar{3}1.3)$

26. The four index Miller–Bravais indices of the direction [100] are:
    (a) $[11\bar{2}0]$
    (b) $[\bar{1}2\bar{1}0]$
    (c) $[2\bar{1}\bar{1}0]$
    (d) $[1\bar{1}00]$

27. The four index Miller–Bravais indices of the direction [010] are:
    (a) $[11\bar{2}0]$
    (b) $[1\bar{1}00]$
    (c) $[2\bar{1}\bar{1}0]$
    (d) $[\bar{1}2\bar{1}0]$

28. The four index Miller–Bravais indices of the direction [110] are:
    (a) $[11\bar{2}0]$
    (b) $[1\bar{1}00]$
    (c) $[2\bar{1}\bar{1}0]$
    (d) $[\bar{1}2\bar{1}0]$

29. In a cubic crystal system, the distance between the consecutive (111) plane is
    2Å. Its lattice parameter is:
    (a) 2.46Å
    (b) 3.46Å
    (c) 4.46Å
    (d) 5.46Å

30. In a tetragonal crystal, the lattice parameters a = b = 2.42Å, and c = 1.74Å. The interplanar spacing between the consecutive (101) plane is:
    (a) 4.41Å
    (b) 3.41Å
    (c) 2.41Å
    (d) 1.41Å

31. The ratio of low index interplanar spacing $d_{100} : d_{110} : d_{111}$ in simple cubic (sc) crystal system is:
    (a) $1 : \frac{1}{\sqrt{2}} : \frac{1}{\sqrt{3}}$
    (b) $1 : \sqrt{2} : \sqrt{3}$
    (c) $1 : \sqrt{2} : \frac{1}{\sqrt{3}}$
    (d) $1 : \frac{1}{\sqrt{2}} : \frac{2}{\sqrt{3}}$

32. The ratio of low index interplanar spacing $d_{100} : d_{110} : d_{111}$ in a body-centered cubic (bcc) crystal system is:
    (a) $1 : \frac{1}{\sqrt{2}} : \frac{1}{\sqrt{3}}$
    (b) $1 : \sqrt{2} : \sqrt{3}$
    (c) $1 : \sqrt{2} : \frac{1}{\sqrt{3}}$
    (d) $1 : \frac{1}{\sqrt{2}} : \frac{2}{\sqrt{3}}$

33. The ratio of low index interplanar spacing $d_{100} : d_{110} : d_{111}$ in a face-centered cubic (fcc) crystal system is:
    (a) $1 : \frac{1}{\sqrt{2}} : \frac{1}{\sqrt{3}}$
    (b) $1 : \sqrt{2} : \sqrt{3}$
    (c) $1 : \sqrt{2} : \frac{1}{\sqrt{3}}$
    (d) $1 : \frac{1}{\sqrt{2}} : \frac{2}{\sqrt{3}}$

34. A tetragonal crystal has the c/a ratio as 1.5. The interplanar spacing between the consecutive (110) plane is:
    (a) 0.707Å
    (b) 0.607Å
    (c) 0.507Å
    (d) 0.407Å

35. A tetragonal crystal has the c/a ratio as 1.5. The interplanar spacing between the consecutive (111) plane is:
    (a) 0.71Å
    (b) 0.61Å
    (c) 0.51Å
    (d) 0.41Å

36. The lattice parameters of an orthorhombic crystal are 2Å, 3Å and 4Å, respectively. The interplanar spacing between the consecutive (110) plane is:
    (a) 3.66Å
    (b) 2.66Å
    (c) 1.66Å
    (d) 0.66Å

37. The lattice parameters of an orthorhombic crystal are 2Å, 3Å and 4Å, respectively. The interplanar spacing between the consecutive (111) plane is:
    (a) 4.54Å
    (b) 3.54Å
    (c) 2.54Å
    (d) 1.54Å

38. The lattice parameters of a hexagonal crystal are $a = b = 4$Å, and $c = 6$Å, respectively. The interplanar spacing between the consecutive $(10\bar{1}0)$ plane is:
    (a) 3.46Å
    (b) 4.46Å
    (c) 5.46Å
    (d) 6.46Å

39. The lattice parameters of a hexagonal crystal are $a = b = 4$Å, and $c = 6$Å, respectively. The interplanar spacing between the consecutive $(\bar{1}100)$ plane is:
    (a) 0.66Å
    (b) 1.66Å
    (c) 2.66Å
    (d) 3.66Å

40. The lattice parameters of a hexagonal crystal are $a = b = 4$Å, and $c = 6$Å, respectively. The interplanar spacing between the consecutive $(11\bar{2}0)$ plane is:
    (a) 4.00Å
    (b) 3.00Å
    (c) 2.00Å
    (d) 1.00Å

41. The lattice parameter of an ideal rhombohedrdron is 3Å. The interplanar spacing between the consecutive (100) plane is:
    (a) 5.45Å
    (b) 4.45Å
    (c) 3.45Å
    (d) 2.45Å

42. The lattice parameter of an ideal rhombohedrdron is 3Å. The interplanar spacing between the consecutive (110) plane is:
    (a) 2.12Å
    (b) 3.12Å
    (c) 4.12Å
    (d) 5.12Å

43. The lattice parameter of an ideal rhombohedrdron is 3Å. The interplanar spacing between the consecutive (111) plane is:
    (a) 1.45Å
    (b) 2.45Å
    (c) 3.45Å
    (d) 4.45Å

44. Show that the planes (321) and $(23\bar{1})$ belong to the zone:
    (a) $[\bar{1}11]$
    (b) $[1\bar{1}1]$
    (c) $[11\bar{1}]$
    (d) $[11\bar{1}]$

45. Show that the zone axes $[11\bar{1}]$ and $[1\bar{2}1]$ are parallel to the plane:
    (a) $(1\bar{2}3)$
    (b) (213)
    (c) (312)
    (d) (123)

46. The zone axis lying at the intersection of the two planes $(21\bar{3})$ and (110) is:
    (a) $[3\bar{3}1]$
    (b) $[3\bar{1}1]$
    (c) $[21\bar{1}]$
    (d) $[21\bar{3}]$

47. The (hkl) plane containing the two directions [131] and $[0\bar{1}1]$ is:
    (a) $(1\bar{3}1)$
    (b) $(4\bar{1}\bar{1})$
    (c) $(3\bar{1}4)$
    (d) $(1\bar{4}3)$

48. The (hkl) plane containing the two directions [102] and [$\bar{1}11$] is:
    (a) $(2\bar{3}1)$
    (b) $(3\bar{1}\bar{1})$
    (c) $(\bar{2}\bar{3}1)$
    (d) $(1\bar{4}2)$
49. The Miller indices (hkl) of the plane sandwiched between two planes (101) and
    (110) is:
    (a) (212)
    (b) (121)
    (c) (122)
    (d) (211)

Answers

 1. (a)
 2. (b)
 3. (c)
 4. (d)
 5. (a)
 6. (b)
 7. (c)
 8. (b)
 9. (d)
10. (c)
11. (a)
12. (b)
13. (c)
14. (d)
15. (b)
16. (d)
17. (c)
18. (a)
19. (a)
20. (c)
21. (c)
22. (b)
23. (d)
24. (a)
25. (b)
26. (c)
27. (d)
28. (a)
29. (b)
30. (d)

31. (a)
32. (c)
33. (d)
34. (a)
35. (b)
36. (c)
37. (d)
38. (a)
39. (b)
40. (c)
41. (d)
42. (a)
43. (b)
44. (c)
45. (d)
46. (a)
47. (b)
48. (c)
49. (d)

Chapter 5
Unit Cell Transformations

5.1 Transformation of Indices of Unit Cell Axes

In order to know the exact relationships (in terms of planes, directions, unit cell volumes, etc. both in direct and reciprocal lattices) between the two sets of unit cells such as one primitive to another, primitive to non-primitive or vice-versa, the transformation of one set of indices to another is carried out simply by the use of vector algebra. It is customary to take one unit cell as the first unit cell (corresponding to the first set of axes) and the other unit cell as the second unit cell (corresponding to the second set of axes). Thus the second set of axes a_2, b_2, c_2 can be defined in terms of the first set of axes a_1, b_1, c_1 by the following simultaneous equations:

$$a_2 = m_{11}a_1 + m_{12}b_1 + m_{13}c_1$$
$$b_2 = m_{21}a_1 + m_{22}b_1 + m_{23}c_1 \qquad (5.1)$$
$$c_2 = m_{31}a_1 + m_{32}b_1 + m_{33}c_1$$

where m_{ij} (i, j =1, 2, 3) are the coefficients (components) of the second set of axes in terms of the first set of axes. In matrix form, Eq. 5.1 becomes

$$\begin{pmatrix} a_2 \\ b_2 \\ c_2 \end{pmatrix} = \begin{pmatrix} m_{11} & m_{12} & m_{13} \\ m_{21} & m_{22} & m_{23} \\ m_{31} & m_{32} & m_{33} \end{pmatrix} \begin{pmatrix} a_1 \\ b_1 \\ c_1 \end{pmatrix} \qquad (5.2)$$

We know that a change of axes will cause a change in the volume of the unit cell. The two unit cell volumes are related through the equation:

© The Author(s), under exclusive license to Springer Nature Singapore Pte Ltd. 2021
M. A. Wahab, *Numerical Problems in Crystallography*,
https://doi.org/10.1007/978-981-15-9754-1_5

$$a_2 \cdot b_2 \times c_2 = \begin{vmatrix} m_{11} & m_{12} & m_{13} \\ m_{21} & m_{22} & m_{23} \\ m_{31} & m_{32} & m_{33} \end{vmatrix} a_1 \cdot b_1 \times c_1 \qquad (5.3)$$

where the matrix of Eq. 5.2 has been changed into a determinant in Eq. 5.3 and is known as the modulus of transformation. The determinant is an integer when the transformation is from a smaller unit cell to a larger one or a simple fraction when the transformation is otherwise. It is interesting to note that Eq. 5.2 can also be used to transform the indices of crystal planes.

5.2   Transformation of Indices of Crystal Planes (Unit Cell)

Solved Example

1. FCC and Rhombohedral

The axial relationships between the translation vectors of the primitive rhombohedron and the fcc unit cells are shown in Fig. 5.1. Let us consider a crystal plane which is referred to as (hkl) in rhombohedral system of axes a_1, b_1, c_1 and (HKL) in fcc system of axes a_2, b_2, c_2. Then, the cubic axes in terms of rhombohedral axes are:

Fig. 5.1 Rhombohedral and cubic axes

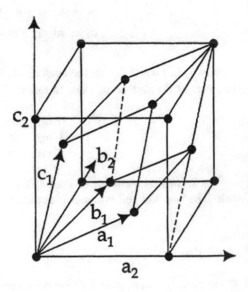

$$a_2 = a_1 + b_1 - c_1$$
$$b_2 = a_1 - b_1 + c_1$$
$$c_2 = - a_1 + b_1 + c_1$$

Now, using these equations, we can write the following transformation equations

$$H = 1\,h + 1\,k - 1\,l$$
$$K = 1\,h - 1\,k + 1\,l \qquad (5.4)$$
$$L = -1\,h + 1\,k + 1\,l$$

The matrix form of these equations is

$$\begin{pmatrix} H \\ K \\ L \end{pmatrix} = \begin{pmatrix} 1 & 1 & -1 \\ 1 & -1 & 1 \\ -1 & 1 & 1 \end{pmatrix} \begin{pmatrix} h \\ k \\ l \end{pmatrix} \qquad (5.5)$$

The determinant of the matrix in Eq. 5.5, $\Delta = 4$. Therefore using Eq. 5.3, we can obtain the volume relationship between the two unit cells

$$V_{FCC} = 4V_{RH}$$

The reverse relationships can be found by determining the inverse of the matrix in Eq. 5.5, that is,

$$\begin{pmatrix} 1 & 1 & -1 \\ 1 & -1 & 1 \\ -1 & 1 & 1 \end{pmatrix}^{-1} = \frac{1}{2} \begin{pmatrix} 1 & 1 & 0 \\ 1 & 0 & 1 \\ 0 & 1 & 1 \end{pmatrix}$$

Hence the inverse relationship (i.e., transformation from fcc to rhombohedral) is

$$\begin{pmatrix} h \\ k \\ l \end{pmatrix} = \frac{1}{2} \begin{pmatrix} 1 & 1 & 0 \\ 1 & 0 & 1 \\ 0 & 1 & 1 \end{pmatrix} \begin{pmatrix} H \\ K \\ L \end{pmatrix}$$

The same can be written as

$$h = \frac{H}{2} + \frac{K}{2} + 0\,L$$
$$k = \frac{H}{2} + 0\,K + \frac{L}{2} \qquad (5.6)$$
$$l = 0\,H + \frac{K}{2} + \frac{L}{2}$$

Let us check the validity of Eqs. 5.4 and 5.6 by taking some examples.

Example 1 Find the equivalent rhombohedral planes for the following fcc planes: (111), (11$\bar{1}$) and (200).

Solution: *Given*: fcc planes: (111), (11$\bar{1}$) and (200).

Let us take them one by one.

Case I: fcc plane: (HKL) $\equiv$ (111).

Substituting the values of H, K and L in Eq. 5.6, we obtain (hkl) $\equiv$ (111) for rhombohedron. Again, substituting the values of h, k and l in Eq. 5.4, we obtain (HKL) $\equiv$ (111) for fcc, this is the same indices with which we started. This also shows that for (111) plane, both crystal systems have identical indices.

Case II: fcc plane: (HKL) $\equiv$ (11$\bar{1}$).

A similar operation with the given H, K and L values will provide us (hkl) $\equiv$ (100) for rhombohedron. Again, substituting the values of h, k and l in Eq. 5.4, we obtain (HKL) $\equiv$ (11$\bar{1}$) for fcc. This confirms the validity of Eqs. 5.4 and 5.6.

Case III: fcc plane: (HKL) $\equiv$ (200).

A similar operation with the given H, K and L values will provide us (hkl) $\equiv$ (110) for rhombohedron. Again, substituting the values of h, k and l in Eq. 5.4, we obtain (HKL) $\equiv$ (200) for fcc. This confirms the validity of Eqs. 5.4 and 5.6. Using these equations, a one to one correspondence of other Miller indices can be obtained.

2. Simple Hexagonal and Orthorhombic

The axial relationships between the translation vectors of simple hexagonal and orthorhombic unit cells are shown in Fig. 5.2. Let us consider a crystal plane which is referred to as (hkil) in a simple hexagonal system of axes a, b, c and (HKL) in orthorhombic system of axes a', b', c'. Then, the orthorhombic axes in terms of the hexagonal axes are:

$$a' = 2a + b, b' = b, c' = c$$

Now, using these equations, we can write the following transformation equations

$$
\begin{aligned}
H &= 2h + 1k + 0l \\
K &= 0h + 1k + 0l \\
L &= 0h + 0k + 1l
\end{aligned}
\qquad (5.7)
$$

The matrix form of these equations is

$$
\begin{pmatrix} H \\ K \\ L \end{pmatrix} = \begin{pmatrix} 2 & 1 & 0 \\ 0 & 1 & 0 \\ 0 & 0 & 1 \end{pmatrix} \begin{pmatrix} h \\ k \\ l \end{pmatrix}
\qquad (5.8)
$$

The determinant of the matrix in Eq. 5.8, $\Delta = 2$. Therefore using Eq. 5.3, we can obtain the volume relationship between the two unit cells

Fig. 5.2 Hexagonal and orthorhombic axes

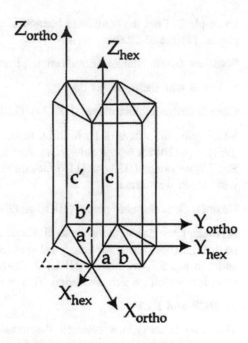

$$V_{ORTH} = 2V_{SH}$$

The reverse relationships can be found by determining the inverse of the matrix in Eq. 5.8, that is,

$$\begin{pmatrix} 2 & 1 & 0 \\ 0 & 1 & 0 \\ 0 & 0 & 1 \end{pmatrix}^{-1} = \frac{1}{2} \begin{pmatrix} 1 & -1 & 0 \\ 0 & 2 & 0 \\ 0 & 0 & 2 \end{pmatrix}$$

Hence the inverse relationship (i.e., transformation from orthorhombic to simple hexagon) is

$$\begin{pmatrix} h \\ k \\ l \end{pmatrix} = \frac{1}{2} \begin{pmatrix} 1 & -1 & 0 \\ 0 & 2 & 0 \\ 0 & 0 & 2 \end{pmatrix} \begin{pmatrix} H \\ K \\ L \end{pmatrix} \tag{5.9}$$

The same can be written as

$$h = \frac{H}{2} - \frac{K}{2} + 0\,L$$
$$k = 0\,H + 1K + 0\,L \tag{5.10}$$
$$l = 0\,H + 0\,K + 1\,L$$

where $i = -(h+k)$

Example 2 Find the equivalent hexagonal planes for the following orthorhombic planes: (110) and (200).

Solution: *Given*: Simple orthorhombic planes: (110) and (200).

Let us take them one by one.

Case I: Orthorhombic plane: (HKL) ≡ (110).

Substituting the values of H, K and L in Eq. 5.10, we obtain (hkl) ≡ (010), so that (hkil) ≡ (01$\bar{1}$0) for hexagonal system. Again, substituting the values of h, k and l in Eq. 5.7, we obtain (HKL) ≡ (110) for orthorhombic system, this is the same indices with which we started.

Case II: Orthorhombic plane: (HKL) ≡ (200).

A similar operation with the given H, K and L values will provide us (hkl) ≡ (100), so that (hkil) ≡ (10$\bar{1}$0) for hexagonal system. Again, substituting the values of h, k and l in Eq. 5.7, we obtain (HKL) ≡ (200) for orthorhombic system, this is the same indices with which we started. This confirms the validity of Eqs. 5.7 and 5.10.

3. HCP and RCP

The axial relationships between the translation vectors of rhombohedral close packing (RCP) and hexagonal close packing (HCP) unit cells are shown in Fig. 5.3. Let us consider a crystal plane which is referred as (hkl) in RCP system of axes a_R, b_R, c_R and (HKIL) in HCP or hexagonal system of axes a_H, b_H, c_H. Then, the hexagonal axes in terms of the rhombohedral axes are:

$$a_H = a_R - b_R, \quad b_H = b_R - c_R, \quad c_H = a_R + b_R + c_R$$

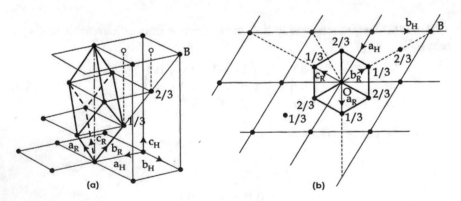

Fig. 5.3 RCP and HCP axes

We can also write two other similar equivalent sets. However, using the the above relationships, we can express hexagonal indices in terms of the rhombohedral indices through the following transformation equations

$$
\begin{aligned}
H &= 1h - 1k + 0l \\
K &= 0h + 1k - 1l \\
L &= 1h + 1k + 1l \\
I &= -(H + K)
\end{aligned}
$$
(5.11)

The matrix form of these equations is

$$
\begin{pmatrix} H \\ K \\ L \end{pmatrix} = \begin{pmatrix} 1 & -1 & 0 \\ 0 & 1 & -1 \\ 1 & 1 & 1 \end{pmatrix} \begin{pmatrix} h \\ k \\ l \end{pmatrix}
$$
(5.12)

The determinant of the matrix in Eq. 5.12, $\Delta = 3$. Therefore using Eq. 5.3, we can obtain the volume relationship between the two unit cells

$$
V_{HCP} = 3V_{RCP}
$$

Now, combining the hexagonal indices algebraically, we obtain

$$
(-H + K + L) = 3k
$$
(5.13)

where k is an integer.

This is an important consequence which tells us that when the planes in rhombohedral lattice are referred to as hexagonal axes, only certain combinations of hexagonal indices given by Eq. 5.13 are allowed. This limits the possible X-ray reflections in the study of such crystals.

The reverse relationships can be found by determining the inverse of the matrix in Eq. 5.12, that is,

$$
\begin{pmatrix} 1 & -1 & 0 \\ 0 & 1 & -1 \\ 1 & 1 & 1 \end{pmatrix}^{-1} = \frac{1}{3} \begin{pmatrix} 2 & 1 & 1 \\ -1 & 1 & 1 \\ -1 & -2 & 1 \end{pmatrix}
$$

Hence the inverse relationship (i.e., transformation from HCP to RCP is

$$
\begin{pmatrix} h \\ k \\ l \end{pmatrix} = \frac{1}{3} \begin{pmatrix} 2 & 1 & 1 \\ -1 & 1 & 1 \\ -1 & -2 & 1 \end{pmatrix} \begin{pmatrix} H \\ K \\ L \end{pmatrix}
$$
(5.14)

The same can be written as

$$h = \frac{2H}{3} + \frac{K}{3} + \frac{L}{3}$$
$$k = -\frac{H}{3} + \frac{K}{3} + \frac{L}{3} \qquad (5.15)$$
$$l = -\frac{H}{3} - \frac{2K}{3} + \frac{L}{3}$$

Example 3 Find the equivalent HCP planes for the following rhombohedral planes:$(10\bar{1}))$ and (200).

Solution: *Given*: Rhombohedral planes: $(10\bar{1}))$ and (200).

Let us take them one by one.

Case I: Rhombohedral plane: $(hkl) \equiv (10\bar{1}))$.

Substituting the values of h, k and l in Eq. 5.11, we obtain $(HKIL) \equiv (11\bar{2}0)$ for HCP. Again, substituting the values of H, K and L in Eq. 5.15, we obtain $(hkl) \equiv (10\bar{1})$ for rhombohedron, this is the same indices with which we started. This confirms the validity of Eqs. 5.11 and 5.15.

Case II: Rhombohedral plane: $(hkl) \equiv (200)$.

A similar operation with the given h, k and l values will provide us $(HKIL) \equiv (20\bar{2}2)$ for HCP. Again, substituting the values of h, k and l in Eq. 5.15, we obtain $(hkl) \equiv (200)$ for rhombohedron. This confirms the validity of Eqs. 5.11 and 5.15. Using these equations, a one to one correspondence of other Miller indices can be obtained.

4. Trigonal and simple Hexagonal

The axial relationships between the translation vectors of trigonal and primitive (simple) hexagonal unit cells are shown in Fig. 5.4. Let us consider a crystal plane which is referred to as (HKL) in trigonal system of axes A_1, A_2, A_3 and (hkil) in hexagonal system of axes a_1, a_2, c.

From Fig. 5.4, we can write the following axial relationships:

$$A_1 = a_1 + c/3, A_2 = a_2 + c/3, A_3 = -a_1 - a_2 + c/3$$

Based on above relationships, we can express trigonal indices in terms of the primitive hexagonal indices through the following transformation equations

$$H = 1h + 0k + 1l/3$$
$$K = 0h + 1k + 1l/3 \qquad (5.16)$$
$$L = -1h - 1k + 1l/3$$

The matrix form of these equations is

Fig. 5.4 Simple hexagon and trigonal axes

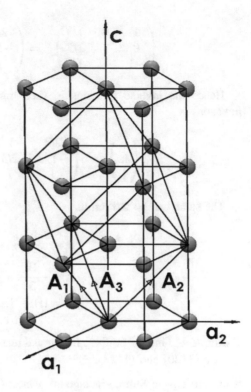

$$\begin{pmatrix} H \\ K \\ L \end{pmatrix} = \begin{pmatrix} 1 & 0 & 1/3 \\ 0 & 1 & 1/3 \\ -1 & -1 & 1/3 \end{pmatrix} \begin{pmatrix} h \\ k \\ l \end{pmatrix} \tag{5.17}$$

The determinant of the matrix in Eq. 5.17, $\Delta = 1$. Therefore using Eq. 5.3, we can obtain the volume relationship between the two unit cells

$$V_{TR} = V_{SH}$$

Now, combining the hexagonal indices algebraically, we obtain

$$(H + K + L) = 1 \tag{5.18}$$

This is an important consequence which tells us that when the planes in a simple hexagonal lattice are referred to as trigonal axes, only certain combinations of trigonal indices given by Eq. 5.18 are allowed. This limits the possible X-ray reflections in the study of such crystals.

The reverse relationships can be found by determining the inverse of the matrix in Eq. 5.17, that is,

$$\begin{pmatrix} 1 & 0 & 1/3 \\ 0 & 1 & 1/3 \\ -1 & -1 & 1/3 \end{pmatrix}^{-1} = \begin{pmatrix} 2/3 & -1/3 & -1/3 \\ -1/3 & 2/3 & -1/3 \\ 1 & 1 & 1 \end{pmatrix}$$

Hence the inverse relationship (i.e., transformation from trigonal to simple hexagon) is

$$\begin{pmatrix} h \\ k \\ l \end{pmatrix} = \begin{pmatrix} 2/3 & -1/3 & -1/3 \\ -1/3 & 2/3 & -1/3 \\ 1 & 1 & 1 \end{pmatrix} \begin{pmatrix} H \\ K \\ L \end{pmatrix} \qquad (5.19)$$

The same can be written as

$$h = \frac{2H}{3} - \frac{K}{3} - \frac{L}{3}$$
$$k = -\frac{H}{3} + \frac{2K}{3} - \frac{L}{3} \qquad (5.20)$$
$$l = 1H + 1K + 1L$$

Example 4 Find the equivalent trigonal planes for the following simple hexagonal planes: $(11\bar{2}0)$ and $(11\bar{2}3)$.

Solution: *Given*: Simple hexagonal planes: $(11\bar{2}0)$ and $(11\bar{2}3)$.

Let us take them one by one.

Case I: Simple hexagonal plane: (hkil) $\equiv (11\bar{2}0)$.

Substituting the values of h, k and l in Eq. 5.16, we obtain (HKL) $\equiv (11\bar{2})$ for the corresponding trigonal plane. Again, substituting the values of H, K and L in Eq. 5.20, we obtain (hkil) $\equiv (11\bar{2}0)$ for simple hexagon, this is the same indices with which we started. This confirms the validity of Eqs. 5.16 and 5.20.

Case II: Simple hexagonal plane: (hkil) $\equiv (11\bar{2}3)$.

A similar operation with the given h, k and l values will provide us (HKL) $\equiv (22\bar{1})$ for the corresponding trigonal plane. Again, substituting the values of H, K and L in Eq. 5.20, we obtain (hkil) $\equiv (11\bar{2}3)$ for hexagonal plane. This confirms the validity of Eqs. 5.16 and 5.20. Using these equations, a one to one correspondence of other Miller indices can be obtained.

5. BCC and Primitive (For All Lattices)

The axial relationships between the translation vectors of body-centered (for simplicity it is taken as cubic) and its primitive unit cells are shown in Fig. 5.5. Let us consider a crystal plane which is referred to as (hkl) in the primitive system of axes a_1, b_1, c_1 and (HKL) in bcc system of axes a_2, b_2, c_2. Writing the second set in terms of the first set, we have:

Fig. 5.5 BCC and its primitive unit cell axes

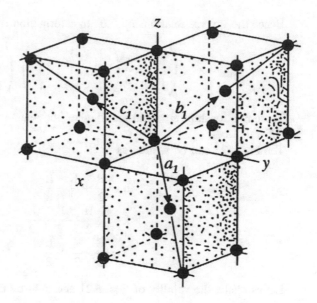

$$a_2 = a_1 + c_1, \; b_2 = a_1 + b_1, \; c_2 = b_1 + c_1$$

Now, using these equations, we can write the following transformation equations

$$H = 1h + 0k + 1l$$
$$K = 1h + 1k + 0l \tag{5.21}$$
$$L = 0h + 1k + 1l$$

The matrix form of these equations is

$$\begin{pmatrix} H \\ K \\ L \end{pmatrix} = \begin{pmatrix} 1 & 0 & 1 \\ 1 & 1 & 0 \\ 0 & 1 & 1 \end{pmatrix} \begin{pmatrix} h \\ k \\ l \end{pmatrix} \tag{5.22}$$

The determinant of the matrix in Eq. 5.22, $\Delta = 2$. Therefore Using Eq. 5.3, we can obtain the volume relationship between the two unit cells

$$V_{BCC} = 2V_P$$

The reverse relationships can be found by determining the inverse of the matrix in Eq. 5.22, that is,

$$\begin{pmatrix} 1 & 0 & 1 \\ 1 & 1 & 0 \\ 0 & 1 & 1 \end{pmatrix}^{-1} = \begin{pmatrix} 1 & 1 & -1 \\ -1 & 1 & 1 \\ 1 & -1 & 1 \end{pmatrix}$$

Hence the inverse relationship (i.e., transformation from bcc to its primitive) is

$$
\begin{pmatrix} h \\ k \\ l \end{pmatrix} = \frac{1}{2} \begin{pmatrix} 1 & 1 & -1 \\ -1 & 1 & 1 \\ 1 & -1 & 1 \end{pmatrix} \begin{pmatrix} H \\ K \\ L \end{pmatrix}
\tag{5.23}
$$

These relationships are valid for all other body-centered and their corresponding primitive lattices.

The Eq. 5.23 can be written as

$$
h = \frac{H}{2} + \frac{K}{2} - \frac{L}{2}
$$
$$
k = -\frac{H}{2} + \frac{K}{2} + \frac{L}{2}
\tag{5.24}
$$
$$
l = \frac{H}{2} - \frac{K}{2} + \frac{L}{2}
$$

Let us check the validity of Eqs. 5.21 and 5.24 by taking some examples.

Example 5 Find the equivalent bcc planes corresponding to its primitive lattice planes: (100), (110) and (111).

Solution: *Given*: Planes of bcc primitive: (100), (110) and (111).

Let us take them one by one.

Case I: Plane of bcc primitive: (hkl) ≡ (100).

Substituting the values of h, k and l in Eq. 5.21, we obtain (HKL) ≡ (110) for bcc plane. Again, substituting the values of H, K and L in Eq. 5.24, we obtain (hkl) ≡ (100) for primitive bcc plane, this is the same indices with which we started. This confirms the validity of Eqs. 5.21 and 5.24.

Case II: Plane of bcc primitive: (hkl) ≡ (110).

A similar operation with the given h, k and l values in Eq. 5.21 will provide us (HKL) ≡ (121) for bcc plane. Again, substituting the values of H, K and L in Eq. 5.24, we obtain (hkl) ≡ (110) for primitive bcc plane. This confirms the validity of Eqs. 5.21 and 5.24.

Case III: Plane of bcc primitive: (hkl) ≡ (111).

A similar operation with the given h, k and l values will provide us (HKL) ≡ (222) for bcc plane. Again, substituting the values of H, K and L in Eq. 5.24, we obtain (hkl) ≡ (111) for primitive bcc plane. This confirms the validity of Eqs. 5.21 and 5.24. Using these equations, a one to one correspondence of other Miller indices can be obtained.

Fig. 5.6 FCC and its
primitive unit cell axes

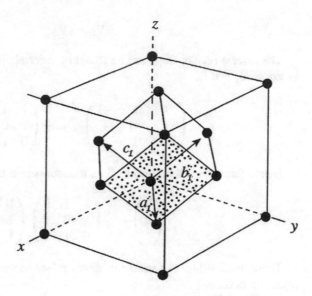

6. FCC and Primitive (For All Lattices)

The axial relationships between the translation vectors of fcc and its primitive unit
cells are shown in Fig. 5.6. Let us consider a crystal plane which is referred to as
(hkl) in the primitive system of axes a_1, b_1, c_1 and (HKL) in fcc system of axes
a_2, b_2, c_2. The relationships between the two sets of axes are:

$$a_2 = a_1 + b_1 - c_1$$
$$b_2 = a_1 - b_1 + c_1$$
$$c_2 = -a_1 + b_1 + c_1$$

Now, using these equations, we can write the following transformation equations

$$H = 1h + 1k - 1l$$
$$K = 1h - 1k + 1l \qquad (5.25)$$
$$L = -1h + 1k + 1l$$

The matrix form of these equations is

$$\begin{pmatrix} H \\ K \\ L \end{pmatrix} = \begin{pmatrix} 1 & 1 & -1 \\ 1 & -1 & 1 \\ -1 & 1 & 1 \end{pmatrix} \begin{pmatrix} h \\ k \\ l \end{pmatrix} \qquad (5.26)$$

The determinant of the matrix in Eq. 5.26, $\Delta = 4$. Therefore using Eq. 5.3, we
can obtain the volume relationship between the two unit cells

$$V_{fcc} = 4V_{rh}$$

The reverse relationships can be found by determining the inverse of the matrix in Eq. 5.26, that is,

$$\begin{pmatrix} 1 & 1 & -1 \\ 1 & -1 & 1 \\ -1 & 1 & 1 \end{pmatrix}^{-1} = \frac{1}{2} \begin{pmatrix} 1 & 1 & 0 \\ 1 & 0 & 1 \\ 0 & 1 & 1 \end{pmatrix}$$

Hence the inverse relationship (i.e., transformation from fcc to primitive) is

$$\begin{pmatrix} h \\ k \\ 1 \end{pmatrix} = \frac{1}{2} \begin{pmatrix} 1 & 1 & 0 \\ 1 & 0 & 1 \\ 0 & 1 & 1 \end{pmatrix} \begin{pmatrix} H \\ K \\ L \end{pmatrix} \qquad (5.27)$$

These relationships are valid for all other face-centered and their corresponding primitive lattices.

The Eq. 5.27 can be written as

$$h = \frac{H}{2} + \frac{K}{2} + 0\,L$$
$$k = \frac{H}{2} + 0\,K + \frac{L}{2} \qquad (5.28)$$

Let us check the validity of Eqs. 5.25 and 5.28 by taking some examples.

Example 6 Find the equivalent fcc planes corresponding to its primitive lattice planes: (100), (110) and (111).

Solution: *Given*: Planes of fcc primitive: (100), (110) and (111).

Let us take them one by one.

Case I: Plane of fcc primitive: (hkl) $\equiv$ (100).

Substituting the values of h, k and l in Eq. 5.25, we obtain (HKL) $\equiv$ $(11\bar{1})$ for fcc plane. Again, substituting the values of H, K and L in Eq. 5.28, we obtain (hkl) $\equiv$ (100) for primitive fcc plane, this is the same indices with which we started. This confirms the validity of Eqs. 5.25 and 5.28.

Case II: Plane of fcc primitive: (hkl) $\equiv$ (110).

A similar operation with the given h, k and l values in Eq. 5.25 will provide us (HKL) $\equiv$ (200) for fcc plane. Again, substituting the values of H, K and L in Eq. 5.28, we obtain (hkl) $\equiv$ (110) for primitive bcc plane. This confirms the validity of Eqs. 5.25 and 5.28.

Case III: Plane of fcc primitive: (hkl) ≡ (111).

A similar operation with the given h, k and l values will provide us (HKL) ≡ (111) for fcc plane. Again, substituting the values of H, K and L in Eq. 5.28, we obtain (hkl) ≡ (111) for primitive fcc plane. This confirms the validity of Eqs. 5.25 and 5.28. This also shows that for (111) plane, both crystal systems have identical indices. Using these equations, a one to one correspondence of other Miller indices can be obtained.

5.3 Transformation of Indices of Direction (Zone Axes)

Unlike the transformation of indices of planes, the transformation of indices of direction is not identical to the transformation axes. In order to find the general equation for transformation of indices of direction, let us assume $[u_1 v_1 w_1]$ and $[u_2 v_2 w_2]$ are the indices of direction (for a vector $\vec{r}$) in the two sets of axes, respectively. Therefore, the vector $\vec{r}$ in terms of components of two sets of unit cell vectors can be written as

$$\vec{r} = u_1 a_1 + v_1 b_1 + w_1 c_1 = u_2 a_2 + v_2 b_2 + w_2 c_2 \tag{5.29}$$

Substituting the values of a_2, b_2 and c_2 from Eq. 5.1 into Eq. 5.29, we obtain

$$\begin{aligned}
\vec{r} &= u_1 a_1 + v_1 b_1 + w_1 c_1 \\
&= u_2(m_{11}a_1 + m_{12}b_1 + m_{13}c_1) \\
&\quad + v_2(m_{21}a_1 + m_{22}b_1 + m_{23}c_1) \\
&\quad + w_2(m_{31}a_1 + m_{32}b_1 + m_{33}c_1)
\end{aligned} \tag{5.30}$$

Now, comparing the coefficients of a_1, b_1 and c_1 in Eq. 5.30, we have

$$\begin{aligned}
u_1 &= m_{11}u_2 + m_{21}v_2 + m_{31}w_2 \\
v_1 &= m_{12}u_2 + m_{22}v_2 + m_{32}w_2 \\
w_1 &= m_{13}u_2 + m_{23}v_2 + m_{33}w_2
\end{aligned} \tag{5.31}$$

Matrix form of Eq. 5.31 is

$$\begin{pmatrix} u_1 \\ v_1 \\ w_1 \end{pmatrix} = \begin{pmatrix} m_{11} & m_{21} & m_{31} \\ m_{12} & m_{22} & m_{32} \\ m_{13} & m_{23} & m_{33} \end{pmatrix} \begin{pmatrix} u_2 \\ v_2 \\ w_2 \end{pmatrix} \tag{5.32}$$

Comparing Eq. 5.32 with Eq. 5.2, we observe the following (which can be used as an instruction):

(i)   The indices of the two sets are interchanged.

(ii) The matrix elements (rows and columns) are interchanged.

The Eq. 5.32 can also be used to transform:

(i)   the reciprocal lattice vectors

(ii) the coordinate positions in the unit cell from one set to another and vice-versa.

Solved Examples

1. **FCC and Rhombohedral**

The axial relationships between the translation vectors of the primitive rhombo-hedral and the fcc lattices are shown in Fig. 5.1. The matrix form of the equation for a plane expressing fcc indices in terms of rhombohedral indices has been given in Eq. 5.5. Now, let us obtain the matrix for transforming the direction [uvw] referred to as rhombohedral lattice in terms of the direction [UVW] referred to as fcc lattice following the above-mentioned instructions. Thus we, have

$$\begin{pmatrix} U \\ V \\ W \end{pmatrix} = \frac{1}{2} \begin{pmatrix} 1 & 1 & 0 \\ 1 & 0 & 1 \\ 0 & 1 & 1 \end{pmatrix} \begin{pmatrix} u \\ v \\ w \end{pmatrix} \tag{5.33}$$

The same can be written as

$$U = \frac{u}{2} + \frac{v}{2} + 0w$$
$$V = \frac{u}{2} + 0v + \frac{w}{2} \tag{5.34}$$
$$W = 0u + \frac{v}{2} + \frac{w}{2}$$

The inverse transformation giving [uvw] in terms of [UVW] is

$$\begin{pmatrix} u \\ v \\ w \end{pmatrix} = \begin{pmatrix} 1 & 1 & -1 \\ 1 & -1 & 1 \\ -1 & 1 & 1 \end{pmatrix} \begin{pmatrix} U \\ V \\ W \end{pmatrix} \tag{5.35}$$

In other words,

$$u = 1U + 1V - 1W$$
$$v = 1U - 1V + 1W \tag{5.36}$$
$$w = -1U + 1V + 1W$$

Let us check the validity of Eqs. 5.34 and 5.36 by taking some examples.

Example 1 Find the equivalent fcc directions for the following rhombohedral directions: [111], [11$\bar{1}$] and [200].

Solution: *Given*: Rhombohedral directions: [111], [11$\bar{1}$] and [200].

Let us take them one by one.

Case I: Rhombohedral direction: [uvw] ≡ [111].

Substituting the values of u, v and w in Eq. 5.34, we obtain [UVW] ≡ [111] for fcc. Again, substituting the values of U, V and W in Eq. 5.36, we obtain [uvw] ≡ [111] for rhombohedron, this is the same indices with which we started. This also shows that for [111] direction, both fcc and rhombohedron have identical indices.

Case II: Rhombohedral direction: [uvw] ≡ [11$\bar{1}$].

A similar operation using Eq. 5.34 with the given u, v and w values will provide us [UVW] ≡ [100] for fcc. Again, substituting the values of U, V and W in Eq. 5.36, we obtain [uvw] ≡ [11$\bar{1}$] for rhombohedron. This confirms the validity of Eqs. 5.34 and 5.36.

Case III: Rhombohedral direction: [uvw] ≡ [200].

A similar operation using Eq. 5.34 with the given u, v and w values will provide us [UVW] ≡ [110] for fcc. Again, substituting the values of U, V and W in Eq. 5.36, we obtain [uvw] ≡ [200] for rhombohedron. This confirms the validity of Eqs. 5.34 and 5.36. Using these equations, a one to one correspondence of other directions can be obtained.

2. Simple Hexagonal and Orthorhombic

The axial relationships between the translation vectors of simple hexagonal and orthorhombic lattices are shown in Fig. 5.2. The matrix form of the equation for a plane expressing orthorhombic indices in terms of hexagonal indices has been given in Eq. 5.8. Now, let us obtain the matrix for transforming the direction [UVW] referred to as orthorhombic lattice in terms of the direction [uviw] referred to as hexagonal lattice following the above-mentioned instructions. Thus, we have

$$\begin{pmatrix} U \\ V \\ W \end{pmatrix} = \frac{1}{2} \begin{pmatrix} 1 & 0 & 0 \\ -1 & 2 & 0 \\ 0 & 0 & 2 \end{pmatrix} \begin{pmatrix} u \\ v \\ w \end{pmatrix} \tag{5.37}$$

The same can be written as

$$U = \frac{1u}{2} + 0\,v + 0\,w$$

$$V = -\frac{1u}{2} + 1\,v + 0\,w \tag{5.38}$$

$$W = 0\,u + 0\,v + 1\,w$$

The inverse transformation giving [uviw] in terms of [UVW] is

$$
\begin{pmatrix} u \\ v \\ w \end{pmatrix} = \begin{pmatrix} 2 & 0 & 0 \\ 1 & 1 & 0 \\ 0 & 0 & 1 \end{pmatrix} \begin{pmatrix} U \\ V \\ W \end{pmatrix}
\tag{5.39}
$$

The same can be written as

$$
\begin{aligned}
u &= 2\,U + 0\,V + 0\,W \\
v &= 1\,U + 1\,V + 0\,W \\
w &= 0\,U + 0\,V + 1\,W
\end{aligned}
\tag{5.40}
$$

where $i = -(u+v)$.

Let us check the validity of Eqs. 5.38 and 5.40 by taking some examples.

Example 2 Find the equivalent orthorhombic directions for the following hexagonal directions: [200], [210] and [211].

Solution: *Given*: Hexagonal directions: [200], [210] and [211].

Let us take them one by one.

Case I: Hexagonal direction: [uvw] $\equiv$ [200].

Substituting the values of u, v and w in Eq. 5.38, we obtain [UVW] $\equiv$ [1$\bar{1}$0] for hexagonal system. Again, substituting the values of U, V and W in Eq. 5.40, we obtain [uvw] $\equiv$ [200] for orthorhombic system, this is the same indices with which we started.

Case II: Hexagonal direction: [uvw] $\equiv$ [210].

A similar operation using Eq. 5.38 with the given u, v and w values will provide us [UVW] $\equiv$ [100] for orthorhombic system. Again, substituting the values of U, V and W in Eq. 5.40, we obtain [uvw] $\equiv$ [210] for hexagonal system. This confirms the validity of Eqs. 5.38 and 5.40.

Case III: Hexagonal direction: [uvw] $\equiv$ [211].

A similar operation using Eq. 5.38 with the given u, v and w values will provide us [UVW] $\equiv$ [101] for orthorhombic system. Again, substituting the values of U, V and W in Eq. 5.40, we obtain [uvw] $\equiv$ [211] for hexagonal system. This confirms the validity of Eqs. 5.38 and 5.40.

3. HCP and RCP

The axial relationships between the translation vectors of rhombohedral close packing (RCP) and hexagonal close packing (HCP) lattices are shown in Fig. 5.3. The matrix form of the equation for a plane expressing hexagonal indices in terms of rhombohedral indices has been given in Eq. 5.12. Now, let us obtain the matrix for transforming the direction [UVW] referred to as HCP lattice in terms of the direction [uvw] referred to as rhombohedral lattice following the above-mentioned instructions. Thus we, have

$$
\begin{pmatrix} U \\ V \\ W \end{pmatrix} = \frac{1}{3} \begin{pmatrix} 2 & -1 & -1 \\ 1 & 1 & -2 \\ 1 & 1 & 1 \end{pmatrix} \begin{pmatrix} u \\ v \\ w \end{pmatrix} \tag{5.41}
$$

The same can be written as

$$
\begin{aligned}
U &= \frac{2u}{3} - \frac{v}{3} - \frac{w}{3} \\
V &= \frac{u}{3} + \frac{v}{3} - \frac{2w}{3} \\
W &= \frac{u}{3} + \frac{v}{3} + \frac{w}{3}
\end{aligned} \tag{5.42}
$$

where I = − (U + V).
The inverse transformation giving [uvw] in terms of [UVW] is

$$
\begin{pmatrix} u \\ v \\ w \end{pmatrix} = \begin{pmatrix} 1 & 0 & 1 \\ -1 & 1 & 1 \\ 0 & -1 & 1 \end{pmatrix} \begin{pmatrix} U \\ V \\ W \end{pmatrix} \tag{5.43}
$$

The same can be written as

$$
\begin{aligned}
u &= 1U + 0V + 1W \\
v &= -1U + 1V + 1W \\
w &= 0U - 1V + 1W
\end{aligned} \tag{5.44}
$$

Example 3 Find the equivalent HCP directions for the following rhombohedral directions: [111] and [330].

Solution: *Given*: Rhombohedral directions: [111] and [330].

Let us take them one by one.
Case I: Rhombohedral direction: [uvw] ≡ [111].

Substituting the values of u, v and w in Eq. 5.42, we obtain [UVIW] ≡ [0001] for HCP. This gives a very important result that [111] rhombohedron direction coincides with the principal axis [0001] of HCP. Again, substituting the values of U, V and W in Eq. 5.44, we obtain [uvw] ≡ [111] for rhombohedron, this is the same indices with which we started. This confirms the validity of Eqs. 5.42 and 5.44.

Case II: Rhombohedral direction: [uvw] ≡ [330].

A similar operation using Eq. 5.42 with the given u, v and w values will provide us [UVIW] ≡ [$1\bar{2}\bar{3}2$] for HCP. Again, substituting the values of U, V and W in

Eq. 5.44, we obtain [uvw] ≡ [330] for rhombohedron. This confirms the validity of
Eqs. 5.42 and 5.44. Using these equations, a one to one correspondence of other
indices of direction can be obtained.

4. **Trigonal and Simple Hexagonal**

The axial relationships between the translation vectors of trigonal and primitive
simple hexagonal unit cells are shown in Fig. 5.4. The matrix form of the equation
for a plane expressing Trigonal indices in terms of hexagonal indices has been
given in Eq. 5.17. Now, let us obtain the matrix for transforming the direction
[UVW] referred to as trigonal lattice in terms of the direction [uviw] referred to as
hexagonal lattice following the above-mentioned instructions. Thus we, have

$$\begin{pmatrix} U \\ V \\ W \end{pmatrix} = \begin{pmatrix} 2/3 & -1/3 & 1 \\ -1/3 & 2/3 & 1 \\ -1/3 & -1/3 & 1 \end{pmatrix} \begin{pmatrix} u \\ v \\ w \end{pmatrix} \tag{5.45}$$

The same can be written as

$$U = \frac{2u}{3} - \frac{v}{3} + 1w$$
$$V = -\frac{u}{3} + \frac{2v}{3} + 1w \tag{5.46}$$
$$W = -\frac{u}{3} - \frac{v}{3} + 1w$$

The inverse transformation giving [uviw] in terms of [UVW] is

$$\begin{pmatrix} u \\ v \\ w \end{pmatrix} = \begin{pmatrix} 1 & 0 & -1 \\ 0 & 1 & -1 \\ 1/3 & 1/3 & 1/3 \end{pmatrix} \begin{pmatrix} U \\ V \\ W \end{pmatrix} \tag{5.47}$$

The same can be written as

$$u = 1U + 0V - 1W$$
$$v = 0U + 1V - 1W \tag{5.48}$$
$$w = 1U/3 + 1V/3 + 1W/3$$

Example 4 Find the equivalent trigonal directions for the following hexagonal
directions: $[21\bar{3}0]$ and $[21\bar{3}3]$.

Solution: *Given*: Hexagonal directions: $[21\bar{3}0]$ and $[21\bar{3}3]$.

Let us take them one by one.

Case I: Hexagonal direction: $[uviw] \equiv [21\bar{3}0]$.

Substituting the values of h, k and l in Eq. 5.46, we obtain $[UVW] \equiv [10\bar{1}]$ for the corresponding trigonal direction. Again, substituting the values of U, V and W in Eq. 5.48, we obtain $[uviw] \equiv [21\bar{3}0]$ for a simple hexagon, this is the same indices with which we started. This confirms the validity of Eqs. 5.46 and 5.48.

Case II: Hexagonal direction: $[uviw] \equiv [11\bar{2}3]$.

A similar operation with the given h, k and l values will provide us $[UVW] \equiv [432]$ for the corresponding trigonal direction. Again, substituting the values of U, V and W in Eq. 5.48, we obtain $[uviw] \equiv [21\bar{3}3]$ for simple hexagon. This confirms the validity of Eqs. 5.46 and 5.48. Using these equations, a one to one correspondence of other Miller indices can be obtained.

5. BCC and Primitive (For All Lattices)

The axial relationships between the translation vectors of body-centered and its primitive unit cells are shown in Fig. 5.5. The matrix form of the equation for a plane expressing body-centered indices in terms of its primitive indices has been given in Eq. 5.22. Now, let us obtain the matrix for transforming the direction [UVW] referred to as body-centered lattice in terms of the direction [uvw] referred to as its primitive lattice following the above-mentioned instructions. Thus we, have

$$\begin{pmatrix} U \\ V \\ W \end{pmatrix} = \frac{1}{2}\begin{pmatrix} 1 & -1 & 1 \\ 1 & 1 & -1 \\ -1 & 1 & 1 \end{pmatrix}\begin{pmatrix} u \\ v \\ w \end{pmatrix} \tag{5.49}$$

These relationships are valid for all other body-centered and their corresponding primitive lattices. The Eq. 5.49 can be written as

$$\begin{aligned} U &= \frac{u}{2} - \frac{v}{2} + \frac{w}{2} \\ V &= \frac{u}{2} + \frac{v}{2} - \frac{w}{2} \\ W &= -\frac{u}{2} + \frac{v}{2} + \frac{w}{2} \end{aligned} \tag{5.50}$$

The inverse transformation giving [uvw] in terms of [UVW] is

$$\begin{pmatrix} u \\ v \\ w \end{pmatrix} = \begin{pmatrix} 1 & 1 & 0 \\ 0 & 1 & 1 \\ 1 & 0 & 1 \end{pmatrix}\begin{pmatrix} U \\ V \\ W \end{pmatrix} \tag{5.51}$$

The same can be written as

$$u = 1U + 1V + 0W$$
$$v = 0U + 1V + 1W \qquad (5.52)$$
$$w = 1U + 0V + 1W$$

Example 5 Find the equivalent bcc directions corresponding to its primitive lattice directions: [200], [220] and [222].

Solution: *Given*: Directions of bcc primitive: [200], [220] and [222].

Let us take them one by one.

Case I: Direction of bcc primitive: [uvw] ≡ [200].

Substituting the values of u, v and w in Eq. 5.50, we obtain [UVW] ≡ $[11\bar{1}]$ for bcc plane. Again, substituting the values of U, V and W in Eq. 5.52, we obtain [uvw] ≡ [200] for primitive bcc direction, this is the same indices with which we started. This confirms the validity of Eqs. 5.50 and 5.52.

Case II: Direction of bcc primitive: [uvw] ≡ [220].

A similar operation with the given u, v and w in Eq. 5.50, we obtain [UVW] ≡ [020] for bcc plane. Again, substituting the values of U, V and W in Eq. 5.52, we obtain [uvw] ≡ [220] for primitive bcc direction, this is the same indices with which we started. This confirms the validity of Eqs. 5.50 and 5.52.

Case III: Direction of bcc primitive: [uvw] ≡ [222].

A similar operation with the given u, v and w in Eq. 5.50, we obtain [UVW] ≡ [111] for bcc plane. Again, substituting the values of U, V and W in Eq. 5.52, we obtain [uvw] ≡ [222] for primitive bcc direction, this is the same indices with which we started. This confirms the validity of Eqs. 5.50 and 5.52. Using these equations, a one to one correspondence of other Miller indices can be obtained.

6. FCC and Primitive (For All Lattices)

The axial relationships between the translation vectors of face-centered and its primitive unit cells are shown in Fig. 5.6. The matrix form of the equation for a plane expressing face-centered indices in terms of its primitive indices has been given in Eq. 5.25. Now, let us obtain the matrix for transforming the direction [UVW] referred to as face-centered lattice in terms of the direction [uvw] referred to as its primitive lattice following the above-mentioned instructions. Thus we, have

$$\begin{pmatrix} U \\ V \\ W \end{pmatrix} = \frac{1}{2} \begin{pmatrix} 1 & 1 & 0 \\ 1 & 0 & 1 \\ 0 & 1 & 1 \end{pmatrix} \begin{pmatrix} u \\ v \\ w \end{pmatrix} \qquad (5.53)$$

These relationships are valid for all other face-centered and their corresponding primitive lattices.

The Eq. 5.53 can be written as

$$U = \frac{u}{2} + \frac{v}{2} + 0\,w$$
$$V = \frac{u}{2} + 0\,v + \frac{w}{2} \tag{5.54}$$
$$W = 0\,u + \frac{v}{2} + \frac{w}{2}$$

The inverse transformation giving [uvw] in terms of [UVW] is

$$\begin{pmatrix} u \\ v \\ w \end{pmatrix} = \begin{pmatrix} 1 & 1 & -1 \\ 1 & -1 & 1 \\ -1 & 1 & 1 \end{pmatrix} \begin{pmatrix} U \\ V \\ W \end{pmatrix} \tag{5.55}$$

The same can be written as

$$u = 1U + 1V - 1W$$
$$v = 1U - 1V + 1W \tag{5.56}$$
$$w = -1U + 1V + 1W$$

Example 6 Find the equivalent fcc directions corresponding to its primitive lattice directions: [200], [220] and [222].

Solution: *Given*: Directions of fcc primitive: [200], [220] and [222].

Let us take them one by one.

Case I: Direction of fcc primitive: [uvw] ≡ [200].

Substituting the values of u, v and w in Eq. 5.54, we obtain [UVW] ≡ [110] for fcc direction. Again, substituting the values of U, V and W in Eq. 5.56, we obtain [uvw] ≡ [200] for primitive fcc direction, this is the same indices with which we started. This confirms the validity of Eqs. 5.54 and 5.56.

Case II: Direction of fcc primitive: [uvw] ≡ [220].

A similar operation with the given u, v and w in Eq. 5.54, we obtain [UVW] ≡ [211] for fcc direction. Again, substituting the values of U, V and W in Eq. 5.56, we obtain [uvw] ≡ [220] for primitive fcc direction, this is the same indices with which we started. This confirms the validity of Eqs. 5.54 and 5.56.

Case III: Direction of fcc primitive: [uvw] ≡ [222].

A similar operation with the given u, v and w in Eq. 5.54, we obtain [UVW] ≡ [222] for fcc direction. Again, substituting the values of U, V and W in Eq. 5.56, we obtain [uvw] ≡ [222] for primitive fcc direction, this is the same indices with which we started. This confirms the validity of Eqs. 5.54 and 5.56. This also shows that for [222] direction, both crystal systems have identical indices. Using these equations, a one to one correspondence of other Miller indices can be obtained.

Multiple Choice Questions (MCQ)

1. The equivalent rhombohedral plane for the (111) fcc plane is:
   (a) (111)
   (b) ($\bar{1}$11)
   (c) (1$\bar{1}$1)
   (d) (11$\bar{1}$)

2. The equivalent rhombohedral plane for the (11$\bar{1}$)fcc plane is:
   (a) (110)
   (b) (100)
   (c) (101)
   (d) (001)

3. The equivalent rhombohedral plane for the (200) fcc plane is:
   (a) (100)
   (b) (101)
   (c) (110)
   (d) (011)

4. The equivalent hexagonal plane for the (110) orthorhombic plane is:
   (a) (1$\bar{1}$00)
   (b) ($\bar{1}$010)
   (c) (10$\bar{1}$0)
   (d) (01$\bar{1}$0)

5. The equivalent hexagonal plane for the (110) orthorhombic plane is:
   (a) (1$\bar{1}$00)
   (b) ($\bar{1}$010)
   (c) (10$\bar{1}$0)
   (d) (01$\bar{1}$0)

6. The equivalent HCP plane for the (10$\bar{1}$) rhombohedral plane is:
   (a) (1$\bar{1}$00)
   (b) (11$\bar{2}$0)
   (c) (1$\bar{2}$10)
   (d) ($\bar{2}$110)

7. The equivalent HCP plane for the $(10\bar{1})$ rhombohedral plane is:
   (a) $(\bar{2}022)$
   (b) $(02\bar{2}2)$
   (c) $(20\bar{2}2)$
   (d) $(\bar{2}202)$

8. The equivalent trigonal plane for the $(11\bar{2}0)$ hexagonal plane is:
   (a) $(\bar{2}11)$
   (b) $(2\bar{2}2)$
   (c) $(1\bar{2}1)$
   (d) $(11\bar{2})$

9. The equivalent trigonal plane for the $(11\bar{2}3)$ hexagonal plane is:
   (a) $(22\bar{1})$
   (b) $(1\bar{2}2)$
   (c) $(2\bar{1}2)$
   (d) $(\bar{1}22)$

10. The equivalent bcc plane corresponding to its primitive (100) plane is:
    (a) (101)
    (b) (110)
    (c) (011)
    (d) (022)

11. The equivalent bcc plane corresponding to its primitive (110) plane is:
    (a) (122)
    (b) (112)
    (c) (121)
    (d) (211)

12. The equivalent bcc plane corresponding to its primitive (111) plane is:
    (a) $(22\bar{2})$
    (b) $(2\bar{2}2)$
    (c) $(\bar{2}22)$
    (d) (222)

13. The equivalent fcc plane corresponding to its primitive (100) plane is:
    (a) $(11\bar{1})$
    (b) (111)
    (c) $(\bar{1}11)$
    (d) $(1\bar{1}1)$

14. The equivalent fcc plane corresponding to its primitive (110) plane is:
    (a) (020)
    (b) (200)
    (c) (002)
    (d) (021)

15. The equivalent fcc plane corresponding to its primitive (111) plane is:
    (a) $(11\bar{1})$
    (b) $(1\bar{1}1)$
    (c) (111)
    (d) $(\bar{1}11)$

16. The equivalent fcc direction for the [111] rhombohedral direction is:
    (a) $[11\bar{1}]$
    (b) $[\bar{1}11]$
    (c) $[1\bar{1}1]$
    (d) [111]

17. The equivalent fcc direction for the $[11\bar{1}]$ rhombohedral direction is:
    (a) [100]
    (b) [010]
    (c) [001]
    (d) [101]

18. The equivalent fcc direction for the [200] rhombohedral direction is:
    (a) [100]
    (b) [110]
    (c) [101]
    (d) [011]

19. The equivalent orthorhombic direction for the [200] hexagonal direction is:
    (a) $[01\bar{1}]$
    (b) $[\bar{1}01]$
    (c) $[1\bar{1}0]$
    (d) $[10\bar{1}]$

20. The equivalent orthorhombic direction for the [210] hexagonal direction is:
    (a) [010]
    (b) [120]
    (c) [210]
    (d) [100]

21. The equivalent orthorhombic direction for the [211] hexagonal direction is:
    (a) [101]
    (b) [011]
    (c) [112]
    (d) [211]

22. The equivalent HCP direction for the [111] rhombohedral direction is:
    (a) [1000]
    (b) [0001]
    (c) [0100]
    (d) [0010]

23. The equivalent HCP direction for the [330] rhombohedral direction is:
    (a) $[21\bar{3}2]$
    (b) $[1\bar{3}22]$
    (c) $[12\bar{3}2]$
    (d) $[\bar{3}212]$

24. The equivalent trigonal direction for the $[21\bar{3}0]$ hexagonal direction is:
    (a) $[011\bar{\ }]$
    (b) $[\bar{1}01]$
    (c) $[1\bar{1}0]$
    (d) $[10\bar{1}]$

25. The equivalent trigonal direction for the $[21\bar{3}3]$ hexagonal direction is:
    (a) [432]
    (b) [423]
    (c) [234]
    (d) [243]

26. The equivalent bcc direction corresponding to its primitive [200] direction is:
    (a) $[1\bar{1}1]$
    (b) $[11\bar{1}]$
    (c) $[\bar{1}11]$
    (d) [111]

27. The equivalent bcc direction corresponding to its primitive [220] direction is:
    (a) [200]
    (b) [220]
    (c) [020]
    (d) [022]

28. The equivalent bcc direction corresponding to its primitive [200] direction is:
    (a) $[1\bar{1}1]$
    (b) $[11\bar{1}]$
    (c) $[\bar{1}11]$
    (d) $[111]$

29. The equivalent fcc direction corresponding to its primitive [200] direction is:
    (a) [110]
    (b) [101]
    (c) [011]
    (d) [111]

30. The equivalent fcc direction corresponding to its primitive [220] direction is:
    (a) [121]
    (b) [211]
    (c) [112]
    (d) [220]

31. The equivalent fcc direction corresponding to its primitive [222] direction is:
    (a) [200]
    (b) [220]
    (c) [222]
    (d) [221]

Answers

 1. (a)
 2. (b)
 3. (c)
 4. (d)
 5. (a)
 6. (b)
 7. (c)
 8. (d)
 9. (a)
10. (b)
11. (c)
12. (d)
13. (a)
14. (b)
15. (c)
16. (d)
17. (a)
18. (b)

19. (c)
20. (d)
21. (a)
22. (b)
23. (c)
24. (d)
25. (a)
26. (b)
27. (c)
28. (d)
29. (a)
30. (b)
31. (c)

Chapter 6
Unit Cell Symmetries and Their Representations

6.1 Unit Cell Symmetry Elements/Operations

Macroscopic symmetry elements/operations exhibited by crystalline solids are:

(i)	Proper rotation	:	$\alpha = \frac{2\pi}{n}$ (through an angle), where n = 1, 2, 3, 4 and 6
(ii)	Mirror (Reflection)	:	m (across a line in 2-D or a plane in 3-D)
(iii)	Inversion center	:	$\bar{1}$ (through a point)
(iv)	Improper rotations	:	(rotoreflection and rotoinversion)

First three symmetry operations are illustrated in Fig. 6.1.

Rotoreflection

It represents a combined operation of rotation followed by a reflection (mirror plane perpendicular to the axis of rotation). The two operations are taking place consecutively. There exists a rotoreflection axis corresponding to each proper rotation axis. The five rotoreflection axes are: $\tilde{1}, \tilde{2}, \tilde{3}, \tilde{4}$ and $\tilde{6}$ (read as one tilde, etc.). Based on the rotoreflection axes, Schoenflies notation has been developed.

Rotoinversion

Like rotoreflection, rotoinversion also represents a combined operation of rotation followed by inversion, consecutively. Also, there exists a rotoinversion axis for each rotation axis. The five rotoinversion axes are: $\bar{1}, \bar{2}, \bar{3}, \bar{4}$, and $\bar{6}$ (read as one bar, etc.). Based on the rotoinversion axes, Hermann-Mauguin (also known as International) notation has been developed.

Equivalence of Rotoreflection and Rotoinversion Axes

When we compare the two improper rotation axes, they are found to be equivalent in pairs. For example:

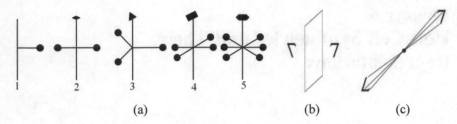

Fig. 6.1 **a** Five proper rotational axes **b** reflection **c** inversion of an object

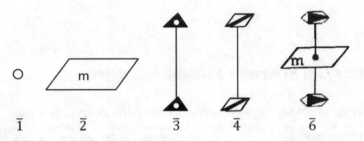

Fig. 6.2 Five rotoreflection and rotoinversion axes

$$\tilde{1} = \frac{1}{m} = m = \bar{2}, \tilde{2} = \bar{1}, \tilde{3} = \frac{3}{m} = \bar{6}, \tilde{4} = \bar{4}, \tilde{6} = \bar{3}$$

They are shown in Fig. 6.2.

On the basis of such observations, the following empirical rules can be formulated which in turn can be used for conversion from one system to another, that is, from rotoreflection to rotoinversion or vice–versa. They are:

$$\tilde{n}_{odd} = \overline{2n} = \frac{n}{m} = n\sigma_h \text{ (mirror perpendicular to the axis)}$$

$$\widetilde{2n} = \tilde{n}_{odd} = ni$$

$$\text{and } \widetilde{4n} = \overline{4n}$$

where n = 1, 2, 3, 4 and 6, m is a mirror plane and i is the center of inversion, respectively. The International/conventional symbol used to represent these axes is provided below in Table 6.1.

Solved Examples

Example 1 Show that the permissible rotational symmetries are limited to only five.

Proof Consider a combination of n-fold axis of rotation (A_n) with a translation (t) as shown in Fig. 6.3 (lower line). Perform two rotational operations of equal magnitude $\left(\alpha = \frac{2\pi}{n}\right)$ but of opposite sense to get new lattice points p and q whose length is an integral multiple of the translation (t). That is

Table 6.1 Conventional symbols for improper axes

Rotoinversion axes	Rotoreflection axes	Conventional symbol	International symbol
$\bar{1}$	$\tilde{2}$	Center of symmetry	$\bar{1}$
$\bar{2}$	$\tilde{1}$	Mirror plane	m
$\bar{3}$	$\tilde{6}$	3 fold rotoinversion	$\bar{3}$
$\bar{4}$	$\tilde{4}$	4 fold rotoinversion	$\bar{4}$
$\bar{6}$	$\tilde{3}$	6 fold rotoinversion	$\bar{6}$

Fig. 6.3 n-fold axis of rotation combined with the translation t in a plane

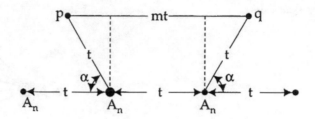

$$pq = mt, \text{ where } m = 0, \pm 1, \pm 2, \ldots \ldots$$

$$\text{or} \quad t + 2t\cos\alpha = mt,$$

$$\text{or} \quad \cos\alpha = \frac{m-1}{t} = \frac{N}{2}, \text{ where } N = 0, \pm 1, \pm 2, \ldots \ldots$$

Since $|\cos\alpha| \leq 1$, the allowed values of N are 2, 1, 0, −1, −2. The corresponding values of α and n are provided in Table 6.2.

Example 2 Show that two mirrors intersecting at an angle θ together produce a rotation of 2θ.

Proof Let us consider two mirror lines (in 3D mirror planes) m_1 and m_2 making an angle θ intersect at a point A. Through reflection, the first mirror m_1 changes (say) a right-handed object 1R into a left-handed object 2L and the second mirror m_2 changes 2L to 3R as shown in Fig. 6.4. For such reflections, the angles

$$\varphi_1 = \varphi_2 \text{ and } \psi_1 = \psi_2$$

Such that

$$\varphi_2 + \psi_1 = \theta$$

The two reflections are equivalent to a rotation about A by the amount

Table 6.2 Allowed
rotational axes

N	cos α	α (°)	n allowed rotational axes
-2	−1	180	2
-1	−1/2	120	3
0	0	90	4
+1	+1/2	60	6
+2	+1	360 or 0	1

Fig. 6.4 Combination of
rotation and reflection
symmetries

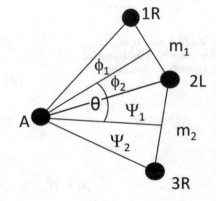

$$\varphi_1 + \varphi_2 + \psi_1 + \psi_2 = 2\theta$$

That is the two mirrors at an angle θ together produce a rotation of 2θ.

The same result can be restated as: When a mirror line in 2-D (in 3-D it becomes a mirror plane) is combined with a proper rotation axis, another mirror line (parallel to the rotation axis) is produced such that the angle between the two mirrors is equal to one-half the throw of the rotation axis. The two mirrors will have different forms. For example:

(i) Combination of 2-fold and m produces a mirror line of second form at 90° away.

(ii) Combination of 4-fold and m produces two mirror lines of second form at 45° away.

(iii) Combination of 6-fold and m produces three mirror lines of second form at 30° away.

They are shown in Fig. 6.5.

Fig. 6.5 Different ways of
combining mirror and
rotations

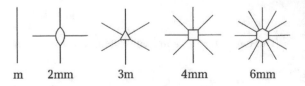

m   2mm   3m   4mm   6mm

Fig. 6.6 a A pentagon, **b** An octagon with their interior angles

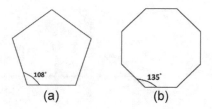

Example 3 Show that neither 5-fold nor 8-fold rotational symmetries can exist in a crystal lattice.

Solution: Construct two-dimensional pentagonal and octagonal unit cells as shown in Fig. 6.6. The interior angle of the pentagon is 108° and that of the octagon is 135°, and none of these angles are quotient of the angle 360°. Therefore, when we try to construct a periodic array using the 5-fold or 8-fold unit cell, we find that the resulting array of pentagons (or octagons) do not fit together neatly and leave empty spaces in between them as shown in Fig. 6.7. Both these considerations indicate that neither 5-fold nor 8-fold symmetries can exist in crystals. In fact, no rotational symmetry greater than 6-fold is possible.

Example 4 Show that $\tilde{1}$ (one tilde) is equivalent to a mirror plane.

Solution: It is a combined operation of rotation and reflection taking place consecutively. In this case, the proper 1-fold rotation will rotate the motif representing all space (here it is taken as 7) through an angle of 0° or 360°, that is, leaving it unchanged. Combining this with a reflection (where mirror is placed perpendicular to the axis of rotation) to produce a configuration as shown in Fig. 6.8a, which is identical to the configuration shown in Fig. 6.1b. Thus, the operation $\tilde{1}$ is equivalent to a reflection through a plane (specifically a mirror plane) .

Example 5 Show that $\tilde{2}$ (two tildes) is equivalent to an inversion center.

Solution: In this case, the proper 2-fold rotation will rotate the motif representing all space (here it is taken as 7) through an angle of 180° and then reflected across an imaginary plane placed perpendicular to the axis of rotation to produce a configuration as shown in Fig. 6.8b, which is identical to the configuration shown in Fig. 6.1c. Thus, the operation $\tilde{2}$ is equivalent to an inversion center.

Fig. 6.7 Gaps shown in (**a**) 5-fold rotation (**b**) 8-fold rotation

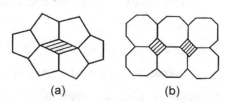

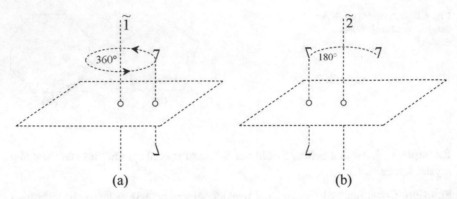

Fig. 6.8 Three-dimensional view of $\tilde{1}$ and $\tilde{2}$

Fig. 6.9 Primitive
Monoclinic

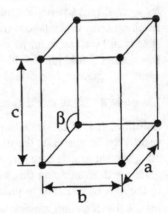

Example 6 Find the total number of symmetry elements that exist in a monoclinic
unit cell.

Solution: Looking at Fig. 6.9, we find that there exist:

(i)   2-fold axis parallel to c-axis and passing through the center of the unit cell.
      There is only one such axis.
(ii) Two mirror planes intersecting at 90° whose axis is parallel to c-axis. There
      are two such planes.
(iii) One center of symmetry at the center of the unit cell.

   Thus, there are:

1-diad + 2 surface planes + 1 center of symmetry = 4

   Therefore, the total number of symmetry elements in a monoclinic unit cell = 4.

Example 7 Find the total number of symmetry elements that exist in an
orthorhombic unit cell.

Fig. 6.10 Primitive
Orthorhombic

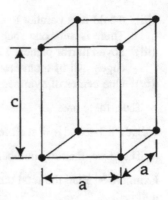

Solution: Looking at Fig. 6.10, we find that there exist:

(i)   2-fold axes parallel to a, b and c-axes and passing through the center of the
      unit cell. There are three such axes.
(ii) Three mirror planes bisecting the parallel faces of the cuboid. All of them are
      passing through the center of the unit cell.
(iii) One center of symmetry at the center of the unit cell.

   Thus, there are:

3-diads + 3 surface planes + 1 center of symmetry = 7

   Therefore, the total number of symmetry elements in an orthorhombic unit cell =
7.

Example 8 Find the total number of symmetry elements that exist in a tetragonal
unit cell.

Solution: Looking at Fig. 6.11, we find that there exist:

(i)   2-fold axes parallel to the face diagonal and passing through the center of the
      unit cell. There are four such axes.

Fig. 6.11 Primitive
tetragonal unit cell

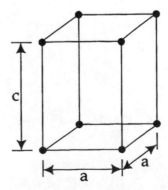

(ii)   4-fold axis parallel to c-axis and passing through the center of the unit cell.
       There is only one such axis.
(iii) Seven mirror planes, 3 bisecting the parallel faces, and 4 connecting diagonal
       edges. All of them are passing through the center of the unit cell.
(iv)   One center of symmetry at the center of the unit cell.

   Thus, there are:

4-diads + 1-tetrad + 7 surface planes + 1 center of symmetry = 13

   Therefore, the total number of symmetry elements in a tetragonal unit cell = 13.

Example 9 Find the total number of symmetry elements that exist in a hexagonal
unit cell.

Solution: Looking at Fig. 6.12, we find that there exist:

 (i)   2-fold axes parallel to the face diagonal and passing through the center of the
       unit cell. There are six such axes.
 (ii) 6-fold axis parallel to c-axis and passing through the center of the unit cell.
       There is only one such axis.
(iii) Seven mirror planes, 3 parallel to the opposite (vertical) faces, 3 parallel to
       opposite vertical edges, all of them are passing through the center of the unit
       cell and parallel to the principal axis. 1 horizontal plane passing through the
       center of the unit cell.
(iv)   One center of symmetry at the center of the unit cell.

   Thus, there are:

6-diads + 1-hexad + 7 surface planes + 1 center of symmetry = 15

   Therefore, the total number of symmetry elements in a hexagonal unit cell = 15.

Fig. 6.12 Simple hexagonal
unit cell

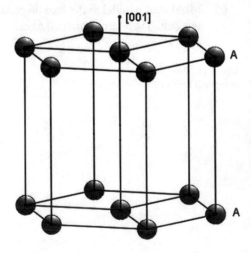

Example 10 Find the total number of symmetry elements that exist in a cube.

Solution: Looking at Fig. 6.13, we find that there exist:

(i)   2-fold axes parallel to the face diagonal and passing through the center of the cube. There are six such axes.

(ii) 3-fold axes passing through the body diagonal. There are four such axes.

(iii) 4-fold symmetry about an axis passing through the center of two opposite faces. There are three such symmetry axes.

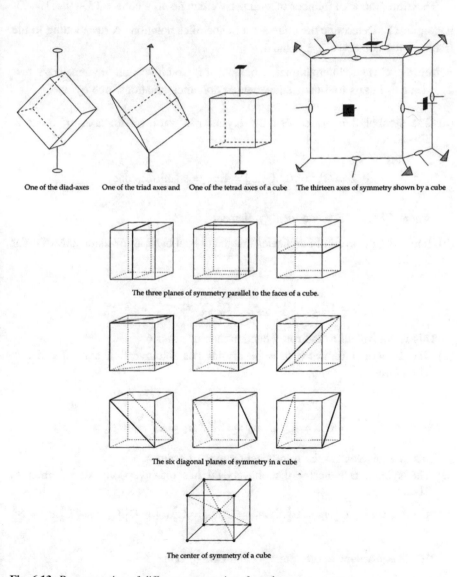

One of the diad-axes   One of the triad axes and   One of the tetrad axes of a cube   The thirteen axes of symmetry shown by a cube

The three planes of symmetry parallel to the faces of a cube.

The six diagonal planes of symmetry in a cube

The center of symmetry of a cube

Fig. 6.13 Representation of different symmetries of a cube

(iv)   Nine mirror planes, 3 bisecting the parallel faces, and 6 connecting diagonal edges. All of them are passing through the center of the cube.
(v)   One center of symmetry at the center of the cube.

Thus, there are:

6-diads + 4-triads + 3-tetrads = 13 axes
3 surface planes + 6 diagonal planes = 9 planes
Center of symmetry = 1

Therefore, the total number of symmetry elements in a cube = 13 + 9 + 1 = 23.

Example 11   Determine the equivalent Schoenflies notations corresponding to the International notations, $\bar{2}, \bar{3}, \bar{4}$ and $\bar{6}$.

Solution: *Given*: International notation of rotoinversion symmetries are: $\bar{2}, \bar{3}, \bar{4}$ and $\bar{6}$. Let us find their equivalent Schoenflies notations, one by one.

(a)   The symbol $\bar{2}$ indicates that the crystal has rotoinversion axis of order 2. Therefore,

$$\bar{2} = C_2^1.i = C_2^1.C_2^1.\sigma_h = E\sigma_h = \sigma \, (\text{mirrorplane})$$

where $C_2^1.C_2^1 = E$ is the identity element.

(b)   Similarly, the symbol $\bar{3}$ indicates that the crystal has rotoinversion axis of order 3. Therefore,

$$\bar{3} = C_3^1.i = C_3^1.C_2^1.\sigma_h = C_6^2.C_6^3.\sigma_h = C_6^5.\sigma_h = S_6^5$$

This is equivalent to the rotoreflection axis of order 6.

(c)   The symbol $\bar{4}$ indicates that the crystal has rotoinversion axis of order 4. Therefore,

$$\bar{4} = C_4^1.i = C_4^1.C_2^1.\sigma_h = C_4^1.C_4^2.\sigma_h = C_4^3.\sigma_h = S_4^3$$

This is equivalent to the rotoreflection axis of order 4.

(d)   The symbol $\bar{6}$ indicates that the crystal has rotoinversion axis of order 6. Therefore,

$$\bar{6} = C_6^1.i = C_6^1.C_2^1.\sigma_h = C_6^1.C_6^3.\sigma_h = C_6^4.\sigma_h = C_3^2.\sigma_h = C_3^3.C_3^2.\sigma_h = C_3^5.\sigma_h = S_3^5$$

This is equivalent to the rotoreflection axis of order 3.

6.2 Matrix Representation of Symmetry Operations

Whenever a symmetry operation is applied to a unit cell, the coordinate axes associated with it change. The resulting change can be conveniently described by the following linear equations.

$$x'_1 = c_{11}x_1 + c_{12}x_2 + c_{13}x_3$$
$$x'_2 = c_{21}x_1 + c_{22}x_2 + c_{23}x_3$$
$$x'_3 = c_{31}x_1 + c_{32}x_2 + c_{33}x_3$$

This can be written in the matrix form as

$$\begin{pmatrix} x'_1 \\ x'_2 \\ x'_3 \end{pmatrix} = \begin{pmatrix} c_{11} & c_{12} & c_{13} \\ c_{21} & c_{22} & c_{23} \\ c_{31} & c_{32} & c_{33} \end{pmatrix} \begin{pmatrix} x_1 \\ x_2 \\ x_3 \end{pmatrix} \qquad (6.1)$$

The matrix elements are given by

$$C_{ij} = \begin{pmatrix} c_{11} & c_{12} & c_{13} \\ c_{21} & c_{22} & c_{23} \\ c_{31} & c_{32} & c_{33} \end{pmatrix}$$

The first script in the symbol C_{ij} refers to the new axes and the second to the old axes. Therefore, the matrix elements may be conveniently recognized as the direction cosines between the new and old axes in the right-handed coordinate system taken in the order old $\rightarrow$ new as shown in the Fig. 6.14. Accordingly, the matrix elements in terms of direction cosines are given as

$$C_{ij} = \begin{pmatrix} cos(x_1x'_1) & cos(x_1x'_2) & cos(x_1x'_3) \\ cos(x_2x'_1) & cos(x_2x'_2) & cos(x_2x'_3) \\ cos(x_3x'_1) & cos(x_3x'_2) & cos(x_3x'_3) \end{pmatrix} \qquad (6.2)$$

Fig. 6.14 Directions cosines

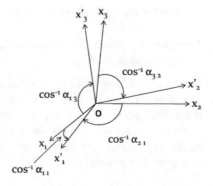

Fig. 6.15 Rotation of axes
about $x_3(x_3')$ axis

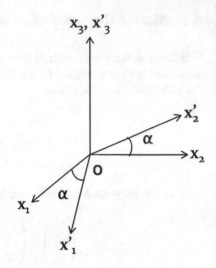

Now, for a given rotation angle α (Fig. 6.15) about any principal axis (say x_3 - axis), above equation will reduce to

$$C_{ij} = \begin{pmatrix} \cos\alpha & \cos(90+\alpha) & \cos 90 \\ \cos(90-\alpha) & \cos\alpha & \cos 90 \\ \cos 90 & \cos 90 & \cos 0 \end{pmatrix}$$

This on simplification gives us

$$C_{ij} = \begin{pmatrix} \cos\alpha & -\sin\alpha & 0 \\ \sin\alpha & \cos\alpha & 0 \\ 0 & 0 & 1 \end{pmatrix} \tag{6.3}$$

From Eq. 6.3, one can easily determine the matrices corresponding to five proper rotational symmetries by replacing $\alpha = \frac{2\pi}{n}$ (where n = 1, 2, 3, 4 and 6).

Solved Examples

(i) *Matrix Using Orthogonal Axes*

Example 1 Obtain the matrix corresponding to $1(\bar{1})$ − fold operation using orthogonal system of axes.

Solution: 1-fold operation is equivalent to a no rotation or a rotation of 360° around any direction in a crystal. This operation does not bring about any change in the axes. This gives us $x_1' = x_1$, $x_2' = x_2$ and $x_3' = x_3$ (Fig. 6.16). The matrix corresponding to this operation is obtained either from Eq. 6.2 by substituting the values of direction cosines or directly from Eq. 6.3 with $\alpha = 0$. This is known as identity matrix and is represented as

Fig. 6.16 Two sets of axes

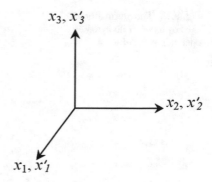

Fig. 6.17 Transformation of
axes by inversion operation

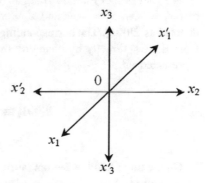

$$1 \equiv \begin{pmatrix} 1 & 0 & 0 \\ 0 & 1 & 0 \\ 0 & 0 & 1 \end{pmatrix} \tag{6.4}$$

On the other hand, $\bar{1}-$ operation inverts the whole space through a point, called the center of symmetry.

This operation gives us $x'_1 = -x_1$, $x'_2 = -x_2$, $x'_3 = -x_3$ as shown in Fig. 6.17. The corresponding matrix is represented as

$$\bar{1} = \begin{pmatrix} -1 & 0 & 0 \\ 0 & -1 & 0 \\ 0 & 0 & -1 \end{pmatrix}$$

This matrix can be obtained by simply changing the sign of the digits of the matrix in Eq. 6.4.

Example 2 Obtain the matrix corresponding to $2(\bar{2})$ − fold operation using orthogonal system of axes.

Solution: In general, 2 and $(\bar{2})$- fold operations are common to all principal crystallographic directions except [111].

Fig. 6.18 Transformation of
axes by a two–fold symmetry
operation parallel to x_3

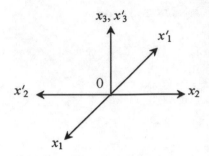

A 2-fold proper rotation along x_3 (or z) axis is shown in Fig. 6.18. This suggests
that $x'_1 = -x_1$, $x'_2 = -x_2$ and $x'_3 = x_3$. In Miller index notation, this operation is
denoted as 2[001]. The corresponding matrix can be obtained either from Eq. 6.2 or
from Eq. 6.4 directly by changing the sign of the first two axes. This operation is
represented as

$$2[001] \equiv \begin{pmatrix} -1 & 0 & 0 \\ 0 & -1 & 0 \\ 0 & 0 & 1 \end{pmatrix}$$

On the other hand, a $\bar{2}-$ operation is an improper rotation equivalent to a mirror
plane, that is, $\bar{2} \equiv m$. A mirror plane perpendicular to the x_3 (z) axis is shown in
Fig. 6.19. This suggests that $x'_1 = x_1$, $x'_2 = x_2$ and $x'_3 = -x_3$.
The corresponding matrix is represented as

$$\bar{2}[001] \equiv m[001] \equiv \begin{pmatrix} 1 & 0 & 0 \\ 0 & 1 & 0 \\ 0 & 0 & -1 \end{pmatrix}$$

Fig. 6.19 Transformation of
axes by a mirror plane
perpendicular to x_3

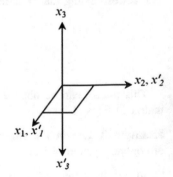

Similarly, the operations 2[010] and m [010] gives us $x_1' = -x_1, x_2' = x_2, x_3' = -x_3$ and $x_1' = x_1, x_2' = -x_2, x_3' = x_3$, respectively. The corresponding matrices are:

$$2[010] \equiv \begin{pmatrix} -1 & 0 & 0 \\ 0 & 1 & 0 \\ 0 & 0 & -1 \end{pmatrix} \text{ and } m[010] \equiv \begin{pmatrix} 1 & 0 & 0 \\ 0 & -1 & 0 \\ 0 & 0 & 1 \end{pmatrix}$$

Example 3 Obtain the matrix corresponding to 3 and 6-fold operation using orthogonal system of axes.

Solution: A 3-fold operation coupled with the orthogonal axes x_1, x_2, x_3 is shown in Fig. 6.20. A proper rotation about x_3-axis transforms x_1 into x_1' and x_2 into x_2', each of which is thrown 120° away from its initial position. Substituting the values α (=120°) in Eq. 6.3, the corresponding matrix can be obtained as

$$3[001] = \begin{pmatrix} \cos 120° & -\sin 120° & 0 \\ \sin 120° & \cos 120° & 0 \\ 0 & 0 & 1 \end{pmatrix} = \begin{pmatrix} -\frac{1}{2} & -\frac{\sqrt{3}}{2} & 0 \\ \frac{\sqrt{3}}{2} & -\frac{1}{2} & 0 \\ 0 & 0 & 1 \end{pmatrix}$$

In a similar manner, the matrix corresponding to a 6-fold rotation is obtained as

$$6[001] = \begin{pmatrix} \cos 60° & -\sin 60° & 0 \\ \sin 60° & \cos 60° & 0 \\ 0 & 0 & 1 \end{pmatrix} = \begin{pmatrix} \frac{1}{2} & -\frac{\sqrt{3}}{2} & 0 \\ \frac{\sqrt{3}}{2} & \frac{1}{2} & 0 \\ 0 & 0 & 1 \end{pmatrix}$$

On the other hand, a proper 3-fold rotation along the body diagonal of a cube, that is, 3[111] transforms x_1 into x_1' and x_2 into x_2' and x_3 into x_3', each of these is thrown by 90° away from its original position. This operation gives us $x_1' = x_2$, $x_2' = x_3$ and $x_3' = x_1$ as shown in Fig 6.21. The corresponding matrix can be obtained by substituting the values of direction cosines in Eq. 6.2, we have

Fig. 6.20 Crystallographic axes chosen for an equilateral triangle

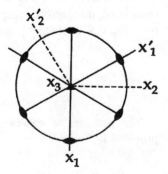

Fig. 6.21 Rotational operation along 3[111] direction

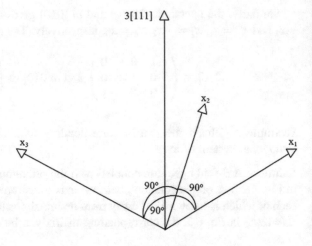

$$3[111] = \begin{pmatrix} \cos 90° & \cos 90° & \cos 0° \\ \cos 0° & \cos 90° & \cos 90° \\ \cos 90° & \cos 0° & \cos 90° \end{pmatrix} = \begin{pmatrix} 0 & 0 & 1 \\ 1 & 0 & 0 \\ 0 & 1 & 0 \end{pmatrix}$$

Example 4 Obtain the matrix corresponding to $4(\bar{4})$ − fold operation using orthogonal system of axes.

Solution: A $4(\bar{4})$−fold operation is possible along the three crystallographic directions [100], [010] and [001]. These directions are equivalent to orthogonal axes. A 4−fold rotation along [001] transforms the axis x_1 into x_1' and x_2 into x_2', each of which is thrown 90° away from its initial position. This gives us $x_1' = x_2$, $x_2' = -x_1$. The corresponding matrix can be represented as

$$4[001] = \begin{pmatrix} 0 & -1 & 0 \\ 1 & 0 & 0 \\ 0 & 0 & 1 \end{pmatrix} \qquad (6.5)$$

This matrix can also be obtained from Eq. 6.3 by substituting $\alpha = 90°$. On the other hand, the matrix corresponding to $\bar{4}[001]$ symmetry operation can be obtained from Eq. 6.5 by changing the sign of the digit 1, that is,

$$\bar{4}[001] = \begin{pmatrix} 0 & 1 & 0 \\ -1 & 0 & 0 \\ 0 & 0 & -1 \end{pmatrix}$$

The matrices of generating elements corresponding to orthogonal axes are provided in Table 6.3.

Table 6.3 Generating elements and their matrices (orthogonal axes)

Identity, $I = \begin{pmatrix} 1 & 0 & 0 \\ 0 & 1 & 0 \\ 0 & 0 & 1 \end{pmatrix}$	Inversion, $\bar{1} = \begin{pmatrix} -1 & 0 & 0 \\ 0 & -1 & 0 \\ 0 & 0 & -1 \end{pmatrix}$
$2[001] = \begin{pmatrix} -1 & 0 & 0 \\ 0 & -1 & 0 \\ 0 & 0 & 1 \end{pmatrix}$	$\bar{2}[001] = m[001] = \begin{pmatrix} 1 & 0 & 0 \\ 0 & 1 & 0 \\ 0 & 0 & -1 \end{pmatrix}$
$2^*[010] = \begin{pmatrix} -1 & 0 & 0 \\ 0 & 1 & 0 \\ 0 & 0 & -1 \end{pmatrix}$	$\bar{2}^*[010] = m[010] = \begin{pmatrix} 1 & 0 & 0 \\ 0 & -1 & 0 \\ 0 & 0 & 1 \end{pmatrix}$
$3[001] = \begin{pmatrix} -\frac{1}{2} & -\frac{\sqrt{3}}{2} & 0 \\ \frac{\sqrt{3}}{2} & -\frac{1}{2} & 0 \\ 0 & 0 & 1 \end{pmatrix}, \bar{3}[001] = \begin{pmatrix} \frac{1}{2} & \frac{\sqrt{3}}{2} & 0 \\ -\frac{\sqrt{3}}{2} & \frac{1}{2} & 0 \\ 0 & 0 & 1 \end{pmatrix}$	$3^*[111] = \begin{pmatrix} 0 & 0 & 1 \\ 1 & 0 & 0 \\ 0 & 1 & 0 \end{pmatrix}$
$4[001] = \begin{pmatrix} 0 & -1 & 0 \\ 1 & 0 & 0 \\ 0 & 0 & 1 \end{pmatrix}$	$\bar{4}[001] = \begin{pmatrix} 0 & 1 & 0 \\ -1 & 0 & 0 \\ 0 & 0 & -1 \end{pmatrix}$
$6[001] = \begin{pmatrix} \frac{1}{2} & -\frac{\sqrt{3}}{2} & 0 \\ \frac{\sqrt{3}}{2} & \frac{1}{2} & 0 \\ 0 & 0 & 1 \end{pmatrix}$	$\bar{6}[001] = \begin{pmatrix} -\frac{1}{2} & \frac{\sqrt{3}}{2} & 0 \\ -\frac{\sqrt{3}}{2} & -\frac{1}{2} & 0 \\ 0 & 0 & -1 \end{pmatrix}$

*Supplementary generating elements

Example 5 Obtain 2-D rotation matrix using Cartesian coordinate system.

Solution: Let P (x, y) be any point on the XY plane to be rotated through an angle θ to get the point Q (x', y') as shown in Fig. 6.22. Let $\hat{i}$ and $\hat{j}$ are the unit vectors along OX and OY axes, respectively, and the line OP makes an angle θ with respect to X-axis. It is to be noted that the vector $\left|\overrightarrow{OP}\right| = \left|\overrightarrow{OQ}\right|$ and $z = z'$. Then from the figure, we can write

Fig. 6.22 Rotation at a point

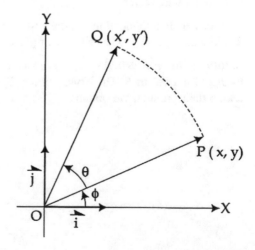

$$\overrightarrow{OP} = x\hat{i} + y\hat{j}$$
$$= \left|\overrightarrow{OP}\right| \cos\theta\hat{i} + \left|\overrightarrow{OQ}\right| \sin\theta\hat{j}$$
$$= \left|\overrightarrow{OQ}\right| \cos\theta\hat{i} + \left|\overrightarrow{OQ}\right| \sin\theta\hat{j}$$

Similarly,

$$\overrightarrow{OQ} = x'\hat{i} + y'\hat{j}$$
$$= \left|\overrightarrow{OQ}\right| \cos(\alpha+\theta)\hat{i} + \left|\overrightarrow{OQ}\right| \sin(\alpha+\theta)\hat{j}$$
$$= \left|\overrightarrow{OQ}\right|(\cos\alpha\cos\theta - \sin\alpha\sin\theta)\hat{i} + \left|\overrightarrow{OQ}\right|(\sin\alpha\cos\theta + \cos\alpha\sin\theta)\hat{j}$$
$$= (x\cos\alpha - y\sin\alpha)\hat{i} + (x\sin\alpha + y\cos\alpha)\hat{j}$$

This gives us the following:

$$x' = x\cos\alpha - y\sin\alpha$$
$$y' = x\sin\alpha + y\cos\alpha$$

In the matrix notation, they can be expressed as

$$\begin{pmatrix} x' \\ y' \end{pmatrix} = \begin{pmatrix} \cos\alpha & -\sin\alpha \\ \sin\alpha & \cos\alpha \end{pmatrix} \begin{pmatrix} x \\ y \end{pmatrix}$$

Here, the matrix $A = \begin{pmatrix} \cos\alpha & -\sin\alpha \\ \sin\alpha & \cos\alpha \end{pmatrix}$ is known as rotation matrix. It is to be noted that the determinant of a rotation matrix is equal to 1. Furthermore, either the row or column vectors of a rotation matrix may be considered to be a pair of orthogonal unit vectors.

Example 6 In a monoclinic crystal system show that the mirror plane parallel to a 2-fold axis (c-axis) produces a mirror plane.

Solution: *Given*: A 2-fold symmetry along c-axis $\equiv 2[001]$, a mirror plane parallel to 2-fold axis $\equiv$ m [010]. Now, writing 2[001] and m[010] in matrix form and taking their product, we obtain

$$2[001] \times m[010] = \begin{pmatrix} -1 & 0 & 0 \\ 0 & -1 & 0 \\ 0 & 0 & 1 \end{pmatrix} \begin{pmatrix} 1 & 0 & 0 \\ 0 & -1 & 0 \\ 0 & 0 & 1 \end{pmatrix}$$

$$= \begin{pmatrix} -1 & 0 & 0 \\ 0 & 1 & 0 \\ 0 & 0 & 1 \end{pmatrix} = m[100]$$

The resulting mirror plane is perpendicular to the x axis.

Example 7 In an orthorhombic crystal system, show that two mutually perpendicular mirror planes produce a 2-fold axis.

Solution: *Given*: Crystal system is orthorhombic, two mutually perpendicular mirror planes. Let us suppose that the two mutually perpendicular mirror planes are: m[100] and m[010]. Now, writing them in the matrix form and taking their product, we obtain

$$m[001] \times m[010] = \begin{pmatrix} -1 & 0 & 0 \\ 0 & 1 & 0 \\ 0 & 0 & 1 \end{pmatrix} \begin{pmatrix} 1 & 0 & 0 \\ 0 & -1 & 0 \\ 0 & 0 & 1 \end{pmatrix}$$

$$= \begin{pmatrix} -1 & 0 & 0 \\ 0 & -1 & 0 \\ 0 & 0 & 1 \end{pmatrix} = 2[001]$$

The resulting 2-fold axis represents the intersection of the two mirrors along c-axis.

Example 8 In an orthorhombic crystal system, show that two mutually perpendicular 2-fold axes produce another 2-fold axis perpendicular to each of the first two axes.

Solution: *Given*: Crystal system is orthorhombic, two mutually perpendicular 2-fold axes. Let us suppose that the two mutually perpendicular axes are: 2[010] and 2[001]. Now, writing 2[010] and 2[001] in the matrix form and taking their product, we obtain

$$2[001] \times 2[010] = \begin{pmatrix} -1 & 0 & 0 \\ 0 & -1 & 0 \\ 0 & 0 & 1 \end{pmatrix} \begin{pmatrix} -1 & 0 & 0 \\ 0 & 1 & 0 \\ 0 & 0 & -1 \end{pmatrix}$$

$$= \begin{pmatrix} 1 & 0 & 0 \\ 0 & -1 & 0 \\ 0 & 0 & -1 \end{pmatrix} = 2[100]$$

The resulting 2-fold axis is perpendicular to the first two axes.

Example 9 In an orthorhombic crystal system, show that three mutually perpendicular mirror planes produce an inversion.

Solution: *Given*: Crystal system is orthorhombic, three mutually perpendicular mirror planes. Let us suppose that that the three mutually perpendicular mirror planes are: m[100], m[010] and m[001]. Now, writing them in the matrix form and taking their product, we obtain

$$
m\,[100] \times m\,[010] \times m\,[001] =
\begin{pmatrix} -1 & 0 & 0 \\ 0 & 1 & 0 \\ 0 & 0 & 1 \end{pmatrix}
\begin{pmatrix} 1 & 0 & 0 \\ 0 & -1 & 0 \\ 0 & 0 & 1 \end{pmatrix}
\begin{pmatrix} 1 & 0 & 0 \\ 0 & 1 & 0 \\ 0 & 0 & -1 \end{pmatrix}
$$

$$
=
\begin{pmatrix} -1 & 0 & 0 \\ 0 & 1 & 0 \\ 0 & 0 & 1 \end{pmatrix}
\begin{pmatrix} 1 & 0 & 0 \\ 0 & -1 & 0 \\ 0 & 0 & -1 \end{pmatrix}
$$

$$
=
\begin{pmatrix} -1 & 0 & 0 \\ 0 & -1 & 0 \\ 0 & 0 & -1 \end{pmatrix}
= i \ \text{(inversion)}
$$

Here, any first two mirror planes could be taken as parallel to 2-fold axis and the third mirror perpendicular to it, which produces an inversion.

(ii) *Matrix Using Crystallographic Axes*

Orthogonal and crystallographic coordinate axes differ for trigonal and hexagonal crystal systems. Let us examine 2-, 3- and 6-fold symmetry operations under crystallographic axes (Fig. 6.23) separately.

Example 10 Obtain the matrix corresponding to 2-fold operation using crystallographic system of axes.

Fig. 6.23 The hexagonal system of axes

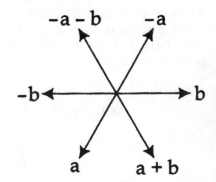

Solution: In a hexagonal system of axes (Fig. 6.23), 2-fold operations are possible along three crystallographic axes: [100], [010], and [001]. A rotation of 180° (π) along [100] axis gives us $x' = x$, $y' = -x - y$ and $z' = -z$. Therefore, the position vector

$$
\begin{aligned}
r' &= x'a + y'b + z'c \\
&= xa + (-x - y)b + (-z)c \\
&= xa - xb - yb - zc \\
&= x(a - b) + y(-b) + z(-c)
\end{aligned}
$$

The corresponding matrix can be represented as

$$
2[100] = \begin{pmatrix} 1 & -1 & 0 \\ 0 & -1 & 0 \\ 0 & 0 & -1 \end{pmatrix}
$$

Similarly, we can obtain other matrices. They are given as

$$
2[010] = \begin{pmatrix} -1 & 0 & 0 \\ 1 & 1 & 0 \\ 0 & 0 & -1 \end{pmatrix} \quad \text{and} \quad 2[001] = \begin{pmatrix} -1 & 0 & 0 \\ 0 & -1 & 0 \\ 0 & 0 & 1 \end{pmatrix}
$$

Example 11 Obtain the matrix corresponding to 3- and 6− fold operations using orthogonal system of axes.

Solution: In a hexagonal system of axes, a 3-fold operation is possible only along [001] axis. A rotation of 120° ($\frac{2\pi}{3}$) along [001] axis gives us $x' = y$, $y' = -x - y$ and $z' = z$. Therefore, the position vector

$$
\begin{aligned}
r' &= x'a + y'b + z'c \\
&= ya + (-x - y)b + zc \\
&= ya - xb - yb + zc \\
&= x(-b) + y(a - b) + zc)
\end{aligned}
$$

The corresponding matrix can be represented as

$$
3[100] = \begin{pmatrix} 0 & -1 & 0 \\ 1 & -1 & 0 \\ 0 & 0 & 1 \end{pmatrix}
$$

Like a 3-fold axis, a 6-fold axis is also possible only along [001] axis. Proceeding in a similar manner, a 6-fold rotation, a rotation of $60°$ $\left(\frac{\pi}{3}\right)$ along [001] axis leads us to obtain the matrix

$$6[001] = \begin{pmatrix} 1 & -1 & 0 \\ 1 & 0 & 0 \\ 0 & 0 & 1 \end{pmatrix}$$

Matrices corresponding to rotoinversion axes can be obtained by simply interchanging the sign of the digits appearing in proper rotation matrices. Table 6.4 provides the matrices of generating elements corresponding to crystallographic axes.

With the help of the generating elements provided in Tables 6.3 and 6.4 and taking into account the group conditions, the matrix representation of 32 point groups can be obtained.

Example 12 Obtain the representative matrices corresponding to five rotoreflection axes, S_1, S_2, S_3, S_4 and S_6.

Solution: *Given*: Five rotoreflecion axes: S_1, S_2, S_3, S_4 and S_6. Determine their matrices.

We know that rotoreflection is a two-step process, that is, a rotation followed by a reflection (perpendicular to the principal c−axis) consecutively.

Therefore,

$$S_n = \sigma_h C_n = \begin{pmatrix} 1 & 0 & 0 \\ 0 & 1 & 0 \\ 0 & 0 & -1 \end{pmatrix} \begin{pmatrix} \cos\theta & -\sin\theta & 0 \\ \sin\theta & \cos\theta & 0 \\ 0 & 0 & 1 \end{pmatrix}$$

$$= \begin{pmatrix} \cos\theta & -\sin\theta & 0 \\ \sin\theta & \cos\theta & 0 \\ 0 & 0 & -1 \end{pmatrix}$$

Now, considering different cases, we have

(i) For $n = 1$, $\theta = 0°$ or $360°$,

$\implies \cos\theta = 1$, $\sin\theta = 0$, and hence,

$$S_1 = \begin{pmatrix} 1 & 0 & 0 \\ 0 & 1 & 0 \\ 0 & 0 & -1 \end{pmatrix} = \sigma_h$$

(ii) For $n = 2$, $\theta = 180°$,

$\implies \cos\theta = -1$, $\sin\theta = 0$, and hence,

Table 6.4 Generating elements and their matrices (crystallographic axes)

Identity, $I = \begin{pmatrix} 1 & 0 & 0 \\ 0 & 1 & 0 \\ 0 & 0 & 1 \end{pmatrix}$	Inversion, $\bar{1} = \begin{pmatrix} -1 & 0 & 0 \\ 0 & -1 & 0 \\ 0 & 0 & -1 \end{pmatrix}$
$2[001] = \begin{pmatrix} -1 & 0 & 0 \\ 0 & -1 & 0 \\ 0 & 0 & 1 \end{pmatrix}$	$\bar{2}[001] = m[001] = \begin{pmatrix} 1 & 0 & 0 \\ 0 & 1 & 0 \\ 0 & 0 & -1 \end{pmatrix}$
$3[001] = \begin{pmatrix} 0 & -1 & 0 \\ 1 & -1 & 0 \\ 0 & 0 & 1 \end{pmatrix}, \bar{3}[001] = \begin{pmatrix} 0 & 1 & 0 \\ -1 & 1 & 0 \\ 0 & 0 & -1 \end{pmatrix}$	$3^*[111] = \begin{pmatrix} 0 & 0 & 1 \\ 1 & 0 & 0 \\ 0 & 1 & 0 \end{pmatrix}$
$4[001] = \begin{pmatrix} 0 & -1 & 0 \\ 1 & 0 & 0 \\ 0 & 0 & 1 \end{pmatrix}$	$\bar{4}[001] = \begin{pmatrix} 0 & 1 & 0 \\ -1 & 0 & 0 \\ 0 & 0 & -1 \end{pmatrix}$
$H \rightarrow 2^*[010] = \begin{pmatrix} -1 & 0 & 0 \\ -1 & 1 & 0 \\ 0 & 0 & -1 \end{pmatrix}$	$H \rightarrow \bar{2}^*[010] = m[010] = \begin{pmatrix} 1 & 0 & 0 \\ 1 & -1 & 0 \\ 0 & 0 & 1 \end{pmatrix}$
$6[001] = \begin{pmatrix} 1 & -1 & 0 \\ 1 & 0 & 0 \\ 0 & 0 & 1 \end{pmatrix}$	$\bar{6}[001] = \begin{pmatrix} -1 & 1 & 0 \\ -1 & 0 & 0 \\ 0 & 0 & -1 \end{pmatrix}$

*Supplementary generating elements

$$S_2 = \begin{pmatrix} -1 & 0 & 0 \\ 0 & -1 & 0 \\ 0 & 0 & -1 \end{pmatrix} = i$$

(iii) For $n = 3$, $\theta = 120°$,

$\Rightarrow \cos\theta = -\frac{1}{2}, \sin\theta = \frac{\sqrt{3}}{2}$, and hence,

$$S_3 = \begin{pmatrix} -\frac{1}{2} & -\frac{\sqrt{3}}{2} & 0 \\ \frac{\sqrt{3}}{2} & -\frac{1}{2} & 0 \\ 0 & 0 & -1 \end{pmatrix}$$

(iv) For $n = 4$, $\theta = 90°$,

$\implies \cos\theta = 0$, $\sin\theta = 1$, and hence,

$$S_4 = \begin{pmatrix} 0 & -1 & 0 \\ 1 & 0 & 0 \\ 0 & 0 & -1 \end{pmatrix}$$

(v) For $n = 6$, $\theta = 60°$,

$\Rightarrow \cos\theta = \frac{1}{2}, \sin\theta = \frac{\sqrt{3}}{2}$, and hence,

$$S_6 = \begin{pmatrix} \frac{1}{2} & -\frac{\sqrt{3}}{2} & 0 \\ \frac{\sqrt{3}}{2} & \frac{1}{2} & 0 \\ 0 & 0 & -1 \end{pmatrix}$$

(iii) *Inverse of a Matrix*

A square matrix $B = A^{-1}$ is called the inverse (multiplicative) of the matrix A if

$$AB = BA = I \tag{6.6}$$

where I is the unit/identity matrix. The inverse of the matrix can be found by determining the minor and co-factor of each matrix element, value of the determinant of the matrix and then use the formula,

$$A^{-1} = \frac{Adj(A)}{|A|} \tag{6.7}$$

where the Adjoint of a given matrix is the transpose of the matrix whose elements are cofactors of elements of the given matrix, that is,

$$Adj(A) = C_{ji} \tag{6.8}$$

and the cofactor,

$$C_{ij} = (-1)^{i+j} M_{ij}(A) \tag{6.9}$$

This suggests that the inverse of a singular matrix does not exist.

Example 13 For a given square matrix $A = (a_{ij})$, find its adjoint matrix and if possible determine its inverse also.

Solution: *Given*: The square matrix $A = (a_{ij})$, $Adj(A) = ?$, $A^{-1} = ?$

Let the given square matrix is a non–singular matrix of order 3×3. This implies, $\Delta = |A| \neq 0$, which means A^{-1} exists.

Now,

$$A = \begin{pmatrix} a_{11} & a_{12} & a_{13} \\ a_{21} & a_{22} & a_{23} \\ a_{31} & a_{32} & a_{33} \end{pmatrix}$$

Let A_{ij} be the cofactor of a_{ij} in $|A|$, then from Eq. 6.9, we have

$$A_{11} = (-1)^{1+1} \begin{vmatrix} a_{22} & a_{23} \\ a_{32} & a_{33} \end{vmatrix}, A_{12} = (-1)^{1+2} \begin{vmatrix} a_{21} & a_{23} \\ a_{31} & a_{33} \end{vmatrix}, A_{13} = (-1)^{1+3} \begin{vmatrix} a_{21} & a_{22} \\ a_{31} & a_{32} \end{vmatrix}$$

$$A_{21} = (-1)^{2+1} \begin{vmatrix} a_{12} & a_{13} \\ a_{32} & a_{33} \end{vmatrix}, A_{22} = (-1)^{2+2} \begin{vmatrix} a_{11} & a_{13} \\ a_{31} & a_{33} \end{vmatrix}, A_{23} = (-1)^{2+3} \begin{vmatrix} a_{11} & a_{12} \\ a_{31} & a_{32} \end{vmatrix}$$

$$A_{31} = (-1)^{3+1} \begin{vmatrix} a_{12} & a_{13} \\ a_{22} & a_{23} \end{vmatrix}, A_{32} = (-1)^{3+2} \begin{vmatrix} a_{11} & a_{13} \\ a_{21} & a_{23} \end{vmatrix}, A_{33} = (-1)^{3+3} \begin{vmatrix} a_{11} & a_{12} \\ a_{21} & a_{22} \end{vmatrix}$$

The matrix of the cofactors, A_{ij} is given by

$$A_{ij} = \begin{pmatrix} A_{11} & A_{12} & A_{13} \\ A_{21} & A_{22} & A_{23} \\ A_{31} & A_{32} & A_{33} \end{pmatrix}$$

Transpose of the matrix A_{ij} is the Adj(A)

$$Adj(A) = A_{ji} = \begin{pmatrix} A_{11} & A_{21} & A_{31} \\ A_{12} & A_{22} & A_{32} \\ A_{13} & A_{23} & A_{33} \end{pmatrix}$$

Hence, the inverse is obtained as

$$A^{-1} = \frac{Adj(A)}{|A|} = \frac{1}{|A|} \begin{pmatrix} A_{11} & A_{21} & A_{31} \\ A_{12} & A_{22} & A_{32} \\ A_{13} & A_{23} & A_{33} \end{pmatrix}$$

Example 14 Find the inverse of the matrix $A = \begin{pmatrix} 0 & 1 & 1 \\ 1 & 0 & 1 \\ 1 & 1 & 0 \end{pmatrix}$ and verify that $A^{-1}A = AA^{-1} = EE$.

Solution: *Given:* the matrix $A = \begin{pmatrix} 0 & 1 & 1 \\ 1 & 0 & 1 \\ 1 & 1 & 0 \end{pmatrix}$

Now, the determinant of the given matrix is

$$|A| = \begin{vmatrix} 0 & 1 & 1 \\ 1 & 0 & 1 \\ 1 & 1 & 0 \end{vmatrix} = 0(0-1) - 1(0-1) + 1(1-0) = 1 + 1 = 2$$

$\Longrightarrow A^{-1}$ exists.

Now, let A_{ij} be the cofactors of the matrix elements a_{ij}, then from Eq. 6.9, we have,

$$A_{11} = (-1)^{1+1}\begin{vmatrix} 0 & 1 \\ 1 & 0 \end{vmatrix} = -1, A_{12} = (-1)^{1+2}\begin{vmatrix} 1 & 1 \\ 1 & 0 \end{vmatrix} = 1,$$

$$A_{13} = (-1)^{1+3}\begin{vmatrix} 1 & 0 \\ 1 & 1 \end{vmatrix} = 1$$

$$A_{21} = (-1)^{2+1}\begin{vmatrix} 1 & 1 \\ 1 & 0 \end{vmatrix} = 1, A_{22} = (-1)^{2+2}\begin{vmatrix} 0 & 1 \\ 1 & 0 \end{vmatrix} = -1,$$

$$A_{23} = (-1)^{2+3}\begin{vmatrix} 0 & 1 \\ 1 & 1 \end{vmatrix} = 1$$

$$A_{31} = (-1)^{3+1}\begin{vmatrix} 1 & 1 \\ 0 & 1 \end{vmatrix} = 1, A_{32} = (-1)^{3+2}\begin{vmatrix} 0 & 1 \\ 1 & 1 \end{vmatrix} = 1,$$

$$A_{33} = (-1)^{3+3}\begin{vmatrix} 0 & 1 \\ 1 & 0 \end{vmatrix} = -1$$

So that the matrix of the cofactors is $A_{ij} = \begin{pmatrix} -1 & 1 & 1 \\ 1 & -1 & 1 \\ 1 & 1 & -1 \end{pmatrix}$

$$\Rightarrow \text{Adj}(A) = A_{ji} = \begin{pmatrix} -1 & 1 & 1 \\ 1 & -1 & 1 \\ 1 & 1 & -1 \end{pmatrix}.$$

Hence,

$$A^{-1} = \frac{\text{Adj}(A)}{|A|} = \frac{1}{2}\begin{pmatrix} -1 & 1 & 1 \\ 1 & -1 & 1 \\ 1 & 1 & -1 \end{pmatrix}$$

Now,

$$A^{-1}A = \frac{1}{2}\begin{pmatrix} -1 & 1 & 1 \\ 1 & -1 & 1 \\ 1 & 1 & -1 \end{pmatrix}\begin{pmatrix} 0 & 1 & 1 \\ 1 & 0 & 1 \\ 1 & 1 & 0 \end{pmatrix} = \begin{pmatrix} 1 & 0 & 0 \\ 0 & 1 & 0 \\ 0 & 0 & 1 \end{pmatrix} = E = I$$

Similarly,

$$AA^{-1} = \begin{pmatrix} 0 & 1 & 1 \\ 1 & 0 & 1 \\ 1 & 1 & 0 \end{pmatrix}\begin{pmatrix} -\frac{1}{2} & \frac{1}{2} & \frac{1}{2} \\ \frac{1}{2} & -\frac{1}{2} & \frac{1}{2} \\ \frac{1}{2} & \frac{1}{2} & -\frac{1}{2} \end{pmatrix} = \begin{pmatrix} 1 & 0 & 0 \\ 0 & 1 & 0 \\ 0 & 0 & 1 \end{pmatrix} = E = I$$

$$\Rightarrow A^{-1}A = AA^{-1} = E.$$

Example 15 General form of the matrix of rotation operation through an angle α about any principal axis (say x_3-axis) is given by

$$A = A_{ij} = \begin{pmatrix} \cos\alpha & -\sin\alpha & 0 \\ \sin\alpha & \cos\alpha & 0 \\ 0 & 0 & 1 \end{pmatrix}$$

Determine its inverse and show that $A^{-1}A = AA^{-1} = E$.

Solution: *Given:* The matrix $A = \begin{pmatrix} \cos\alpha & -\sin\alpha & 0 \\ \sin\alpha & \cos\alpha & 0 \\ 0 & 0 & 1 \end{pmatrix}$

Now, the determinant of the given matrix is

$$|A| = \begin{vmatrix} \cos\alpha & -\sin\alpha & 0 \\ \sin\alpha & \cos\alpha & 0 \\ 0 & 0 & 1 \end{vmatrix} = \cos^2\alpha + \sin^2\alpha + 0 = 1$$

$\Longrightarrow$ Inverse of the matrix A exists

Let A_{ij} be the cofactors of a_{ij} in $|A|$, then

$A_{11} = \cos\alpha,$	$A_{12} = -\sin\alpha,$	$A_{13} = 0$
$A_{21} = \sin\alpha,$	$A_{22} = \cos\alpha,$	$A_{23} = 0$
$A_{31} = 0,$	$A_{32} = 0,$	$A_{33} = 1$

The matrix of the cofactors, A_{ij} is given by

$$A_{ij} = \begin{pmatrix} \cos\alpha & -\sin\alpha & 0 \\ \sin\alpha & \cos\alpha & 0 \\ 0 & 0 & 1 \end{pmatrix}$$

$$\Rightarrow \text{Adj}(A) = \begin{pmatrix} \cos\alpha & \sin\alpha & 0 \\ -\sin\alpha & \cos\alpha & 0 \\ 0 & 0 & 1 \end{pmatrix}$$

Hence,

$$A^{-1} = \frac{\text{Adj}(A)}{|A|} = \frac{1}{1}\begin{pmatrix} \cos\alpha & \sin\alpha & 0 \\ -\sin\alpha & \cos\alpha & 0 \\ 0 & 0 & 1 \end{pmatrix} = \begin{pmatrix} \cos\alpha & \sin\alpha & 0 \\ -\sin\alpha & \cos\alpha & 0 \\ 0 & 0 & 1 \end{pmatrix}$$

$$\text{Now, } A^{-1}A = \begin{pmatrix} \cos\alpha & \sin\alpha & 0 \\ -\sin\alpha & \cos\alpha & 0 \\ 0 & 0 & 1 \end{pmatrix} \begin{pmatrix} \cos\alpha & \sin\alpha & 0 \\ -\sin\alpha & \cos\alpha & 0 \\ 0 & 0 & 1 \end{pmatrix}$$

$$= \begin{pmatrix} \cos^2\alpha + \sin^2\alpha & -\cos\alpha\sin\alpha + \cos\alpha\sin\alpha & 0 \\ -\sin\alpha\cos\alpha + \sin\alpha\cos\alpha & \sin^2\alpha + \cos^2\alpha & 0 \\ 0 & 0 & 1 \end{pmatrix}$$

$$= \begin{pmatrix} 1 & 0 & 0 \\ 0 & 1 & 0 \\ 0 & 0 & 1 \end{pmatrix} = E$$

$$\text{Similarly, } AA^{-1} = \begin{pmatrix} \cos\alpha & -\sin\alpha & 0 \\ \sin\alpha & \cos\alpha & 0 \\ 0 & 0 & 1 \end{pmatrix} \begin{pmatrix} \cos\alpha & \sin\alpha & 0 \\ -\sin\alpha & \cos\alpha & 0 \\ 0 & 0 & 1 \end{pmatrix}$$

$$= \begin{pmatrix} \cos^2\alpha + \sin^2\alpha & \cos\alpha\sin\alpha - \cos\alpha\sin\alpha & 0 \\ \sin\alpha\cos\alpha - \sin\alpha\cos\alpha & \sin^2\alpha + \cos^2\alpha & 0 \\ 0 & 0 & 1 \end{pmatrix}$$

$$= \begin{pmatrix} 1 & 0 & 0 \\ 0 & 1 & 0 \\ 0 & 0 & 1 \end{pmatrix} = E$$

$$\Rightarrow A^{-1}A = AA^{-1} = E$$

(iv) *Orthogonal Matrix*

An orthogonal matrix is a unitary matrix whose all elements are real. One can immediately see that $A^{-1} = A^{T}$ if A is orthogonal.

Since $AA^{T} = I$, it follows that

$$\sum_{k=0}^{n} A_{ik}A_{jk} = \delta_{ij}$$

$$= 1, if\ i = j$$
$$= 0, if\ i \neq j$$

This can be stated as:

(i) the sum of the squares of the elements in any row (column) is equal to 1, that is, $A_{11}^2 + A_{12}^2 + A_{13}^2 = 1$, etc. and

(ii) the sum of the products of elements from one row (column) and the corresponding element of another row (column) is equal to zero, that is,

$A_{11}A_{21} + A_{12}A_{22} + A_{13}A_{23} = 0$, etc.

Example 16 Obtain the most general form of an orthogonal matrix of order 2.

Solution: Let us start with an arbitrary matrix of order 2, which can be written as

$$A = \begin{pmatrix} a & b \\ c & d \end{pmatrix}$$

where a, b, c and d are any scalars, real or complex. If the matrix A is to be an orthogonal matrix, then its elements must satisfy the following orthogonal condition, that is,

	$a^2 + b^2 = 1$	(x)
	$c^2 + d^2 = 1$	(y)
and	$ac + bd = 0$	(z)

The general solution of Eq. (x) is $a = \cos\theta$ and $b = \sin\theta$, where θ is real or complex. Similarly, the solution of Eq. (y) can be $c = \cos\phi$ and $d = \sin\phi$, where ϕ is a scalar. In order to satisfy Eq. (z), we can find that θ and ϕ must be related through

$$\cos\theta \cos\phi + \sin\theta \sin\phi = 0$$
$$\text{or } \cos(\theta - \phi) = 0$$
$$\Rightarrow \theta - \phi$$

Therefore, the most general form of the orthogonal matrix of order 2 becomes

$$A = \begin{pmatrix} \cos\theta & \sin\theta \\ \pm\sin\theta & \pm\cos\theta \end{pmatrix}$$

Choosing upper signs, we get $\Delta = \pm 1$. On the other hand, the lower signs will give, $\Delta = -1$. Thus, the most general form is

$$A = \begin{pmatrix} \cos\theta & \sin\theta \\ -\sin\theta & \cos\theta \end{pmatrix}$$

Example 17 Show that the matrix A corresponding to a 6-fold rotation is an orthogonal matrix. Obtain its inverse. The matrix A is:

$$A = \begin{pmatrix} \frac{1}{2} & -\frac{\sqrt{3}}{2} & 0 \\ \frac{\sqrt{3}}{2} & \frac{1}{2} & 0 \\ 0 & 0 & 1 \end{pmatrix}$$

Solution: Let the matrix A is an orthogonal matrix. For this,

$$A^{-1} = A^T \text{ and } AA^{-1} = AA^T = E$$

so that, $A^T = \begin{pmatrix} \frac{1}{2} & \frac{\sqrt{3}}{2} & 0 \\ -\frac{\sqrt{3}}{2} & \frac{1}{2} & 0 \\ 0 & 0 & 1 \end{pmatrix}$

and, $AA^T = \begin{pmatrix} \frac{1}{2} & -\frac{\sqrt{3}}{2} & 0 \\ \frac{\sqrt{3}}{2} & \frac{1}{2} & 0 \\ 0 & 0 & 1 \end{pmatrix} \begin{pmatrix} \frac{1}{2} & \frac{\sqrt{3}}{2} & 0 \\ -\frac{\sqrt{3}}{2} & \frac{1}{2} & 0 \\ 0 & 0 & 1 \end{pmatrix}$

$$= \begin{pmatrix} \frac{1}{4}+\frac{3}{4} & \frac{\sqrt{3}}{4}-\frac{\sqrt{3}}{4} & 0 \\ \frac{\sqrt{3}}{4}-\frac{\sqrt{3}}{4} & \frac{3}{4}+\frac{1}{2} & 0 \\ 0 & 0 & 1 \end{pmatrix} = \begin{pmatrix} 1 & 0 & 0 \\ 0 & 1 & 0 \\ 0 & 0 & 1 \end{pmatrix} = E$$

$\Rightarrow$ The given matrix A is an orthogonal matrix.

6.3 Group (Point) Representation of Symmetry Operations

(i) Elements of Group Theory

We know that there is an intimate connection between the symmetry operations (of a point group) and the mathematical group. It is, therefore, we shall discuss certain elementary aspects of the group theory in this section.

Group

A set of symmetry elements/operations forms a group if and only if the following group conditions are satisfied.

1. A product of two symmetry elements A and B in a group is equivalent to a symmetry element C, also an element of the same group, such that

$$AB = C \qquad (6.10)$$

where the product AB means the operation B followed by the operation A. In general, we observe that the product AB $\neq$ BA. However, if AB = BA the group is said to be commutative or abelian.

2. Every element of the group obeys the associative law of combination. If A, B and C are the elements of the group, then

$$(AB)C = A(BC) \qquad (6.11)$$

3. Every group contains one element called the identity element E, such that

$$AE = EA = A \qquad (6.12)$$

4. Every element in the group has its inverse also in the group, such that

$$AA^{-1} = A^{-1}A = E \qquad (6.13)$$

Solved Examples

Example 1 Show that the symmetry elements E, C_3, C_3^2 constitute a group. What is its order?

Solution: Given: Three symmetry elements are: E, C_3, C_3^2 order of the group = ?

Since the three symmetry elements are independent, hence the order of the group is 3. Now let us check the following conditions.

 (i) To check the closure property, let us take products, that is,

$$E\,C_3 = C_3$$
$$E\,C_3^2 = C_3^2$$
$$C_3 C_3^2 = C_3^1 C_3^2 = C_3^3 = E$$

 (ii) To check the associative property, let us consider the triple products, that is,

$$C_3\left(C_3 C_3^2\right) = (C_3 C_3) C_3^2$$
$$\text{LHS} = C_3\left(C_3 C_3^2\right) = C_3 C_3^3 = C_3\,E = C_3$$
$$\text{RHS} = (C_3 C_3) C_3^2 = \left(C_3^2\right) C_3^2 = C_3$$
$$\Rightarrow \text{LHS} = \text{RHS}$$

 (iii) The given symmetry elements contain one identify element, E which leaves the other members unchanged, that is,

$$E\,C_3 = C_3\,E = C_3$$
$$E\,C_3^2 = C_3^2\,E = C_3^2$$

(iv) To check the inverse, we observe that the inverse of E is E itself, while that of

$$C_3^1 \text{ is } C_3^{-1} = C_3^{-1}C_3^3 = C_3^2$$
$$\text{and } C_3^2 \text{ is } C_3^{-2} = C_3^{-2}C_3^3 = C_3$$

Since the three given symmetry elements satisfy all the group conditions, hence they form a group of order 3.

Example 2 Show that the symmetry elements E, C_2, σ_h, i constitute a group. What is its order? Name the point group?

Solution: Given: Four symmetry elements are: E, C_2, σ_h, i; order of group = ? point group = ?

Since the four given symmetry elements are independent, hence the order of the group is 4

Now let us check that they follow the group conditions

(i) To check the closure property, let us take their products

$$C_2\sigma_h = i$$
$$C_2\, i = \sigma_h$$
$$\sigma_h\, i = C_2$$

(ii) To cheek the associative property, let us consider the triple product

$$C_2(\sigma_h\, i) = (C_2\sigma_h)i$$
$$\text{LHS} = C_2(\sigma_h\, i) = C_2\, C_2 = E$$
$$\text{RHS} = (C_2\sigma_h)i = i \cdot i = E$$
$$\Rightarrow \text{LHS} = \text{RHS}$$

(iii) The given symmetry elements contain one identify elements, E which leaves the other members unchanged, that is,

$$E\, C_2 = C_2\, E = C_2$$
$$E\, \sigma_h = \sigma_h\, E = \sigma_h$$
$$E\, i = i\, E = i$$

(iv) To check the inverse, we observe that the inverse of E is E itself, while that of

$$C_2 \text{ is } C_2^{-1} = C_2^{-1}C_2^2 = C_2$$
$$\sigma_h \text{ is } \sigma_h^{-1} = \sigma_h$$
$$\text{and } i \text{ is } i^{-1} = i$$

$\Rightarrow$ Every symmetry element has its own inverse.

Since the four given symmetry elements are independent and satisfy all the group conditions, hence they form a group of order 4. Further, there is one horizontal mirror plane perpendicular to 2 fold rotation axis, therefore the point group is C_{2h} (2/m).

Example 3 Show that the symmetry elements E, C_2, σ_{xz}, σ_{yz} constitute a group. What is its point group?

Solution: Given: Four symmetry elements are: E, C_2, σ_{xz}, σ_{yz}; point group = ?

Now, let us check that they follow the group conditions.
To check the closure property let us take their products, that is,

$$C_2\sigma_{xz} = \sigma_{yz}$$
$$C_2\sigma_{yz} = \sigma_{xz}$$
$$\sigma_{xz}\sigma_{yz} = C_2$$

To check the associative property, let us consider the triple product, that is,

$$C_2(\sigma_{xz}\sigma_{yz}) = (C_2\sigma_{xz})\sigma_{yz}$$
$$\text{LHS} = C_2(\sigma_{xz}\sigma_{yz}) = C_2\, C_2 = E$$
$$\text{RHS} = (C_2\sigma_{xz})\sigma_{yz} = \sigma_{yz}\sigma_{yz} = E$$
$$\Rightarrow \text{LHS} = \text{RHS}$$

The given symmetry elements contain one identify elements E, which leaves the other members unchanged, that is,

$$E\, C_2 = C_2\, E = C_2$$
$$E\, \sigma_{xz} = \sigma_{xz}\, E = \sigma_{xz}$$
$$E\, \sigma_{yz} = \sigma_{yz}\, E = \sigma_{yz}$$

Every given symmetry elements are found to have its have own inverse, that is,

$$C_2\, C_2^{-1} = C_2\, C_2 = E$$
$$\sigma_{xz}\sigma_{xz^{-1}} = \sigma_{xz}\sigma_{xz} = E$$
$$\sigma_{yz}\sigma_{yz^{-1}} = \sigma_{yz}\sigma_{yz} = E$$

Since the four given symmetry elements are independent and satisfy all the group conditions, therefore they form a group of order 4. Further, there are two vertical mirror planes and one 2-fold pure rotation axis, therefore the point group is C_{2v} (mm2).

(ii) **Important Properties of a Group**

Cyclic Group

A group is said to be cyclic if all elements of the group can be generated by one element, such as $A, A^2, A^3, \ldots A^n \,(=E)$. The element A is called the generator of the group, where n refers to the total number of elements in the group and is called the order of the group. Cyclic groups are abelian but the converse is not true. The point groups C_2 (2), C_3 (3), C_4 (4) are C_6 (6) are example of cyclic groups.

Abelian Group

A group is said to be abelian if all its elements commute with one another. Further, two elements A and B are said to commute with one another if AB = BA. In abelian groups, each element is in a class by itself, since

$$XAX^{-1} = AXX^{-1} = AE = A$$

Order of the Group

In general, it is the number of non-equivalent symmetry elements in the group. For example, the point group C_{2h} (2/m) has four non-equivalent symmetry elements E (1), C_2 (2), σ_h (m) and S_2 ($\bar{1}$). Hence, the order of this point group 4.

Example 4 Out of 32 point groups how many belong to cyclic group. Write the member of each group in their increasing order.

Solution: Given: Thirty-two point groups; No. of cyclic groups and their members = ?

Ten cyclic groups of order,

$$h = 1 \text{ is} : C_1$$
$$h = 2 \text{ are} : C_2, Ci \text{ and } C_s$$
$$h = 3 \text{ is} : C_3$$
$$h = 4 \text{ are} : C_4 \text{ and } S_4$$
$$h = 6 \text{ are} : C_6, \; S_3 = C_{3h}(\bar{6}) \text{ and } S_6 = C_{3i}(\bar{3})$$

Members of C_1: E

$$C_2 : E, C_2$$
$$C_i : E, i$$
$$C_s : E, \sigma_h$$
$$C_3 : E, C_3, C_3^2$$
$$C_4 : E, C_4, C_4^2 = (C_2), C_4^3$$
$$C_6 : E, C_6, C_6^2 = (C_3), C_6^3 = (C_2), C_6^4 = (C_3^2), C_6^5$$

$$S_3 : S_3^6(= E), S_3^5, S_3^4(= C_3), S_3^3(= \sigma_h), S_3^2, (= C_3^2), S_3$$
$$S_4 : S_4^4(= E), S_4^3, S_4^2(= C_2), S_4$$
$$S_6 : S_6^6(= E), S_6^5, S_6^4(= C_3^2), S_6^3(= i), S_6^2(= C_3), S_6$$

Further, we know that all cyclic groups are abelian, hence all ten cyclic groups mentioned above are abelian.

Example 5 Generate the symmetry elements corresponding to five improper rotation axes: S_1, S_2, S_3, S_4 and S_6.

Solution: Given: improper rotation axes are: S_1, S_2, S_3, S_4 and S_6.

Based on the empirical rules formulated in Sect. 6.1, the odd and even improper rotation axes can be symbolized as $S_n^{k_1}$ and $S_n^{k_2}$, respectively, where $k_1 = 2n$ and $k_2 = n$. Therefore, we can write, S_n^{2n} for n_{odd} and S_n^n for n_{even} rotation axes, respectively. Now, the symmetry elements for various improper axes are:

For n = 1, $S_n^{2n} = S_1^2$ gives S_1^1 and S_1^2 symmetry elements, where

$$S_1^1 = C_1^1 \sigma_h = E \; \sigma_h = \sigma_h$$
$$S_1^2 = (C_1^1 \sigma_h)^2 = (C_1^1)^2 (\sigma_h)^2 = E.E = E$$

$\Rightarrow$The symmetry elements are: E, σ_h
For n = 2, $S_n^n \equiv S_2^2$ gives S_2^1 and S_2^2 symmetry elements, where

$$S_2^1 = C_2^1 \sigma_h = i$$
$$S_2^2 = C_2^2 (\sigma_h)^2 = E.E = E$$

$\Longrightarrow$ The symmetry elements are: E, i
For n = 3, $S_n^{2n} = S_3^6$ gives S_3^1, S_3^2, S_3^3, S_3^4, S_3^5 and S_3^6 symmetry elements, where

$$S_3^1 = S_3$$
$$S_3^2 = C_3^2 (\sigma_h)^2 = C_3^2 \; E = C_3^2$$
$$S_3^3 = C_3^3 (\sigma_h)^3 = E \; (\sigma_h)^2 \sigma_h = E.E.\sigma_h = \sigma_h$$
$$S_3^4 = C_3^4 (\sigma_h)^4 = C_3^3 \; C_3 \cdot (\sigma_h^2)^2 = E.C_3(E)^2 = C_3.E = C_3$$
$$S_3^5 = S_3^5$$
$$S_3^6 = C_3^6 (\sigma_h)^6 = (C_3^3)^2 (\sigma_h^2)^3 = (E)^2 .(E)^3 = E$$

$\Longrightarrow$ The symmetry elements are: E, $C_3, C_3^2, \sigma_h, S_3$ and S_3^5
For n = 4, $S_n^n = S_4^4$ gives S_4^1, S_4^2, S_4^3 and S_4^4 symmetry elements, where

$$S_4^1 = S_4$$
$$S_4^2 = C_4^2(\sigma_h)^2 = C_2^1 E = C_2$$
$$S_4^3 = S_4^3$$
$$S_4^4 = C_4^4(\sigma_h)^4 = E(\sigma_h^2)^2 = (E)^2 = E$$

$\implies$ The symmetry elements are: E, C_2, S_4 and S_4^3.

For n = 6, $S_n^n = S_6^6$ gives S_6^1, S_6^2, S_6^3, S_6^4, S_6^5 and S_6^6 symmetry elements, where

$$S_6^1 = S_6$$
$$S_6^2 = C_6^2(\sigma_h)^2 = C_3^1\ E = C_3$$
$$S_6^3 = C_6^3(\sigma_h)^3 = C_2^1(\sigma_h)^2\sigma_h = C_2.E.\sigma_h = C_2\sigma_h = i$$
$$S_6^4 = C_6^4(\sigma_h)^4 = C_3^2(\sigma_h^2)^2 = C_3^2(E)^2 = C_3^2$$
$$S_6^5 = S_6^5$$
$$S_6^6 = C_6^6(\sigma_h)^6 = E(\sigma_h^2)^3 = E.(E)^3 = E$$

$\implies$ The symmetry elements are: E, C_3, C_3^2, i, S_6 and S_6^5.

Isomorphic Group

The two or more groups are said to be isomorphic if they obey the same group multiplication table. This means that there is a one -to-one correspondence between elements A, B, … of one group and those A', B', …. Of the other, such that AB = C implies A'B' = C' and vice versa.

Example 6 Show that the point groups C_2, C_s, C_i and a group containing 1, −1 as members are isomorphic

Solution: Given: Three point groups are C_2, C_s, C_i; one mathematical group containing 1, -1 as members. Let us write their group multiplication tables and check the one-to-one correspondence between their elements.

Group multiplication tables of C_2, C_s, C_i and the elements 1, −1 are:

C_2	E	C_2
E	E	C_2
C_2	C_2	E

C_s	E	σ_h
E	E	σ_h
σ_h	σ_h	E

C_i	E	I
E	E	I
I	I	E

$\bar{1}$	1	-1
1	1	-1
-1	-1	1

Here, $1 \leftrightarrow E$, $C_2 \leftrightarrow \sigma_h$, $\sigma_h \leftrightarrow i$, $i \leftrightarrow -1$, $-1 \leftrightarrow C_2$ are related.

Elements of any one group show one-to-one correspondence with elements of any others groups. Hence, they are isomorphic.

Example 7 Show that the point group C_4 is isomorphic with a group containing 1, i,-1,-i as members.

Solution: Given: point group C_4, one mathematical group containing 1, i, -1, $-i$ as members. Let us write the group multiplication tables and check one-to-one correspondence between their elements.

Group multiplication tables of C_4 and the elements 1, i, -1, $-i$ are:

C_4	E	C_4	$C_4^2 = (C_2)$	C_4^3
E	E	C_4	C_4^2	C_4^3
C_4	C_4	C_4^2	C_4^3	E
C_4^2	C_4^2	C_4^3	E	C_4
C_4^3	C_4^3	E	C_4	C_4^2

	1	i	-1	$-i$
1	1	i	-1	$-i$
i	i	-1	$-i$	1
-1	-1	$-i$	1	$-i$
$-i$	$-i$	1	I	-1

Comparing their elements shows that they follow the one-to-one citation, that is,

C_4	C_4^2	C_4^3	$C_4^4 (= E)$
i	i^2	i^3	i^4
i	-1	$-i$	1

$\implies$ They are isomorphic

Example 8 Show that points groups C_{2h}, D_2 and C_{2v} are isomorphic.

Solution: Given: Three points groups are: C_{2h}, D_2 and C_{2v}.

Let us write the group multiplication tables and check the one-to-one correspondence between their elements. Group multiplication tables are:

C_{2h}	E	C_2	σ_h	$-i$
E	E	C_2	σ_h	$-i$
C_2	C_2	E	$-i$	σ_h
σ_h	σ_h	$-i$	E	C_2
$-i$	$-i$	σ_h	C_2	E

D_2	E	C_{2x}	C_{2y}	C_{2z}
E	E	C_{2x}	C_{2y}	C_{2z}
C_{2x}	C_{2x}	E	C_{2z}	C_{2y}
C_{2y}	C_{2y}	C_{2z}	E	C_{2x}
C_{2z}	C_{2z}	C_{2y}	C_{2x}	E

C_{2v}	E	C_2	σ_{v1}	σ_{v2}
E	E	C_2	σ_{v1}	σ_{v2}
C_2	C_2	E	σ_{v2}	σ_{v1}
σ_{v1}	σ_{v1}	σ_{v2}	E	C_2
σ_{v2}	σ_{v2}	σ_{v1}	C_2	E

By a close look at the group multiplication tables, a one-to-one correspondence is observed clearly. Hence, they are isomorphic.

Finite Group

A group containing finite number of elements is called a finite group. For example, crystallographic point groups and space groups are finite groups.

Generators of a Finite Group

It is possible to generate all elements of a group by starting from a certain set of elements (at the most three) and taking their power and products. However, it is to be noted that the definition of generator is not always unique. For example, in the point group (D_2) 222, the possible generator are 2[100] and 2[010] or 2[100] and 2 [001] or 2[010] and 2[001].

Subgroups and Super Groups

A set of symmetry elements is said to be a subgroup of a bigger group (called super group) if the set itself forms a group and satisfies all group conditions. In general, every group has two trivial subgroups, the identity element and the group itself. However, in the simplest term, it can be said that the addition of symmetry elements to a point group produces super groups while the suppression of the symmetry elements from point group produces subgroup. For example, we know that the point group E(1) is the least symmetric and is the subgroup of all other 31 point groups. On the other hand, the point groups D_{6h} (6/mmm) and O_h (m3m) can have no super group because no symmetry elements could be added to them to obtain any new point group. Subgroups, super groups and order of the point groups are illustrated in Fig. 6.24.

A group is called proper group if there are symmetry elements of the super group not contained in the subgroup. For example, the set of point E (1), C_2 (2), σ_h (m) and S_2 ($\bar{1}$) is a proper subgroup of the point group C_{2h} (2/m).

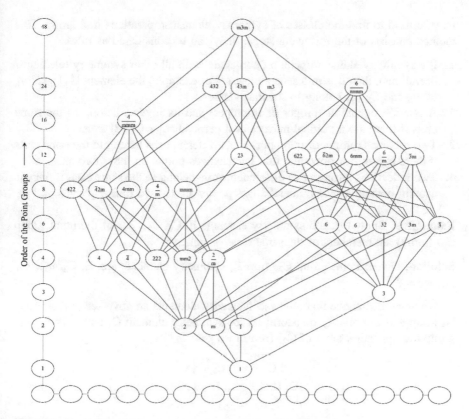

Fig. 6.24 Illustration of subgroups, super groups and order of the point groups

Classes

Two elements A and B in a group belong to the same class if there is an element X
within group such that

$$X^{-1}AX = B \tag{6.14}$$

where X^{-1} is the inverse of X. From Eq. 6.14, we can say that B is a similarity
transform of A by X, or that A and B are conjugate to one another. Making use of
the similarity transform of one element by other elements one can determine
whether a set of elements from classes or not.

The order of a class (c) of the group must be an integral factor of the order of the
group (g), that is,

$$g = mc \tag{6.15}$$

The similarity transformation method is sometimes too elaborate to find the
classes, particularly in high symmetry systems. Therefore, an alternative method

may be used to find the classes of symmetry elements/operations in a group. This method consists of the following steps. They can be considered as rules.

1. If a symmetry element/operation commutes with all other symmetry (elements/operations), then it is in a separate class. For example, the element E (1), C_s, σ_h or m_h and C_2 (2) belong to separate classes.
2. A rotation operation (proper or improper) and its inverse belong to the same class if there are n vertical mirrors or n perpendicular C_2 (2) axes.
3. Two rotations (proper or improper) about different axes belong to the same class if there is a mirror operation that interchanges points on these two axes.
4. Two reflections through two different mirrors belong to the same class if there is another operation which interchanges points on the two mirror planes.

Example 9 Show that the symmetry elements E, C_2, σ_{xz}, σ_{yz} of the point group C_{2v} (mm2) are members of the same class or not.

Solution: Given: Point group C_{2v} (mm2), symmetry elements are: E, C_2, σ_{xz}, σ_{yz} classes = ?

We have seen above that all these symmetry elements are inverses of their own. Now, applying similarity transform on the symmetry element C_2 and referring the group multiplication table of C_{2v} (mm2) we get

$$E\ C_2\ E = E\ C_2 = C_2$$
$$C_2\ C_2\ C_2 = C_2\ E = C_2$$
$$\sigma_{xz}\ C_2\ \sigma_{xz} = \sigma_{xz}\ \sigma_{yz} = C_2$$
$$\sigma_{yz}\ C_2\ \sigma_{yz} = \sigma_{yz}\ \sigma_{xz} = C_2$$

$\Longrightarrow$ All similarity transforms generated the same symmetry element C_2 and hence it forms a class of its own.

Similarly, applying similarity transforms on other symmetry operations, we obtain

$$E\ \sigma_{xz}\ E = E\ \sigma_{xz} = \sigma_{xz}$$
$$C_2\sigma_{xz}\ C_2 = C_2\sigma_{xz} = \sigma_{xz}$$
$$\sigma_{xz}\sigma_{xz}\sigma_{xz} = \sigma_{xz}\ E = \sigma_{xz}$$
$$\sigma_{yz}\sigma_{xz}\sigma_{yz} = \sigma_{xz}\ C_2 = \sigma_{xz}$$

and

$$E\sigma_{yz}\ E = E\sigma_{yz} = \sigma_{yz}$$
$$C_2\sigma_{yz}C_2 = C_2\sigma_{xz} = \sigma_{yz}$$
$$\sigma_{xz}\sigma_{yz}\sigma_{xz} = \sigma_{xz}C_2 = \sigma_{yz}$$
$$\sigma_{yz}\sigma_{yz}\sigma_{yz} = \sigma_{yz}E = \sigma_{yz}$$

Like C_2, we find σ_{xz} and σ_{yz} operations also form their own classes. Hence, all the four members of the point group C_{2v} (mm2) belong to separate classes. Thus, the numbers of classes are 4.

Example 10 Show that the symmetry elements E, C_3, C_3^2 of the point group C_3 (3) belong to the same class or not.

Solution: Given: Point group C_3 (3), symmetry elements E, C_3, C_3^2, classes = ?

We have seen above that the symmetry elements E is the inverse of itself while C_3 and C_3^2 are the inverses of each other. Now, let us apply similarity transforms on the given symmetry elements to check their classes by referring the group multiplication table of the point group $C_3(3)$ we get

$$E\,C_3\,E = E\,C_3 = C_3$$
$$C_3 C_3 C_3^2 = C_3\,E = C_3$$
$$C_3^2 C_3 C_3 = C_3^2 C_3^2 = C_3^3 C_3 = C_3$$

$$E\,E\,E = E\,E = E$$
$$C_3\,E\,C_3^2 = C_3\,C_3^2 = E$$
$$C_3^2\,E\,C_3 = C_3^2\,C_3 = E$$

$$E\,C_3^2\,E = E\,C_3^2 = C_3^2$$
$$C_3\,C_3^2\,C_3^2 = C_3\,C_3 = C_3^2$$
$$C_3^2\,C_3^2\,C_3 = C_3^2\,E = C_3^2$$

From this, we observe that different similarly transform gives different symmetry operation/element. This implies that each element forms its own class. Since, the point group C_3 is cyclic, hence it is abelian. Accordingly, each element is in a class by itself. Thus, the numbers of classes are 3.

Example 11 Determine the number of classes corresponding to the point group D_3 (32) whose symmetry elements are E, C_3, C_3^2, C_{2x}, C_{2y}, C_{2xy}.

Solution: *Given*: Point group D_3 (32), symmetry elements are E, C_3, C_3^2, C_{2x}, C_{2y}, C_{2xy}.

Number of classes = ?

We know that E is the inverse of itself. C_3 and C_3^2 are the inverses of each other. C_{2x}, C_{2y}, C_{2xy} are inverses of their own. Further, the identity element, E, belongs to a class of its own. Now, applying similarly transforms on C_3 and C_3^2 as given below.

$$E \ C_3 \ E = E \ C_3 = C_3$$
$$C_3 \ C_3 \ C_3^2 = C_3 \ E = C_3$$
$$C_3^2 \ C_3 \ C_3 = C_3^2 \ C_3^2 = C_3$$
$$C_{2x} \ C_3 \ C_{2x} = \ C_{2x} \ C_{2xy} = C_3^2$$
$$C_{2y} \ C_3 \ C_{2y} = C_{2y} \ C_{2x} = C_3^2$$
$$C_{2xy} \ C_3 \ C_{2xy} = C_{2xy} \ C_{2y} = C_3^2$$

$$E \ C_3^2 \ E = E \ C_3^2 = C_3^2$$
$$C_3 \ C_3^2 \ C_3^2 = C_3 C_3 = C_3^2$$
$$C_3^2 \ C_3^2 \ C_3 = C_3^2 \ E = C_3^2$$
$$C_{2x} \ C_3^2 \ C_{2x} = C_{2x} \ C_{2y} = C_3$$
$$C_{2y} \ C_3^2 \ C_{2y} = C_{2y} \ C_{2xy} = C_3$$
$$C_{2xy} \ C_3^2 \ C_{2xy} = C_{2xy} \ C_{2x} = C_3$$

We observe that they are generated either C_3 or C_3^2 which means that two operations are the members of the same class. Further, applying similarity transform on C_{2x}, we obtain

$$E \ C_{2x} \ E = E \ C_{2x} = C_{2x}$$
$$C_3 \ C_{2x} \ C_3^2 = C_3 \ C_{2xy} = C_{2y}$$
$$C_3^2 \ C_{2x} \ C_3 = C_3^2 \ C_{2y} = C_{2xy}$$
$$C_{2x} \ C_{2x} \ C_{2x} = C_{2x} \ E = C_{2x}$$
$$C_{2y} \ C_{2x} \ C_{2y} = C_{2y} \ C_3 = C_{2xy}$$
$$C_{2xy} \ C_{2x} \ C_{2xy} = C_{2xy} \ C_3^2 = C_{2y}$$

This generates C_{2x}, C_{2y} and C_{2xy}. Similar results can be obtained when similarity transforms are taken on C_{2y} and C_{2xy}. The results imply that the operations C_{2x}, C_{2y} and C_{2xy} belong to the same class. Therefore, the three classes are: E, $2C_3$, $3C_2$.

Example 12 Determine the number of classes corresponding to the point group C_{4v} (4mm) whose symmetry elements are: E, $C_4, C_4^2, C_4^3, \sigma_x, \sigma_y, \sigma_1$ and σ_2.

Solution: Given: Point group C_{4v} (4mm), symmetry elements are: E, C_4, C_4^2, $C_4^3, \sigma_x, \sigma_y, \sigma_1$ and σ_2.

No. of classes = ?

We know that E is the inverse of itself. Similarly, C_2 is the inverse of itself. C_4 and C_4^3 are the inverses of each other's. All mirror operations are inverses of their own.

Since the number of symmetry elements is large so let us use the rules instead of similarity transforms to obtain the number of classes. Thus, using the rules we have

1. According to rule 1, the identity element E forms a class of its own.
2. According to rule 2, C_2 belongs to a separate class.
3. According to rule 2, both C_4 and its inverse C_4^3 belong to one class (due to the presence of a vertical C_2 axis).
4. According to rule 4, the first two mirror symmetries σ_x and σ_y belong to one class while the other mirror symmetries σ_1 and σ_2 belong to another class.

Hence, in all there are five separate classes. All eight symmetry elements can be written in five classes as: E, C_2, $2C_4$, $2\sigma_x$ and $2\sigma_1$.

Example 13 Determine the number of classes corresponding to the point group D_{2d} ($\bar{4}$2m) whose symmetry elements are E, S_4, S_3^2, S_4^3, C_{2x}, C_{2y}, σ_1 and σ_2.

Solution: *Given*: Point group D_{2d} ($\bar{4}$2m), symmetry elements are: E, S_4, S_3^2, S_4^3, C_{2x}, C_{2y}, σ_1 and σ_2.

No. of classes = ?
We know that E and C_2 are inverses of their own. Similarly, S_4 and S_4^3 are the inverses of each other. C_{2x}, C_2, σ_1, and σ_2 are inverses of their own.
Let us obtain the number of classes using the above-mentioned rules.

1. According to rule 1, the identity element E forms a class of its own.
2. According to rule 2, C_2 belongs to a separate class.
3. According to rule 2, S_4 and S_4^3 belong to one class (due to vertical C_2 axis).
4. According to rule 3, C_{2x} and C_{2y} belong to another separate class (due to mirror plane σ_1 or σ_2 which interchanges points on the two axes).
5. According to rule 4, σ_1 and σ_2 belong to the same class (due to S_4 axis which interchanges points on the two mirrors) .

Hence, in all there are five separate classes. All eight symmetry elements can be written in five classes as: E, C_2, $2S_4$, $2C_{2x}$, $2\sigma_1$.

Multiple Choice Questions (MCQ)

1. Permissible rotational symmetries in a crystalline solid are limited to:
   (a) 5
   (b) 4
   (c) 3
   (d) 2

2. Two mirrors at an angle θ together produce a rotation of:
   (a) θ
   (b) 2θ
   (c) 3θ
   (d) 4θ

3. $\tilde{1}$(one tilde) is equivalent to:
   (a)   3m
   (b)   3/m
   (c)   m
   (d)   1/m

4. $\tilde{2}$ (two tilde) is equivalent to:
   (a) $\bar{4}$
   (b) $\bar{3}$
   (c) $\bar{2}$
   (d) $\bar{1}$

5. $\tilde{3}$ (three tilde) is equivalent to:
   (a) $\bar{6}$
   (b) $\bar{4}$
   (c) $\bar{3}$
   (d) $\bar{2}$

6. $\tilde{4}$ (four tilde) is equivalent to:
   (a) $\bar{6}$
   (b) $\bar{4}$
   (c) $\bar{3}$
   (d) $\bar{2}$

7. $\tilde{6}$ (six tilde) is equivalent to:
   (a) $\bar{6}$
   (b) $\bar{4}$
   (c) $\bar{3}$
   (d) $\bar{2}$

8. Total number of symmetry elements exist in a monoclinic unit cell is:
   (a)   9
   (b)   7
   (c)   5
   (d)   3

9. Total number of symmetry elements exist in a orthorhombic unit cell is:
   (a)   7
   (b)   5
   (c)   3
   (d)   1

10. Total number of symmetry elements exist in a tetragonal unit cell is:
    (a) 11
    (b) 13
    (c) 15
    (d) 17

11. Total number of symmetry elements exist in a hexagonal unit cell is:
    (a) 11
    (b) 13
    (c) 15
    (d) 17

12. Total number of symmetry elements exist in a cubic unit cell is:
    (a) 17
    (b) 19
    (c) 21
    (d) 23

13. The matrix corresponding to a 2 − fold rotation along z-axis is:

(a) $\begin{pmatrix} -1 & 0 & 0 \\ 0 & -1 & 0 \\ 0 & 0 & 1 \end{pmatrix}$

(b) $\begin{pmatrix} -1 & 0 & 0 \\ 0 & 1 & 0 \\ 0 & 0 & -1 \end{pmatrix}$

(c) $\begin{pmatrix} 1 & 0 & 0 \\ 0 & -1 & 0 \\ 0 & 0 & -1 \end{pmatrix}$

(d) $\begin{pmatrix} -1 & 0 & 0 \\ 0 & -1 & 0 \\ 0 & 0 & -1 \end{pmatrix}$

14. The matrix corresponding to a 2-fold rotation along y-axis is:

(a) $\begin{pmatrix} -1 & 0 & 0 \\ 0 & -1 & 0 \\ 0 & 0 & 1 \end{pmatrix}$

(b) $\begin{pmatrix} -1 & 0 & 0 \\ 0 & 1 & 0 \\ 0 & 0 & -1 \end{pmatrix}$

(c) $\begin{pmatrix} 1 & 0 & 0 \\ 0 & -1 & 0 \\ 0 & 0 & -1 \end{pmatrix}$

(d) $\begin{pmatrix} -1 & 0 & 0 \\ 0 & -1 & 0 \\ 0 & 0 & -1 \end{pmatrix}$

15. The matrix corresponding to a 2 − fold rotation along x-axis is:

(a) $\begin{pmatrix} -1 & 0 & 0 \\ 0 & -1 & 0 \\ 0 & 0 & 1 \end{pmatrix}$

(b) $\begin{pmatrix} -1 & 0 & 0 \\ 0 & 1 & 0 \\ 0 & 0 & -1 \end{pmatrix}$

(c) $\begin{pmatrix} 1 & 0 & 0 \\ 0 & -1 & 0 \\ 0 & 0 & -1 \end{pmatrix}$

(d) $\begin{pmatrix} -1 & 0 & 0 \\ 0 & -1 & 0 \\ 0 & 0 & -1 \end{pmatrix}$

16. The matrix corresponding to a 3-fold rotation along z-axis using orthogonal system is:

(a) $\begin{pmatrix} 1/2 & -\sqrt{3}/2 & 0 \\ \sqrt{3}/2 & 1/2 & 0 \\ 0 & 0 & 1 \end{pmatrix}$

(b) $\begin{pmatrix} -1/2 & \sqrt{3}/2 & 0 \\ -\sqrt{3}/2 & -1/2 & 0 \\ 0 & 0 & 1 \end{pmatrix}$

(c) $\begin{pmatrix} 1/2 & \sqrt{3}/2 & 0 \\ -\sqrt{3}/2 & 1/2 & 0 \\ 0 & 0 & 1 \end{pmatrix}$

(d) $\begin{pmatrix} -1/2 & -\sqrt{3}/2 & 0 \\ \sqrt{3}/2 & -1/2 & 0 \\ 0 & 0 & 1 \end{pmatrix}$

17. The matrix corresponding to a $\bar{3}$-fold rotation along z-axis using orthogonal system is:

(a) $\begin{pmatrix} 1/2 & -\sqrt{3}/2 & 0 \\ \sqrt{3}/2 & 1/2 & 0 \\ 0 & 0 & 1 \end{pmatrix}$

(b) $\begin{pmatrix} -1/2 & \sqrt{3}/2 & 0 \\ -\sqrt{3}/2 & -1/2 & 0 \\ 0 & 0 & 1 \end{pmatrix}$

(c) $\begin{pmatrix} 1/2 & \sqrt{3}/2 & 0 \\ -\sqrt{3}/2 & 1/2 & 0 \\ 0 & 0 & 1 \end{pmatrix}$

(d) $\begin{pmatrix} -1/2 & -\sqrt{3}/2 & 0 \\ \sqrt{3}/2 & -1/2 & 0 \\ 0 & 0 & 1 \end{pmatrix}$

18. The matrix corresponding to a 6-fold rotation along z-axis using orthogonal system is:

(a) $\begin{pmatrix} 1/2 & -\sqrt{3}/2 & 0 \\ \sqrt{3}/2 & 1/2 & 0 \\ 0 & 0 & 1 \end{pmatrix}$

(b) $\begin{pmatrix} -1/2 & \sqrt{3}/2 & 0 \\ -\sqrt{3}/2 & -1/2 & 0 \\ 0 & 0 & 1 \end{pmatrix}$

(c) $\begin{pmatrix} 1/2 & \sqrt{3}/2 & 0 \\ -\sqrt{3}/2 & 1/2 & 0 \\ 0 & 0 & 1 \end{pmatrix}$

(d) $\begin{pmatrix} -1/2 & -\sqrt{3}/2 & 0 \\ \sqrt{3}/2 & -1/2 & 0 \\ 0 & 0 & 1 \end{pmatrix}$

19. The matrix corresponding to a $\bar{6}$-fold rotation along z-axis using orthogonal system is:

(a) $\begin{pmatrix} 1/2 & -\sqrt{3}/2 & 0 \\ \sqrt{3}/2 & 1/2 & 0 \\ 0 & 0 & 1 \end{pmatrix}$

(b) $\begin{pmatrix} -1/2 & \sqrt{3}/2 & 0 \\ -\sqrt{3}/2 & -1/2 & 0 \\ 0 & 0 & 1 \end{pmatrix}$

(c) $\begin{pmatrix} 1/2 & \sqrt{3}/2 & 0 \\ -\sqrt{3}/2 & 1/2 & 0 \\ 0 & 0 & 1 \end{pmatrix}$

(d) $\begin{pmatrix} -1/2 & -\sqrt{3}/2 & 0 \\ \sqrt{3}/2 & -1/2 & 0 \\ 0 & 0 & 1 \end{pmatrix}$

20. The matrix corresponding to a 3-fold rotation along z-axis using crystallographic system is:

(a) $\begin{pmatrix} -1 & 1 & 0 \\ -1 & 0 & 0 \\ 0 & 0 & -1 \end{pmatrix}$

(b) $\begin{pmatrix} 1 & -1 & 0 \\ 1 & 0 & 0 \\ 0 & 0 & 1 \end{pmatrix}$

(c) $\begin{pmatrix} 0 & 1 & 0 \\ -1 & 1 & 0 \\ 0 & 0 & -1 \end{pmatrix}$

(d) $\begin{pmatrix} 0 & -1 & 0 \\ 1 & -1 & 0 \\ 0 & 0 & 1 \end{pmatrix}$

21. The matrix corresponding to a $\bar{3}$-fold rotation along z-axis using crystallographic system is:

(a) $\begin{pmatrix} -1 & 1 & 0 \\ -1 & 0 & 0 \\ 0 & 0 & -1 \end{pmatrix}$

(b) $\begin{pmatrix} 1 & -1 & 0 \\ 1 & 0 & 0 \\ 0 & 0 & 1 \end{pmatrix}$

(c) $\begin{pmatrix} 0 & 1 & 0 \\ -1 & 1 & 0 \\ 0 & 0 & -1 \end{pmatrix}$

(d) $\begin{pmatrix} 0 & -1 & 0 \\ 1 & -1 & 0 \\ 0 & 0 & 1 \end{pmatrix}$

22. The matrix corresponding to a 6-fold rotation along z-axis using crystallographic system is:

(a) $\begin{pmatrix} -1 & 1 & 0 \\ -1 & 0 & 0 \\ 0 & 0 & -1 \end{pmatrix}$

(b) $\begin{pmatrix} 1 & -1 & 0 \\ 1 & 0 & 0 \\ 0 & 0 & 1 \end{pmatrix}$

(c) $\begin{pmatrix} 0 & 1 & 0 \\ -1 & 1 & 0 \\ 0 & 0 & -1 \end{pmatrix}$

(d) $\begin{pmatrix} 0 & -1 & 0 \\ 1 & -1 & 0 \\ 0 & 0 & 1 \end{pmatrix}$

23. The matrix corresponding to a $\bar{6}$-fold rotation along z-axis using crystallographic system is:

(a) $\begin{pmatrix} -1 & 1 & 0 \\ -1 & 0 & 0 \\ 0 & 0 & -1 \end{pmatrix}$

(b) $\begin{pmatrix} 1 & -1 & 0 \\ 1 & 0 & 0 \\ 0 & 0 & 1 \end{pmatrix}$

(c) $\begin{pmatrix} 0 & 1 & 0 \\ -1 & 1 & 0 \\ 0 & 0 & -1 \end{pmatrix}$

(d) $\begin{pmatrix} 0 & -1 & 0 \\ 1 & -1 & 0 \\ 0 & 0 & 1 \end{pmatrix}$

24. The matrix corresponding to a 4-fold rotation along z-axis using orthogonal system is:

(a) $\begin{pmatrix} -1 & 0 & 0 \\ 0 & -1 & 0 \\ 0 & 0 & -1 \end{pmatrix}$

(b) $\begin{pmatrix} 0 & -1 & 0 \\ 1 & 0 & 0 \\ 0 & 0 & 1 \end{pmatrix}$

(c) $\begin{pmatrix} 0 & 1 & 0 \\ -1 & 0 & 0 \\ 0 & 0 & -1 \end{pmatrix}$

(d) $\begin{pmatrix} 1 & 0 & 0 \\ 0 & 1 & 0 \\ 0 & 0 & 1 \end{pmatrix}$

25. The matrix corresponding to a $\bar{4}$-fold rotation along z-axis using orthogonal system is:

(a) $\begin{pmatrix} -1 & 0 & 0 \\ 0 & -1 & 0 \\ 0 & 0 & -1 \end{pmatrix}$

(b) $\begin{pmatrix} 0 & -1 & 0 \\ 1 & 0 & 0 \\ 0 & 0 & 1 \end{pmatrix}$

(c) $\begin{pmatrix} 0 & 1 & 0 \\ -1 & 0 & 0 \\ 0 & 0 & -1 \end{pmatrix}$

(d) $\begin{pmatrix} 1 & 0 & 0 \\ 0 & 1 & 0 \\ 0 & 0 & 1 \end{pmatrix}$

26. The matrix corresponding to a 1-fold rotation along z-axis using orthogonal system is:

(a) $\begin{pmatrix} -1 & 0 & 0 \\ 0 & -1 & 0 \\ 0 & 0 & -1 \end{pmatrix}$

(b) $\begin{pmatrix} 0 & -1 & 0 \\ 1 & 0 & 0 \\ 0 & 0 & 1 \end{pmatrix}$

(c) $\begin{pmatrix} 0 & 1 & 0 \\ -1 & 0 & 0 \\ 0 & 0 & -1 \end{pmatrix}$

(d) $\begin{pmatrix} 1 & 0 & 0 \\ 0 & 1 & 0 \\ 0 & 0 & 1 \end{pmatrix}$

27. The matrix corresponding to a 1bar (Place bar symbol over 1)-fold rotation along z-axis using orthogonal system is:

(a) $\begin{pmatrix} -1 & 0 & 0 \\ 0 & -1 & 0 \\ 0 & 0 & -1 \end{pmatrix}$

(b) $\begin{pmatrix} 0 & -1 & 0 \\ 1 & 0 & 0 \\ 0 & 0 & 1 \end{pmatrix}$

(c) $\begin{pmatrix} 0 & 1 & 0 \\ -1 & 0 & 0 \\ 0 & 0 & -1 \end{pmatrix}$

(d) $\begin{pmatrix} 1 & 0 & 0 \\ 0 & 1 & 0 \\ 0 & 0 & 1 \end{pmatrix}$

28. When a mirror plane is combined with a 2-fold axis (perpendicular to mirror) produces the point group:
    (a) 2
    (b) 2/m
    (c) 3m
    (d) 3/m

29. Two mutually perpendicular 2-fold axes produce:
    (a) 4-fold
    (b) $\bar{4}$-fold
    (c) 2-fold
    (d) $\bar{2}$-fold

30. A combination of 2-fold and mirror produces a mirror line of second from away from the first:
    (a) 30°
    (b) 45°
    (c) 60°
    (d) 90°

31. A combination of 4-fold and mirror produces a pair of mirror lines of second from away from the first:
    (a) 30°
    (b) 45°
    (c) 60°
    (d) 90°

32. A combination of 6-fold and mirror produces three mirror lines of second from away from the first:
    (a) 30°
    (b) 45°
    (c) 60°
    (d) 90°

33. The order of monoclinic point group $C_{2h}(2/m)$ is:
    (a) 1
    (b) 2
    (c) 4
    (d) 6

34. The order of orthorhombic point group $D_{2h}(mmm)$ is:
    (a) 1
    (b) 2
    (c) 4
    (d) 6

35. The order of tetragonal point group $D_{4h}(4/mmm)$ is:
    (a) 16
    (b) 12
    (c) 8
    (d) 4

36. The order of trigonal point group $D_{3d}(\bar{3}m)$ is:
    (a) 16
    (b) 12
    (c) 8
    (d) 4

37. The order of hexagonal point group $C_{6h}(6/m)$ is:
    (a) 4
    (b) 8
    (c) 12
    (d) 16

38. The order of hexagonal point group $D_{6h}(6/mmm)$ is:
    (a) 6
    (b) 12
    (c) 18
    (d) 24

39. The order of cubic point group $T_h(m3)$ is:
    (a) 24
    (b) 18
    (c) 12
    (d) 6

40. The order of cubic point group $O_h(m3m)$ is:
    (a) 12
    (b) 24
    (c) 36
    (d) 48

41. The symmetry elements E, C_2, σ_{xz} and σ_{yz} belong to the point group:
    (a) C_{2v} (mm2)
    (b) C_{2h} (2/m)
    (c) D_2 (222)
    (d) D_{2h} (mmm)

42. The symmetry elements E, C_2, σ_h and i belong to the point group:
    (a) C_{2v} (mm2)
    (b) C_{2h} (2/m)
    (c) D_2 (222)
    (d) D_{2h} (mmm)

43. The symmetry elements E, C_{2x}, C_{2y} and C_{2z} belong to the point group:
    (a) C_{2v} (mm2)
    (b) C_{2h} (2/m)
    (c) D_2 (222)
    (d) D_{2h} (mmm)

44. The symmetry elements E, C_{2x}, C_{2y}, C_{2z}, i, σ_x, σ_y, and σ_z belong to the point group:
    (a) C_{2v} (mm2)
    (b) C_{2h} (2/m)
    (c) D_2 (222)
    (d) D_{2h} (mmm)

45. The symmetry elements E, C_4, C_4^2, C_4^3, σ_x, σ_y, σ_1 and σ_2 belong to the point group:
    (a) C_{4v} (4 mm)
    (b) C_{4h} (4/m)
    (c) D_4 (422)
    (d) D_{2d} ($\bar{4}$2m)

46. The symmetry elements E, $C_4, C_4^2, C_4^3, i, S_4, S_4^3$ and σ_h belong to the point group:
    (a) C4v (4 mm)
    (b) C_{4h} (4/m)
    (c) D_4 (422)
    (d) D_{2d} ($\bar{4}$2m)

47. The symmetry elements E, $C_4, C_4^2, C_4^3, C_{2x}, C_{2y}, C_2(1)$ and $C_2(2)$ belong to the point group:
    (a) C_{4v} (4 mm)
    (b) C_{4h} (4/m)
    (c) D_4 (422)
    (d) D_{2d} ($\bar{4}$2m)

48. The symmetry elements belong to the point group:
    (a) C_{4v} (4 mm)
    (b) C_{4h} (4/m)
    (c) D_4 (422)
    (d) Same as option d of 47

Answers

   1. (a)
   2. (b)
   3. 3.(c)
   4. (d)
   5. (a)
   6. (b)
   7. (c)
   8. (d)
   9. (a)
  10. (b)
  11. (c)
  12. (d)
  13. (a)
  14. (b)
  15. (c)
  16. (d)
  17. (c)
  18. (a)
  19. (b)
  20. (d)
  21. (c)
  22. (b)
  23. (a)

24. (b)
25. (c)
26. (d)
27. (a)
28. (b)
29. (c)
30. (d)
31. (b)
32. (a)
33. (c)
34. (d)
35. (a)
36. (b)
37. (c)
38. (d)
39. (a)
40. (d)
41. (a)
42. (b)
43. (c)
44. (d)
45. (a)
46. (b)
47. (c)
48. (d)

Chapter 7
Diffraction of Waves and Particles by Crystal

7.1 Production of X-Rays

(a) X-rays are produced when a beam of highly accelerated charged particles such as electrons is allowed to strike a metal target. In the process, electrons suffer rapid deceleration and the energy lost by them is emitted in the form of "continuous X-rays," that is,

$$\Delta E_{max} = eV_{max} = \frac{1}{2} mv^2 = h\nu_{max} = \frac{hc}{\lambda_{min}}$$

$$\text{or } \lambda_{min} = \frac{hc}{eV_{max}} = \frac{6.626 \times 10^{-34}\text{Js} \times 3 \times 10^8 \text{ms}^{-1}}{1.6 \times 10^{-19}\text{C} \times V} = \frac{1.24 \times 10^{-6}}{V} \text{ V - m}$$

$$(7.1)$$

where m is the mass of electron, e the charge and V the electrode potential. The continuous X-ray spectra of Tungsten at various accelerating potentials are shown in Fig. 7.1.

The above equation also gives the maximum frequency of X-ray corresponding to λ_{min} as

$$\nu_{max} = \frac{c}{\lambda_{min}}$$

(b) The characteristic X-rays (K_α, K_β, etc.) are produced when the bombarded electrons eject some of the orbital electrons of the target atom and the corresponding vacancies are filled by the less energetic outer shell electrons (Fig. 7.2). The energy level diagram and allowed transitions are shown in Fig. 7.3. The allowed transitions are:

M. A. Wahab, *Numerical Problems in Crystallography*,
https://doi.org/10.1007/978-981-15-9754-1_7

Fig. 7.1 X-ray spectra of
tungsten at various
accelerating potentials

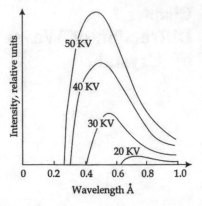

$E_{K\alpha_1} = W_K - W_{L_{III}}$ due to K to L_{III} transition

and $E_{K\alpha_2} = W_K - W_{L_{II}}$ due to K to L_{II} transition

where $\lambda_K(\alpha_1) = 1.54051\,\text{Å}$

and $\lambda_K(\alpha_2) = 1.54433\,\text{Å}$

Thus to excite K radiation from the target material, for example, the bombarding electrons must have the energy equal to W_K. Therefore,

Fig. 7.2 Electron orbits in an
atom, showing transitions
associated with the production
of characteristics X-rays

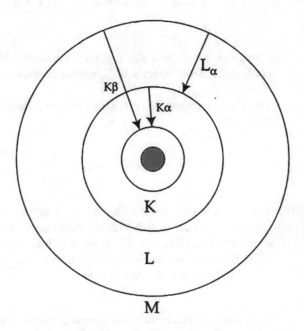

Fig. 7.3 Energy level diagram showing the allowed transitions

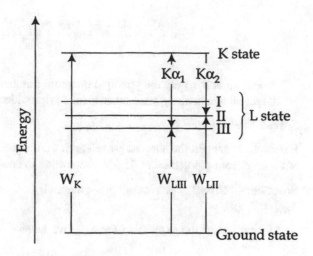

$$eV_K = W_K = h\nu_K = \frac{hc}{\lambda_K}$$

$$\text{or } \lambda_K(\alpha) = \frac{hc}{eV_K} = \frac{hc}{E_L - E_K}$$

where V_K is the K excitation voltage and λ_K is the K absorption edge wavelength (in angstroms). On a similar line one can obtain

$$\lambda_K(\beta) = \frac{hc}{eV_K} = \frac{hc}{E_M - E_K}$$

where E_K, E_L and E_M respectively, are energies corresponding to K, L and M levels.

(c) According to Moseley, the characteristic frequency emitted by an element (in each K, L, M and N series) is found to depend on its atomic number according to the equation

$$\sqrt{\nu} \propto a(Z - b)$$

$$\text{or } \nu = [a(Z - b)]^2$$

where a and b are constants for a particular series (their values are different for different series). The constant a is related to Rydberg constant, b is a small number called the nuclear screening constant and (Z-b) is the effective mass number of the element. For K-spectrum, b = 1.

(d) The Moseley's law deduced from Bohr's atomic model (taking into account the two energy levels) is given by

$$v = \frac{\Delta E}{h} = \frac{E_2 - E_1}{h} = \frac{m\ e^4(Z-b)^2}{8\varepsilon_0^2 h^3}\left(\frac{1}{n_1^2} - \frac{1}{n_2^2}\right)$$

where n_1 and n_2 are the principal quantum numbers. A list of commonly used target elements along with other details is provided in Table 7.1.

Solved Examples

Example 1 Obtain the shortest wavelength that is present in the X-rays produced at an accelerating potential of 45 kV. Determine the corresponding frequency also.

Solution: *Given*: Accelerating potential, $V = 45$ kV $= 45 \times 10^3$ V, $\lambda_{min} = ?, v_{max} = ?$

Making use of the expression for λ_{min}, we have

$$\lambda_{min} = \frac{hc}{eV_{max}} = \frac{6.626 \times 10^{-34} \times 3 \times 10^8}{1.6 \times 10^{-19} \times 45 \times 10^3}$$
$$= 2.76 \times 10^{-11} \text{m} = 0.276\,\text{Å}$$

Corresponding frequency is given by

$$v_{max} = \frac{c}{\lambda_{min}} = \frac{3 \times 10^8}{2.76 \times 10^{-11}}$$
$$= 1.08 \times 10^{19}\ \text{Hz}.$$

Example 2 If the potential difference across an X-ray tube is 40 kV and the current passing through it is 30 mA, determine the number of electrons striking the target per second and the speed at which they strike. Also determine the minimum wavelength of the X-ray produced.

Solution: *Given*: $V = 40$ kV, $I = 30$ mA $= 3 \times 10^{-2}$A, $n = ?, v = ?, \lambda_{min} = ?$.

We know that the number of electrons striking the target will be given by

Table. 7.1 Target elements, corresponding X-ray wavelengths and filter elements

Elements with at. no		X-ray Wavelengths		Filter Element	
Target	Filter	$K_{\alpha 1}$	$K_{\alpha 2}$	Density (g/cm^2)	Optimum thickness (mm)
Cr (24)	V (23)	2.2896	2.2935	0.009	0.016
Fe (26)	Mn (25)	1.9360	1.9399	0.012	0.016
Cu (29)	Ni (28)	1.5405	1.5405	0.019	0.021
Mo (42)	Nb (41)	0.7093	0.7135	0.069	0.108
Au (47)	Pd (46)	0.5594	0.5638	0.030	0.030

$$n = \frac{Q}{e} = \frac{I \times t}{e} = \frac{3 \times 10^{-2} \times 1}{1.6 \times 10^{-19}} = 1.87 \times 10^{17}/s$$

Further, we have

$$eV = \frac{1}{2}mv^2 \text{ or } v^2 = \frac{2eV}{m} = \frac{2 \times 1.6 \times 10^{-19} \times 4 \times 10^4}{9.1 \times 10^{-31}} = 1.406 \times 10^{16}$$

or $v = 1.19 \times 10^8 \text{m/s}$

and $\lambda_{min} = \frac{hc}{eV_{max}} = \frac{6.626 \times 10^{-34} \times 3 \times 10^8}{1.6 \times 10^{-19} \times 4 \times 10^4}$

$\quad = 3.11 \times 10^{-11}\text{m} = 0.311 \text{ Å}$

Example 3   Electrons bombarding the anode of a Coolidge tube produce X-rays of wavelength 1 Å. Determine the energy of each electron at the time of impact.

Solution: *Given*: $\lambda = 1$ Å $= 10^{-10}\text{m}$, energy of each electron at the time of impact = ?

Energy of an electron at the time of impact is given by

$$\Delta E = \frac{hc}{e\lambda} = \frac{6.626 \times 10^{-34} \times 3 \times 10^8}{1.6 \times 10^{-19} \times 10^{-10}} = 1.24 \times 10^4 eV$$

Example 4   Calculate Planck's constant when an X-ray tube operating at 30 kV emits a continuous X-ray spectrum which has $\lambda_{min} = 0.414$ Å.

Solution: *Given*: Operating voltage of the X-ray tube = 30 kV = 30 $\times$ $10^3 V$, $\lambda_{min} = 0.414$ Å $= 0.414 \times 10^{-10}\text{m}$, h = ?

We know that

$$eV_{max} = h\nu_{max} = \frac{hc}{\lambda_{min}}$$

Therefore,

$$h = \frac{eV_{max} \times \lambda_{min}}{c} = \frac{1.6 \times 10^{-19} \times 30 \times 10^3 \times 0.414 \times 10^{-10}}{3 \times 10^8}$$

$\quad = 6.624 \times 10^{-34} \text{J - s}$

Example 5   An X-ray tube is operating at 40 kV and tube current 25 mA, deduce the power input of the tube. If this power is maximum achievable in this tube, then determine the maximum permissible tube current at 50 kV.

Solution: *Given*: First operating voltage = 40 kV = 40×10^3V, tube current = 25 mA =25×10^{-3}A, second operating voltage = 50 kV = 50×10^3V, power (input) = ?, tube current = ?

We know that power is defined as

$$P = VI = 40 \times 10^3 \times 25 \times 10^{-3} = 1000\text{watt}$$

In second case, P = 1000 W and V = 50 kV = 50×10^3V. Therefore,

$$I = \frac{P}{V} = \frac{1000}{50 \times 10^3} = 0.02\text{A} = 20 \text{ mA}$$

Example 6 For platinum metal, the energy levels K, L and M lie roughly at 78, 12 and 3 eV, respectively. Determine the approximate wavelength of characteristic K_α and K_β lines.

Solution: *Given*: Energy levels (Fig. 7.4) of platinum lie at $E_1 = 78$ keV, $E_2 = 12$keV, $E_3 = 3$keV, $K_\alpha = ?$, $K_\beta = ?$

Draw the energy level diagram as shown in Fig. 7.4. From this, we can obtain

$$(h\nu)_{K_\alpha} = E_L - E_K = -12 + 78 = 66 \text{ keV} \tag{i}$$

$$\text{and } (h\nu)_{K_\beta} = E_M - E_K = -3 + 78 = 75 \text{ keV} \tag{ii}$$

Equation (i) can be written as:

Fig. 7.4 Energy level diagram for platinum

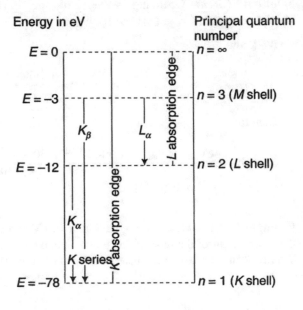

$$\frac{hc}{\lambda_K(\alpha)} = E_L - E_K$$

or

$$\lambda_K(\alpha) = \frac{hc}{E_L - E_K} = \frac{6.626 \times 10^{-34} \times 3 \times 10^8}{1.6 \times 10^{-19} \times 66 \times 10^3} = 0.19\,\text{Å}$$

Similarly, from Eq. (ii), we can obtain $\lambda_K(\beta) = 0.17\,\text{Å}$.

Example 7 An unknown element is found to emit L_α X-rays of wavelength 4.174 Å, while the wavelength of L_α (X-ray line) of platinum (atomic number 78) is 1.321 Å. Determine the atomic number of unknown element where the value of b for L_α is 7.4.

Solution: *Given*: Wavelength of the unknown element (say x) $\lambda_\alpha = 4.174\,\text{Å}$, $\lambda_L(\alpha)$ of Pt = 1.321 Å, At. No. (Pt) = 78, b = 7.4, At. No. of element (x) = ?

We know that the characteristic frequency emitted by an element and its atomic number is related through

$$\nu = [a(Z - b)]^2$$

where a, b are constants.
Thus, for Pt, this equation can be written as

$$\nu_{Pt} = \frac{c}{\lambda_{Pt}} = [a(78 - 7.4)]^2 \tag{i}$$

Similarly, for unknown element

$$\nu_x = \frac{c}{\lambda_x} = [a(x - 7.4)]^2 \tag{ii}$$

Dividing Eqs. (ii) from (i) and solving for x, we obtain

$$x = 7.4 + \left(\frac{\lambda_{Pt}}{\lambda_x}\right)^{1/2} \times 70.6$$

$$= 7.4 + \left(\frac{1321}{4174}\right)^{1/2} \times 70.6 = 47.1$$

Thus the unknown element is silver.

Example 8 The wavelength of K_α (X-ray line) of tungsten is 210 Å. Calculate its value for copper if the atomic number of tungsten is 74 and that of copper is 29.

Solution: *Given*: $\lambda_K(\alpha)$ of W = 210 Å, At. No. (W) = 74, At. No. (Cu) = 29, for K-series of X-ray b = 1, $\lambda_K(\alpha)$ of copper = ?

We know that the characteristic frequency emitted by an element and its atomic number is related through

$$v = [a(Z - b)]^2$$

where a, b are constants.

Thus, for tungsten, this equation can be written as

$$v_W = \frac{c}{\lambda_W} = [a(74 - 1)]^2 = (a \times 73)^2 \tag{i}$$

Similarly, for copper

$$v_{Cu} = \frac{c}{\lambda_{Cu}} = (a \times 28)^2 \tag{ii}$$

From these two equations, we can obtain

$$\lambda_{Cu} = \left(\frac{73}{28}\right)^2 \times 210 = 1427\,\text{Å}$$

Example 9 The wavelengths of K_α (X-ray lines) of iron and platinum are 1.93 Å and 0.19 Å, respectively. Determine the wavelengths of K_α lines for tin and barium.

Solution: *Given*: $\lambda_{Fe} = 1.93\,\text{Å}, \lambda_{Pt} = 0.19\,\text{Å}, \lambda_{Sn} = ?, \lambda_{Ba} = ?$

We know that the atomic number $Z_{Fe} = 26$, $Z_{Pt} = 78$, $Z_{Sn} = 50$ and $Z_{Ba} = 56$, also for K_α line b = 1

Now, for K_α line, the Moseley's law is given by

$$v \propto (Z - 1)^2, \text{ or } \frac{c}{\lambda} \propto (Z - 1)^2, \text{ or } \lambda \propto \frac{1}{(Z - 1)^2}$$

So that

$$\lambda_{Fe} \propto \frac{1}{(25)^2} \quad \text{and} \quad \lambda_{Sn} \propto \frac{1}{(49)^2}$$

and

$$\frac{\lambda_{Sn}}{\lambda_{Fe}} = \frac{(25)^2}{(49)^2}$$

or

$$\lambda_{Sn} = \frac{(25)^2}{(49)^2} \times \lambda_{Fe}$$

$$= \frac{(25)^2}{(49)^2} \times 1.93 = 0.5\,\text{Å}$$

Making a similar calculation by combining platinum and barium, we can obtain $\lambda_{Ba} = 0.37\,\text{Å}$.

Example 10 Wavelength of X-ray K_α line of an unknown element is 0.7185 Å. Find the element.

Solution: *Given*: $\lambda_K(\alpha) = 0.7185\,\text{Å} = 0.7185 \times 10^{-10}\text{m}$, element = ?

Let us use Moseley's law given by

$$\nu = \frac{c}{\lambda} = \frac{m\,c^4(Z-b)^2}{8\varepsilon_0^2\,h^3}\left(\frac{1}{n_1^2} - \frac{1}{n_2^2}\right)$$

Since, K_α transition takes place from $n = 2$ to $n = 1$ and $b = 1$. Therefore, substituting $n_1 = 1, n_2 = 2$ in the above equation, we obtain

$$\lambda_K(\alpha) = \frac{4}{3}\left(\frac{8\varepsilon_0^2 ch^3}{me^4(Z-1)^2}\right)$$

or

$$(Z-1)^2 = \frac{32 \times \varepsilon_0^2 \times c \times h^3}{3m \times e^4 \times \lambda_K(\alpha)}$$

$$= \frac{32 \times (8.85 \times 10^{-12})^2 \times 3 \times 10^8 \times (6.626 \times 10^{-34})^3}{3 \times 9.1 \times 10^{-31} \times (1.6 \times 10^{-19})^4 \times 0.7185 \times 10^{-10}} = 1701.54$$

or

$$Z = 1 + 41.25 \cong 42$$

$\implies$ The required element is molybdenum.

Example 11 Determine the wavelengths of K_α, K_β and L_α lines for copper (Z = 29).

Solution: *Given*: Z (copper) = 29, $K_\alpha = ?, K_\beta = ?, L_\alpha = ?$

We know that the transitions corresponding to K_α, K_β and L_α are:

$$K_\alpha : (n = 2) \rightarrow (n = 1), \quad K_\beta : (n = 3) \rightarrow (n = 1) \text{ and } L_\alpha : (n = 3) \rightarrow (n = 2)$$

Let us find the required wavelengths, one by one by using the equation

$$\bar{v} = \frac{1}{\lambda} = \frac{v}{c} = \frac{me^4 (Z - b)^2}{8\varepsilon_0^2 h^3} \left(\frac{1}{n_1^2} - \frac{1}{n_2^2} \right)$$

or

$$\lambda = \frac{8\varepsilon_0^2 ch^3}{me^4 \, Z^2 \times \left(\frac{1}{n_1^2} - \frac{1}{n_2^2} \right)}$$

Substituting different values of constants, we obtain

$$\lambda_K(\alpha) = \frac{8 \times (8.85 \times 10^{-12})^2 \times 3 \times 10^8 \times (6.626 \times 10^{-34})^3}{9.1 \times 10^{-31} \times (1.6 \times 10^{-19})^4 \times (29)^2 \times \left(\frac{1}{1^2} - \frac{1}{2^2} \right)}$$
$$= 1.22 \times 10^{-10} \text{m} = 1.22 \, \text{Å}$$

Similarly,

$$\lambda_K(\beta) = \frac{8 \times (8.85 \times 10^{-12})^2 \times 3 \times 10^8 \times (6.626 \times 10^{-34})^3}{9.1 \times 10^{-31} \times (1.6 \times 10^{-19})^4 \times (29)^2 \times \left(\frac{1}{1^2} - \frac{1}{3^2} \right)}$$
$$= 1.45 \times 10^{-10} \text{m} = 1.45 \, \text{Å}$$

and

$$\lambda_L(\alpha) = \frac{8 \times (8.85 \times 10^{-12})^2 \times 3 \times 10^8 \times (6.626 \times 10^{-34})^3}{9.1 \times 10^{-31} \times (1.6 \times 10^{-19})^4 \times (29)^2 \times \left(\frac{1}{2^2} - \frac{1}{3^2} \right)}$$
$$= 7.84 \times 10^{-10} \text{m} = 7.84 \, \text{Å}$$

7.2 X-Ray Diffraction by Crystals

(a) The well-known Bragg's equation is derived on the basis of the following:

   (i) A crystal is made up of various sets of equidistant parallel planes without taking into consideration the actual distribution of atoms in them.

Fig. 7.5 Bragg's reflection of X-rays from the atomic planes

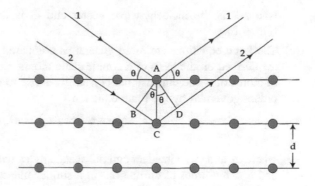

(ii) For a constructive interference between the rays scattered from any two consecutive crystal planes, the path difference must be an integral multiple of the wavelength (Fig. 7.5).
The resulting Bragg's equation is

$$2d\sin\theta = n\lambda \qquad (7.2)$$

where d is the interplanar spacing and n = 0, 1, 2 ....., represents the order of reflection. The highest possible order can be determined by the condition

$$(\sin\theta)_{max} = 1, \quad \text{so that} \quad \frac{n\lambda}{2d} \le 1$$

This indicates that λ must not be greater than twice the interplanar spacing, otherwise, no diffraction will occur.

(b) The vector form of the Bragg's law is given by

$$G^2 + 2k.G = 0$$

Fig. 7.6 Ewald construction

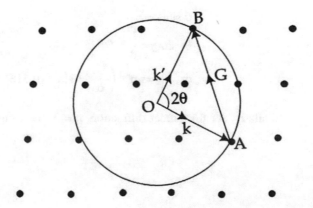

where k is the incident wave vector and G is the reciprocal lattice vector (Fig. 7.6).

(c) The Laue equations are more general as compared to Bragg's equation. They are derived on the basis of a simple static atomic model of a crystal. Each row of atoms of a one-dimensional crystal gives rise to diffraction cones of various orders governed by the Laue equation

$$a(\cos \alpha - \cos \alpha_0) = e\lambda$$

where α_0 and α are incident and diffracted angles with respect to a row of atoms and e is the order of diffraction. In a similar manner, for a space lattice three such equations (for three crystallographic axes) can be obtained. They are:

$$a(\cos \alpha - \cos \alpha_0) = e\lambda$$
$$b(\cos \beta - \cos \beta_0) = f\lambda \quad\quad\quad (7.3)$$
$$c(\cos \gamma - \cos \gamma_0) = g\lambda$$

where $\beta_0, \beta, \gamma_0, \gamma$, f and g have the same meaning for other rows of atoms.

Solved Examples

Example 1 When a crystal is subjected to a monochromatic X-ray beam, the first-order diffraction is observed at an angle of 15°. Determine the angles for second and third orders when the same X-ray beam is used.

Solution: *Given*: n = 1, $\theta_1 = 15°$, λ = fixed, θ_2 = ?, and θ_3 = ?

For first-order diffraction, the Bragg's equation is

$$2d \sin \theta_1 = \lambda$$

$$\text{or} \quad \frac{\lambda}{d} = 2 \sin \theta_1 = 2 \sin 15° = 0.518$$

For second-order diffraction, n = 2, the Bragg's equation becomes

$$d \sin\theta_2 = \lambda$$

$$\text{or } \sin\theta_2 = \frac{\lambda}{d} = 0.518$$

$$\text{or} \quad \theta_2 = \sin^{-1}\left(\frac{\lambda\lambda}{d}\right) = \sin^{-1}(0.518) = 31.2°$$

Similarly, for third-order diffraction, n = 3, the Bragg's equation becomes

$$2d \sin \theta_3 = 3\lambda$$

$$\text{or } \sin \theta_3 = \frac{3\lambda}{2d} = \frac{3}{2} \times 0.518 = 0.777$$

$$\text{or } \theta_3 = \sin^{-1}\left(\frac{3\lambda}{2d}\right) = \sin^{-1}(0.777) = 51°$$

Example 2 When a Cu crystal is subjected to an X-ray beam the first-order (111) plane appears at an angle of 21.7°. The lattice parameter of copper (fcc) is found to be 3.61 Å. Find the wavelength of the X-ray used.

Solution: *Given*: Crystal is fcc, $a = 3.61$Å, n = 1, plane $\equiv (111), \theta_1 = 21.7°$, $\lambda = ?$

For a cubic crystal the interplanar spacing is given by

$$d = \frac{a}{\left(h^2 + k^2 + l^2\right)^{1/2}} = \frac{3.61}{\sqrt{3}}$$

Further, the first-order Bragg's equation is

$$2d \sin \theta_1 = \lambda$$

$$\text{or } \lambda = 2d \sin \theta_1 = \frac{2 \times 3.61 \times \sin 21.7}{\sqrt{3}} = 1.54 \, \text{Å}$$

Example 3 When a nickel crystal (fcc) is exposed to an X-ray of wavelength 1.54 Å, Bragg's angle corresponding to (220) reflection is 38.2°. Determine its lattice parameter.

Solution: Given: $(hkl) \equiv (220)$, structure is fcc, $\theta = 38.2°, \lambda = 1.54 \, \text{Å}, a = ?$

For a cubic crystal the interplanar spacing is given by

$$d = \frac{a}{\left(h^2 + k^2 + l^2\right)^{1/2}} = \frac{a}{\left(2^2 + 2^2 + 0^2\right)^{1/2}} = \frac{a}{2\sqrt{2}}$$

Further, from Bragg's equation, we have

$$2d_{hkl} \sin \theta = \lambda$$

$$\text{or } 2d_{hkl} = \frac{\lambda}{\sin \theta}$$

$$\text{or } \frac{2a}{2\sqrt{2}} = \frac{1.54}{\sin 38.2}$$

$$\text{or } a = \frac{\sqrt{2} \times 1.54}{\sin 38.2} = 3.52 \, \text{Å}$$

Example 4 X-rays of wavelength $1.54\,\text{Å}$ are used to calculate the spacing between (200) planes in aluminum. The first-order Bragg's angle corresponding to this reflection is $22.4°$. Determine the lattice parameter of aluminum crystal.

Solution: *Given*: $\lambda = 1.54\,\text{Å}, n = 1, (hkl) \equiv (200)$, crystal is Al, $\theta = 22.4°, a = ?$

From the first-order Bragg's equation, we can write

$$d = \frac{\lambda}{2\sin\theta} = \frac{1.54}{2 \times \sin 22.4}$$

Further, from the relationship between interplanar spacing and lattice parameter, we can write

$$a = d\left(h^2 + k^2 + l^2\right)^{1/2}$$
$$= d\left(2^2 + 0^2 + 0^2\right)^{1/2}$$
$$\text{or } a = 2d = \frac{2 \times 1.54}{2 \times \sin 22.4} = 4.041 \cong 4.05 \text{ for Al}$$

Example 5 A bcc molybdenum sample was studied with the help of X-rays of wavelength $1.543\,\text{Å}$. Diffraction from $\{200\}$ planes was obtained at $2\theta = 58.618°$. Assuming this to be of first order, determine the lattice parameter.

Solution: *Given*: The crystal structure of molybdenum is bcc, $\lambda = 1.543\,\text{Å}, \{hkl\} \equiv \{200\}, 2\theta = 58.618°$, so that $\theta = 29.309°$, a = ?

For $\{200\}$ planes in a cubic crystal, the interplanar spacing is given by

$$d_{200} = \frac{a}{\left(h^2 + k^2 + l^2\right)^{1/2}} = \frac{a}{\left(2^2 + 0^2 + 0^2\right)^{1/2}} = \frac{a}{2}$$

Further, from Bragg's equation, we have

$$a = \frac{\lambda}{\sin\theta} = \frac{1.543}{\sin 29.309} = 3.15\,\text{Å}$$

Example 6 X-rays of unknown wavelength are diffracted from an iron sample. First peak was observed for (110) planes at $2\theta = 44.70°$. If the lattice parameter of bcc iron is $2.87\,\text{Å}$, determine the wavelength of the X-ray used.

Solution: Given: The crystal structure of iron is bcc $(hkl) \equiv (110)$, $2\theta = 44.70°$, so that $\theta = 22.35°$ a $= 2.87\,\text{Å}, \lambda = ?$

For (110) planes in a cubic crystal, the interplanar spacing is given by

$$d_{110} = \frac{a}{\left(h^2 + k^2 + l^2\right)^{1/2}} = \frac{a}{\left(1^2 + 1^2 + 0^2\right)^{1/2}} = \frac{2.87}{\sqrt{2}}$$

Further, from Bragg's equation, we have

$$\lambda = 2d \sin \theta = \frac{2 \times 2.87 \times \sin(22.35)}{\sqrt{2}} = 1.543 \,\text{Å}$$

Example 7   Show that the greater is the angle of diffraction, greater is the accuracy in determining the lattice parameters.

Solution: Let us start with the Bragg's equation with interplanar spacing $d = a$ (also lattice parameter as in a simple cubic crystal) as

$$2a \sin \theta = n\lambda$$

$$\text{or } a \sin \theta = \frac{n\lambda}{2}$$

Differentiating this equation, we obtain

$$a \cos \theta \, d\theta + \sin \theta \, da = 0$$

$$\text{or } \frac{da}{a} = -\cot \theta \, d\theta$$

It is evident that as $\theta \to \frac{\pi}{2}$, $\cot\theta \to 0$, that is, the error in the lattice parameter approaches zero. This proves the above statement.

Example 8   Determine the longest wavelength that can be analyzed by a rock-salt crystal with interplanar spacing 2.82 Å in the first and second orders of X-ray diffraction.

Solution: *Given*: Interplanar spacing, $d = 2.82$ Å, $n = 1, 2$, $\lambda_{max} = ?$

We know that the Bragg's equation is given by

$$2d \sin \theta = n\lambda$$

Here, for the λ to be maximum, $\sin \theta$ must be maximum, that is, $(\sin \theta)_{max} = 1$
This gives us

$$n\lambda_{max} = 2d$$

$$\text{or } \lambda_{max} = \frac{2d}{n} = \frac{2 \times 2.82}{1} = 5.64 \,\text{Å for n} = 1$$

$$\text{and } \lambda_{max} = \frac{2d}{n} = \frac{2 \times 2.82}{2} = 2.82 \,\text{Å for n} = 2$$

Example 9 Calculate the glancing angle on the plane (110) of a cube of rock-salt (a = 2.81 Å) corresponding to second order maximum for the X-rays of wavelength 0.71 Å.

Solution: *Given*: Crystal plane $(hkl) \equiv (110), a = 2.81 \,\text{Å} = 2.81 \times 10^{-10}\text{m, n} = 2, \lambda = 0.71 \,\text{Å} = 0.71 \times 10^{-10}\text{m}, \theta = ?$

For (110) plane the interplanar spacing is given by

$$d_{110} = \frac{a}{\left(h^2 + k^2 + l^2\right)^{1/2}} = \frac{2.81}{\left(1^2 + 1^2 + 0^2\right)^{1/2}} = \frac{2.81}{\sqrt{2}} = 1.99 \,\text{Å}$$

Now, using Bragg's equation, the glancing angle for n = 2 is obtained as

$$\theta = \sin^{-1}\left(\frac{\lambda}{d}\right) = \sin^{-1}\left(\frac{0.71}{1.99}\right) = 20.9°$$

Example 10 A beam of X-ray is incident on NaCl crystal $(a = 2.81 \,\text{Å})$. The first-order reflection is observed at a glancing angle 8° 35'. Determine the wavelength of the X-ray used and the second-order Bragg's angle.

Solution: *Given*: $d = 2.81 \,\overset{\circ}{A} = 2.81 \times 10^{-10}\text{m}, \theta_1 = 8°35' = 8.6°, \lambda = ?, \theta_2 = ?$

For n = 1, the Bragg's equation is

$$\lambda = 2d \sin \theta = 2 \times 2.81 \times \sin(8.6) = 0.84 \,\text{Å}$$

Now, for n = 2, $\theta = ?$

$$\theta = \sin^{-1}\left(\frac{\lambda}{d}\right) = \sin^{-1}\left(\frac{0.84}{2.81}\right) = 17.4°$$

Example 11 An X-ray of wavelength 0.440 Å is reflected from the cube face of a rock-salt crystal (d = 2.814 Å). Determine reflected angles.

Solution: *Given*: $\lambda = 0.440 \,\text{Å}, d = 2.814 \,\text{Å}$. Using Bragg's equation, angles for various reflections (i.e., θ_1 for n = 1, θ_2 for n = 2, etc.) can be determined.

We know that

$$\sin\theta = \frac{n\lambda}{2d} \text{ or } \theta = \sin^{-1}\left(\frac{n\lambda}{2d}\right)$$

Therefore,

$$\theta_1 = \sin^{-1}\left(\frac{1\lambda}{2d}\right) = \sin^{-1}\left(\frac{0.440}{5.628}\right) = \sin^{-1}(0.0782) = 4°29'$$

Similarly,

$$\theta_2 = \sin^{-1}\left(\frac{2\lambda}{2d}\right) = \sin^{-1}(2 \times 0.0782) = 8°59'$$

$$\theta_3 = \sin^{-1}\left(\frac{3\lambda}{2d}\right) = \sin^{-1}(3 \times 0.0775) = 13°34'$$

$$\theta_4 = \sin^{-1}\left(\frac{4\lambda}{2d}\right) = \sin^{-1}(4 \times 0.0775) = 18°13'$$

$$\theta_5 = \sin^{-1}\left(\frac{5\lambda}{2d}\right) = \sin^{-1}(5 \times 0.0775) = 23°$$

Hence, the reflected beam will be observed at the following angles: $4°29', 8°59', 13°34', 18°13'$ and $23°$, etc.

Example 12 A beam of X-rays having wavelengths in the range $0.2\,\text{Å}$ to $1\,\text{Å}$ is allowed to incident at an angle of $9°$ with the cube face of rock-salt crystal $(d = 2.814\,\text{Å})$. Determine the wavelengths of the diffracted beam.

Solution: *Given*: $d = 2.814\,\text{Å}$, Bragg's angle, $\theta = 9°$.

Using the Bragg's equation $2d\sin\theta = n\lambda$, we have

$$1\lambda_1 = 5.628\sin 9° \text{ or } \lambda_1 = 0.8804\,\text{Å}$$

$$2\lambda_2 = 5.628\sin 9° \text{ or } \lambda_2 = 0.4402\,\text{Å}$$

$$3\lambda_3 = 5.628\sin 9° \text{ or } \lambda_3 = 0.2935\,\text{Å}$$

$$4\lambda_4 = 5.628\sin 9° \text{ or } \lambda_4 = 0.2201\,\text{Å}$$

$$5\lambda_5 = 5.628\sin 9° \text{ or } \lambda_5 = 0.1760\,\text{Å}$$

The last one is less than 0.2 Å and hence the diffracted beam contains the first four wavelengths, that is, $0.8804\,\text{Å}, 0.4402\,\text{Å}, 0.2935\,\text{Å}$, and $0.02201\,\text{Å}$, in the ratio $1 : \frac{1}{2} : \frac{1}{3} : \frac{1}{4}$.

Example 13 The angles for the first-order reflection from (100), (110) and (111) face of sodium chloride crystals using monochromatic X-rays are 5.9°, 8.4° and 5.2°, respectively. Determine the nature of the given crystal.

Solution: *Given*: Crystal planes are $(100), (110)$ and $(111), \theta_{(100)} = 5.9°, \theta_{(110)} = 8.4°, \theta_{(111)} = 5.2°, n = 1$. Making use of the Bragg's equation, let us calculate the ratio of interplanar distances corresponding to the planes (100), (110) and (111). The Bragg's equation for n = 1 is

$$2d \sin \theta = \lambda \text{ or } d = \frac{\lambda}{2 \sin \theta}$$

Therefore,

$$
\begin{aligned}
d_{(100)} : d_{(110)} : d_{(111)} &= \frac{\lambda}{2 \sin 5.9} : \frac{\lambda}{2 \sin 8.4} : \frac{\lambda}{2 \sin 5.2} \\
&= \frac{1}{\sin 5.9} : \frac{1}{\sin 8.4} : \frac{1}{\sin 5.2} \\
&= \frac{1}{0.103} : \frac{1}{0.146} : \frac{1}{0.0906} \\
&= 1 : 0.705 : 1.14 = 1 : \frac{1}{\sqrt{2}} : \frac{2}{\sqrt{3}}
\end{aligned}
$$

$\implies$ Nature of the crystal is fcc.

Example 14 A crystal plane is mounted on an X-ray spectrometer. The glancing angles of incidence beam for three reflections are 5° 58′, 12° 01′ and 18° 12′. Show that these are successive orders of reflections from the same plane. Also find the interplanar spacing, the wavelength of the X-rays used is 0.586 Å.

Solution: *Given*: The glancing angles $\theta_1 = 5°58′ = 5.966°, \theta_2 = 12°01′ = 12.01°$ and $\theta_3 = 18°11′ = 18.18°, \lambda = 0.586\,\text{Å} = 0.586 \times 10^{-10}\text{m}$, angles representing I, II, III order reflections = ?, d = ?

In order to show that the given angles represent I, II and III orders of reflections from the same crystal plane, let us check the ratio of $\sin \theta_1, \sin \theta_2$ and $\sin \theta_3$.

$$\sin\theta_1 : \sin\theta_2 : \sin\theta_3 = \sin 5.966 : \sin 12.01 : \sin 18.18 = 0.1040 : 0.2080$$
$$: 0.3120 = 1 : 2 : 3$$

$\Longrightarrow$ These angles of incidence correspond to I, II and III orders of reflections, respectively.

Now, from Bragg's equation $2d\sin\theta = n\lambda$, we obtain

$$\text{For I order, } d = \frac{1\lambda}{2\sin\theta_1} = \frac{0.586 \times 10^{-10}}{2 \times 0.1040} = 2.817 \times 10^{-10}\text{m}$$

$$\text{For II order, } d = \frac{2\lambda}{2\sin\theta_2} = \frac{2 \times 0.586 \times 10^{-10}}{2 \times 0.2080} = 2.817 \times 10^{-10}\text{m}$$

$$\text{For III order, } d = \frac{3\lambda}{2\sin\theta_3} = \frac{3 \times 0.586 \times 10^{-10}}{2 \times 0.3120} = 2.817 \times 10^{-10}\text{m}$$

Hence the mean value of the interplanar spacing, $d = 2.817 \times 10^{-10}\text{m}$.

7.3   Absorption of X-Rays

(i) The intensity of X-ray beam transmitted through a given material decreases exponentially as

$$I = I_0 \exp(-\mu)x$$

where μ is the linear attenuation coefficient, and I_0 is the maximum intensity of X-ray at $x = x_0$ as shown in Fig. 7.7.

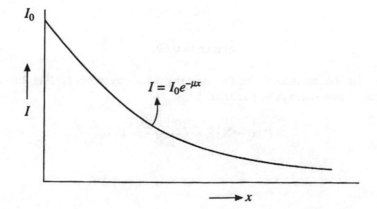

Fig. 7.7 Intensity as a function of thickness of the absorbing material

(ii) Thickness of the absorbing material also called as half-value layer is defined as

$$x_H = \frac{0.693}{\mu}$$

Solved Examples

Example 1 Calculate the percentage of intensity of the X-ray that will pass through 0.005 m thick of an aluminum sheet. The linear attenuation coefficient for aluminum for K_α line emitted by a tungsten target in is 139 m^{-1}.

Solution: *Given:* $\mu = 139 m^{-1}$, thickness x = 0.005 m, percentage of transmitted intensity = ?

Let X% of the intensity is being transmitted through the aluminum sheet, then we have

$$X\% = \frac{I}{I_0}$$

$$\text{or } \frac{X}{100} = \exp(-\mu x)$$

or

$$X = 100 \exp(-\mu x)$$
$$= 100 \exp(-139 \times 0.005) = 50\%$$

Example 2 Determine the attenuation percentage in an X-ray beam after passing through a thickness of a material equal to four-half value thickness.
Solution: *Given:* $x = 4x_H$, attenuation % = ?

We know that the attenuation coefficient in terms of half value layer is given by

$$\mu = \frac{0.693}{x_H}$$
$$\text{or } \mu x_H = 0.693$$

Now, let the attenuation in the intensity of X-ray after passing through a thickness of material is X%, therefore

$$\frac{I}{I_0} = (100 - X)\% = \frac{100 - x}{100} = 1 - \frac{X}{100}$$

Also,

$$\frac{I}{I_0} = \exp(-\mu x) = \exp(-4\mu x_H)$$

or

$$1 - \frac{X}{100} = \exp(-4 \times 0.693) = 0.0625368$$
$$\text{or } X = (1 - 0.0625368) \times 100 = 93.7\%$$

Example 3 A beam of X-ray consists of equal intensities of wavelengths 0.064 Å and 0.098 Å. When they pass through a piece of lead, their attenuated beam intensity is in the ratio 3:1. The mass absorption coefficients are 0.164 m^2/kg for harder component and 0.35 m^2/kg for softer component, respectively. Calculate the thickness of lead if its density is 11340 kg/m^3.

Solution: *Given*: $\lambda_1 = 0.064$ Å ,$\lambda_2 = 0.098$ Å ,$\mu_{m1} = 0.164 \frac{m^2}{kg}$ for λ_1, $\mu_{m2} = 0.35 \frac{m^2}{kg}$ for λ_2, $I(\lambda_1) = I(\lambda_2)$, $\rho(Pb) = 11340 \frac{kg}{m^3}$, also according to question $I_1 : I_2 = 3 : 1, x = ?$

Here, μ_1 and μ_2 are :

$$\mu_1 = \mu_{m1} \times \rho = 0.164 \times 11340 = 1859.76 m^{-1}$$
$$\mu_2 = \mu_{m2} \times \rho = 0.35 \times 11340 = 3969.00 m^{-1}$$

Now, making use of the exponential equation for intensity variation and taking the ratio, we obtain

$$3 = \frac{\exp(-1859.76)}{\exp(-3969.00)}$$
$$= \exp(2109.24)x$$
$$\text{or} \quad 2109.24x = \ln 3$$
$$\text{or} \quad x = \frac{\ln 3}{2109.24} = 5.22 \times 10^{-4} \text{ m}$$

Example 4 A copper sheet of 1.05 mm thick can reduce the intensity of the X-ray beam to 7.5% of its original value. Determine the value of its mass absorption coefficient of copper if its density is 8930 kg/m^3.

Solution: *Given*: x = 1.05 mm = 1.05×10^{-3} m, $\frac{I}{I_0} = 7.5\%$, $\rho(Cu) = 8930 \frac{kg}{m^3}$, $\mu_{m=?}$

We know that

$$I = I_0 \exp(-\mu)x$$

or

$$\frac{I}{I_0} = I_0\exp(-\mu)x = 7.5\% = \frac{7.5}{100} = 0.075$$

$$\text{or} \quad \exp(\mu x) = \frac{1}{0.075} = 13.33$$

$$\text{or} \quad \mu x = \ln(13.33)$$

or

$$\mu = \frac{\ln(13.33)}{x}$$

$$= \frac{\ln(13.33)}{1.05 \times 10^{-3}} = 2466.7\text{m}^{-1}$$

Further, the mass absorption coefficient is given by

$$\mu_m = \frac{\mu}{\rho} = \frac{2466.7}{8930}$$

$$= 0.276 \text{ m}^2/\text{kg}$$

7.4   Other Diffraction Methods

(i) Neutron Diffraction

Depending upon the energy of the neutron from the pile of a reactor, the wavelength of a neutron beam is given by

$$\lambda = \frac{h}{\sqrt{2mE}} = \frac{0.28}{\sqrt{E}}$$

where mass of the neutron, $m = 1.675 \times 10^{-27}\text{kg}$.

(ii) Electron Diffraction

Ignoring the relativistic correction, the wavelength associated with the moving electron beam is given by

$$\lambda = \frac{h}{\sqrt{2mE}} = \frac{h}{\sqrt{2meV}} = \frac{12.24}{\sqrt{V}}$$

where mass of the neutron, $m = 9.1 \times 10^{-31}\text{kg}$.

Solved Examples

Example 1 Determine the angle between the incident beam and the crystal so that the reflected neutrons will have a kinetic energy of 0.1 eV, when a beam of neutron with energies ranging from zero to several electron volts is directed at a crystal with interplanar spacing 3.03 Å.

Solution: *Given*: E = 0.1 eV = 0.1 × 1.6 × 10⁻¹⁹J = 16 × 10⁻²¹J, M_n = 1.67 × 10⁻²⁷kg, d = 3.03 Å = 3.03 ×10⁻¹⁰m, θ = ?

Wavelength associated with this energy is

$$\lambda = \frac{h}{(2M_nE)^{1/2}}$$

$$= \frac{6.626 \times 10^{-34}}{(2 \times 1.67 \times 10^{-27} \times 16 \times 10^{-21})^{1/2}}$$

$$= 0.906 \times 10^{-10}m = 0.906 \,\text{Å}$$

Now, according to Bragg's equation $2d \sin\theta = n\lambda$ (for n = 1), the glancing angle is given by

$$\theta = \sin^{-1}\left(\frac{\lambda}{2d}\right) = \sin^{-1}\left(\frac{0.906}{6.06}\right) = 8°, 36'$$

Example 2 A neutron beam of kinetic energy 0.04 eV is diffracted by the plane (100) of Sylvine crystal (d_{100} = 3.14 Å). Determine the glancing angle θ at which the first-order Bragg's spectrum will be observed. Given neutron rest mass = 1.67 × 10⁻²⁷kg.

Solution: *Given*: K. E of neutron = 0.04 eV = 0.04 × 1.6 × 10⁻¹⁹J = 64 × 10⁻²²J, M_n = 1.67 × 10⁻²⁷kg,   crystal   plane   ≡ (100), d_{100} = 3.14 Å = 3.14 × 10⁻¹⁰m, θ = ?

We know that the wavelength associated with a moving neutron is given by

$$\lambda = \frac{h}{(2M_nE)^{1/2}} = \frac{6.626 \times 10^{-34}}{(2 \times 1.67 \times 10^{-27} \times 64 \times 10^{-22})^{1/2}}$$

$$= 1.43 \times 10^{-10}m$$

Now, according to Bragg's equation $2d \sin\theta = n\lambda$ (for n = 1), the glancing angle is given by

$$\theta = \sin^{-1}\left(\frac{\lambda}{2d}\right) = \sin^{-1}\left(\frac{1.43 \times 10^{-10}}{2 \times 3.14 \times 10^{-10}}\right)$$

$$= \sin^{-1}\left(\frac{1.43}{2 \times 3.14}\right) = \sin^{-1}(0.2277)$$

$$= 13.16° = 13°, 10'$$

Example 3 Calculate the energy (in eV) associated with an electron of wavelength 3×10^{-2}m.

Solution: *Given*: $\lambda = 3 \times 10^{-2}$m, E = ?

We know that the wavelength associated with a moving electron is given by

$$\lambda = \frac{h}{\sqrt{2mE}}$$

or

$$E = \frac{h^2}{2m\lambda^2} = \frac{(6.626 \times 10^{-34})^2}{2 \times 9.1 \times 10^{-31} \times 9 \times 10^{-4}} \text{ J}$$

$$= \frac{(6.626 \times 10^{-34})^2}{2 \times 9.1 \times 10^{-31} \times 9 \times 10^{-4} \times 1.6 \times 10^{-19}} \text{ eV}$$

$$= 1.68 \times 10^{-15} \text{eV}$$

Example 4 Calculate the de Broglie wavelength of electrons and their velocity when the first Bragg's maximum of electron diffraction in a nickel crystal (d = 0.4086 Å) is found to occur at a glancing angle of 65°.

Solution: *Given*: n = 1, d = 0.4086 Å = 0.4086 $\times 10^{-10}$m, θ=65°, λ = ?, v = ?

The Bragg's equation for n = 1 is

$$\lambda = 2d\sin\theta$$

$$= 2 \times 0.4086 \times 10^{-10} \times \sin 65°$$

$$= 2 \times 0.4086 \times 10^{-10} \times 0.9063$$

$$= 0.74 \times 10^{-10}\text{m}$$

Now, for the de Broglie equation, the velocity of the electron is given by

$$v = \frac{h}{m\lambda} = \frac{6.626 \times 10^{-34}}{2 \times 9.1 \times 10^{-31} \times 0.74 \times 10^{-10}}$$

$$= 9.84 \times 10^6 \text{m/s}$$

Example 5 Through what potential should a beam of electrons be accelerated so that their wavelength becomes 1.54 Å. Assuming that an electron beam can undergo diffraction by crystal.

Solution: *Given*: n = 1, $\lambda = 1.54\,\text{Å} = 1.54 \times 10^{-10}\text{m}, V = ?$

We know that the wavelength associated with the electrons is given by

$$\lambda = \frac{h}{mv} \text{ or } v = \frac{h}{m\lambda}$$

$$\text{Also, } eV = \frac{1}{2}\,mv^2 = \frac{1}{2}m\left(\frac{h}{m\lambda}\right)^2$$

or

$$V = \frac{h^2}{2m\,\lambda^2 e}$$

$$= \frac{(6.626 \times 10^{-34})^2}{2 \times 9.1 \times 10^{-31} \times (1.54 \times 10^{-10})^2 \times 1.6 \times 10^{-19}} = 63.6 \text{ volts}$$

Example 6 Electrons are accelerated to 344 V and are reflected from a crystal. The first reflection occurs when the glancing angle is 60°. Determine the interplanar spacing of the crystal.

Solution: *Given*: V = 344 V, $\theta = 60°$, interpanar spacing d = ?

We know that the wavelength associated with the electrons is given by

$$\lambda = \frac{h}{\sqrt{2mE}} = \frac{h}{\sqrt{2meV}}$$

$$= \frac{6.626 \times 10^{-34}}{(2 \times 9.1 \times 10^{-31} \times 1.6 \times 10^{-19} \times 344)^{1/2}}$$

$$= 0.662 \times 10^{-10} = 0.662\,\text{Å}$$

Now, the Bragg's equation for n = 1 is

$$2d \sin\theta = \lambda$$

$$\text{or } d = \frac{\lambda}{2 \sin\theta} = \frac{0.662}{2 \sin 60} = 0.38\,\text{Å}$$

Multiple Choice Questions (MCQ)

1. The shortest wavelength present in the X-rays produced at an accelerating potential of 45 kV is:
   (a) 0.276 Å
   (b) 1.276 Å
   (c) 2.276 Å
   (d) 3.276 Å

2. The frequency corresponding to the wavelength present in the X-rays produced at an accelerating potential of 45 kV is:
   (a) 1.08×10^{16} Hz
   (b) 1.08×10^{17} Hz
   (c) 1.08×10^{18} Hz
   (d) 1.08×10^{19} Hz

3. The potential difference across an X-ray tube is 40 kV and the current passing through it is 30 mA, the number of electrons striking the target per second will be:
   (a) 1.87×10^{16} /s
   (b) 1.87×10^{17} /s
   (c) 1.87×10^{18} /s
   (d) 1.87×10^{19} /s

4. The potential difference across an X-ray tube is 40 kV and the current passing through it is 30 mA, the speed at which the electrons will strike the target will be:
   (a) 1.19×10^{6} m/s
   (b) 1.19×10^{7} m/s
   (c) 1.19×10^{8} m/s
   (d) 1.19×10^{9} m/s

5. The potential difference across an X-ray tube is 40 kV and the current passing through it is 30 mA, the minimum wavelength of the X-ray produced will be:
   (a) 0.311 Å
   (b) 1.311 Å
   (c) 2.311 Å
   (d) 3.311 Å

6. Electrons bombarding the anode of a Coolidge tube produce X-rays of wavelength 1 Å. The energy of each electron at the time of impact is:
   (a) 1.24×10^{3} eV
   (b) 1.24×10^{4} eV
   (c) 1.24×10^{5} eV
   (d) 1.24×10^{6} eV

7. An X-ray tube operating at 30 kV emits continuous X-rays with $\lambda_{min} =$ 0.414 Å, the Planck's constant can found to be:
   (a) 6.324×10^{-34} J-s
   (b) 6.424×10^{-34} J-s
   (c) 6.524×10^{-34} J-s
   (d) 6.624×10^{-34} J-s

8. An X-ray tube is operating at 40 kV and tube current 25 mA. The power input in the tube is:
   (a) 800 W
   (b) 900 W
   (c) 1000 W
   (d) 1100 W

9. An X-ray tube is operating at 40 kV and tube current 25 mA. If the power is maximum achievable in this tube, then the maximum permissible tube current at 50 kV is:
   (a) 20 mA
   (b) 30 mA
   (c) 40 mA
   (d) 50 mA

10. For platinum metal, the energy levels K, L and M lie roughly at 78, 12 and 3 eV, respectively. The approximate wavelength of characteristic K_α line is:
    (a) 0.09 Å
    (b) 0.19 Å
    (c) 0.29 Å
    (d) 0.39 Å

11. For platinum metal, the energy levels K, L and M lie roughly at 78, 12 and 3 eV, respectively. The approximate wavelength of characteristic K_β line is:
    (a) 0.37 Å
    (b) 0.27 Å
    (c) 0.17 Å
    (d) 0.07 Å

12. The wavelength of L_α (X-ray line) of platinum (atomic number 78) is 1.321 Å. An unknown element emits L_α X-rays of wavelength 4.174 Å. If the value of b for L_α is 7.4, the atomic number of unknown element is:
    (a) 17.1
    (b) 27.1
    (c) 37.1
    (d) 47.1

13. The wavelength of K_α (X-ray line) of tungsten is 210 Å. If the atomic numbers of tungsten and copper are, respectively, 74 and 29, the wavelength of copper is:
    (a) 1427 Å
    (b) 1527 Å
    (c) 1627 Å
    (d) 1727 Å

14. The wavelengths of K_α (X-ray lines) of iron and platinum are 1.93 Å and 0.19 Å, respectively. The wavelengths of K_α lines for tin is:
    (a) 0.4 Å
    (b) 0.5 Å
    (c) 0.6 Å
    (d) 0.7 Å

15. The wavelengths of K_α (X-ray lines) of iron and platinum are 1.93 Å and 0.19 Å, respectively. The wavelengths of K_α lines for barium is:
    (a) 0.17 Å
    (b) 0.27 Å
    (c) 0.37 Å
    (d) 0.47 Å

16. The atomic number of the element which has K_α (X-ray line) of wavelength 0.7185 Å is:
    (a) 39
    (b) 40
    (c) 41
    (d) 42

17. The wavelengths of K_α line for copper (Z = 29) is:
    (a) 1.22 Å
    (b) 2.22 Å
    (c) 3.22 Å
    (d) 4.22 Å

18. The wavelengths of K_β line for copper (Z = 29) is:
    (a) 0.45 Å
    (b) 1.45 Å
    (c) 2.45 Å
    (d) 3.45 Å

19. The wavelengths of L_α line for copper (Z = 29) is:
    (a) 5.84 Å
    (b) 6.84 Å
    (c) 7.84 Å
    (d) 8.84 Å

20. When a crystal is subjected to monochromatic X-ray beam, the first-order diffraction is observed at an angle of 15°. If the same X-ray beam is used, the angle for second-order diffraction is:
    (a) 61.2°
    (b) 51.2°
    (c) 41.2°
    (d) 31.2°

21. When a crystal is subjected to monochromatic X-ray beam, the first-order diffraction is observed at an angle of 15°. If the same X-ray beam is used, the angle for third-order diffraction is:
    (a) 51°
    (b) 41°
    (c) 31°
    (d) 21°

22. The lattice parameter of copper (fcc) is 3.61 Å. The first-order (111) plane appears at an angle of 21.7°. The wavelength of the X-ray used is:
    (a) 0.54 Å
    (b) 1.54 Å
    (c) 2.54 Å
    (d) 3.54 Å

23. Nickel (fcc) has the Bragg's angle for its (220) reflection 38.2°. If the wavelength of the X ray used is 1.54 Å, the lattice parameter of nickel will be:
    (a) 5.52 Å
    (b) 4.52 Å
    (c) 3.52 Å
    (d) 2.52 Å

24. X-rays of wavelength 1.54 Å are used to calculate the spacing between (200) planes in aluminum. If the first-order Bragg's angle corresponding to this reflection is 22.4°, the lattice parameter of aluminum will be:
    (a) 1.05 Å
    (b) 2.05 Å
    (c) 3.05 Å
    (d) 4.05 Å

25. A bcc molybdenum sample was studied with the help of X-ray diffraction using X-rays of wavelength 1.543 Å. Diffraction from {200} planes was obtained at $2\theta = 58.618°$. Assuming the diffraction to be of first order, the value of lattice parameter will be:
    (a) 3.15 Å
    (b) 4.15 Å
    (c) 5.15 Å
    (d) 6.15 Å

26. X-rays of unknown wavelength are diffracted from an iron sample. First peak was observed for (110) planes at $2\theta = 44.70°$. If the lattice parameter of bcc iron is 2.87 Å, the wavelength of the X-ray used is:
    (a) 0.543 Å
    (b) 1.543 Å
    (c) 2.543 Å
    (d) 3.543 Å

27. The longest wavelength that can be analyzed by a rock-salt crystal with interplanar spacing 2.82 Å in the first order of X-ray diffraction is:
    (a) 3.64 Å
    (b) 4.64 Å
    (c) 5.64 Å
    (d) 6.64 Å

28. The longest wavelength that can be analysed by a rock-salt crystal with interplanar spacing 2.82 Å in the second order of X-ray diffraction is:
    (a) 5.82 Å
    (b) 4.82 Å
    (c) 3.82 Å
    (d) 2.82 Å

29. The glancing angle on the plane (110) of a cube of rock-salt ($a = 2.81$ Å) corresponding to second order maximum for the X-rays of wavelength 0.71 Å is:
    (a) 20.9°
    (b) 30.9°
    (c) 40.9°
    (d) 50.9°

30. A beam of X-ray is incident on a sodium chloride crystal $(a = 2.81\,\text{Å})$. The first-order reflection is observed at a glancing angle 8° 35'. The wavelength of the X-ray used is:
    (a) 3.84 Å
    (b) 2.84 Å
    (c) 1.84 Å
    (d) 0.84 Å

31. A beam of X-ray is incident on a sodium chloride crystal $(a = 2.81\,\text{Å})$. The first-order reflection is observed at a glancing angle 8° 35'. The angle for the second order of reflection is:
    (a) 7.4°
    (b) 17.4°
    (c) 27.4°
    (d) 37.4°

32. A beam of X-ray of wavelength 0.440 Å is used to study the cube face of a rock-salt crystal (d = 2.814 Å). The first-order reflection is observed at a glancing angle of:
    (a) 4° 29'
    (b) 6° 29'
    (c) 8° 29'
    (d) 10° 29'

33. A beam of X-ray of wavelength 0.440 Å is used to study the cube face of a rock-salt crystal (d = 2.814 Å). The third-order reflection is observed at a glancing angle of:
    (a) 33° 34'
    (b) 23° 34'
    (c) 13° 34'
    (d) 3° 34'

34. A beam of X-ray having wavelengths in the range 0.2 Å to 1 Å is allowed to incident at an angle of 9° with the cube face of rock-salt crystal $(d = 2.814\,\text{Å})$. For the diffracted beam, the maximum allowed value of the wavelength is:
    (a) 0.88 Å
    (b) 0.77 Å
    (c) 0.66 Å
    (d) 0.55 Å

35. A beam of X-ray having wavelengths in the range 0.2 Å to 1 Å is allowed to incident at an angle of 9° with the cube face of rock-salt crystal $(d = 2.814 \text{ Å})$. For the diffracted beam, the minimum allowed value of the wavelength is:
    (a) 0.12 Å
    (b) 0.22 Å
    (c) 0.32 Å
    (d) 0.42 Å

36. The angles for the first-order reflection from (100), (110) and (111) face of sodium chloride crystals using monochromatic X-rays are 5.9°, 8.4° and 5.2°, respectively. Nature of the given crystal is:
    (a) S C
    (b) B C C
    (c) F C C
    (d) D C

37. A crystal plane is mounted on an X-ray spectrometer. The glancing angles of incidence for three reflections are 5° 58′, 12° 01′ and 18° 12′, respectively. If the wavelength of the X-rays used is 0.586 Å, the average value of the inter-planar spacing will be:
    (a) 5.817 Å
    (b) 4.817 Å
    (c) 3.817 Å
    (d) 2.817 Å

38. The linear attenuation coefficient for aluminum is 139 m^{-1} for K$_\alpha$ line emitted by a tungsten target in an X-ray tube. The percentage of intensity of this X-ray will pass through 0.005 m thick aluminum sheet is:
    (a) 40%
    (b) 50%
    (c) 60%
    (d) 70%

39. The attenuation percentage in an X-ray beam after passing through a thickness of a material equal to four-half value thickness:
    (a) 93.7%
    (b) 83.7%
    (c) 73.7%
    (d) 63.7%

40. A beam of X-ray consists of equal intensities of wavelengths 0.064 Å and 0.098 Å. When they pass through a piece of lead, their attenuated beam intensity is in the ratio 3:1. The mass absorption coefficients are 0.164 m^2/kg for harder component and 0.35 m^2/kg for softer component, respectively. If the density of lead piece is 11340 kg/m^3, its thickness will be:
    (a) 7.22×10^{-4} m
    (b) 6.22×10^{-4} m
    (c) 5.22×10^{-4} m
    (d) 4.22×10^{-4} m

41. A copper sheet of 1.05 mm thick can reduce the intensity of the X-ray beam to 7.5% of its original value. If the density of copper sheet is 8930 kg/m^3, its mass absorption coefficient will be:
    (a) 3.276 m^2 /kg
    (b) 2.276 m^2 /kg
    (c) 1.276 m^2 /kg
    (d) 0.276 m^2 /kg

42. A beam of neutron with energies ranging from zero to several electron volts is directed at a crystal with interplanar spacing 3.03 Å. In order that the reflected neutrons have a kinetic energy of 0.1 eV, the angle between the incident beam and the crystal is:
    (a) 8° 36′
    (b) 6° 36′
    (c) 4° 36′
    (d) 2° 36'

43. A neutron beam (with neutron rest mass = 1.67×10^{-27} kg) of kinetic energy 0.04 eV is diffracted by the plane (100) of Sylvine crystal ($d_{100} = 3.14$ Å). The glancing angle θ at which the first-order Bragg's spectrum is observed will be:
    (a) 12.16°
    (b) 13.16°
    (c) 14.16°
    (d) 15.16°

44. The energy (in eV) associated with an electron of wavelength 3×10^{-2} m is:
    (a) 3.68×10^{-15} eV
    (b) 2.68×10^{-15} eV
    (c) 1.68×10^{-15} eV
    (d) 0.68×10^{-15} eV

45. The first Bragg's maximum of electron diffraction in a nickel crystal
    (d = 0.4086 Å) is found to occur at a glancing angle of 65°. The de Broglie
    wavelength of the electrons is:
    (a) 6.74 Å
    (b) 4.74 Å
    (c) 2.74 Å
    (d) 0.74 Å

46. The first Bragg's maximum of electron diffraction in a nickel crystal
    (d = 0.4086 Å) is found to occur at a glancing angle of 65°. The velocity of
    electrons will be:
    (a) 9.84×10^6 m/s
    (b) 8.84×10^6 m/s
    (c) 7.84×10^6 m/s
    (d) 6.84×10^6 m/s

47. An electron beam can undergo diffraction by crystal. In order to have its
    wavelength equal to 1.54 Å, the potential through which the beam of electrons
    should be accelerated is:
    (a) 62.6 V
    (b) 63.6 V
    (c) 64.6 V
    (d) 65.6 V

48. Electrons are accelerated to 344 V and are reflected from a crystal. The first
    reflection occurs when the glancing angle is 60°. The interplanar spacing of the
    crystal is:
    (a) 6.38 Å
    (b) 4.38 Å
    (c) 2.38 Å
    (d) 0.38 Å

Answers

1. (a)
2. (d)
3. (b)
4. (c)
5. (a)
6. (b)
7. (d)
8. (c)
9. (a)
10. (b)

11. (c)
12. (d)
13. (a)
14. (b)
15. (c)
16. (d)
17. (a)
18. (b)
19. (c)
20. (d)
21. (a)
22. (b)
23. (c)
24. (d)
25. (a)
26. (b)
27. (c)
28. (d)
29. (a)
30. (d)
31. (b)
32. (a)
33. (c)
34. (a)
35. (b)
36. (c)
37. (d)
38. (b)
39. (a)
40. (c)
41. (d)
42. (a)
43. (b)
44. (c)
45. (d)
46. (a)
47. (b)
48. (d)

Chapter 8
Structure Factor Calculations

8.1 The Structure Factor

We know that the intensity of the diffracted beams depends on atomic scattering factor and the position of each atom in the unit cell. Since, X-rays are scattered by electrons of atom only and not by the nucleus, therefore X-rays scattered from one part of an atom interfere with those scattered from other parts at all angles of scattering (at $2\theta = 0$, all atoms scatter in phase). Further, the diffracted beams are obtained as a result of combination of the scattered waves from the electrons of all the atoms in the unit cell. This process involves two distinct contributions:

Scattering from electrons of the same atom (called the atomic scattering factor or atomic form factor f).

The summation of this scattering from all atoms in the unit cell (called geometrical structure factor F).

The atomic scattering factor is a measure of the efficiency of an atom in scattering X-rays. It is defined as the ratio:

$$f = \frac{\text{amplitude of the X - ray wave scattered by an atom}}{\text{amplitude of the X - ray wave scattered by one electron}}$$

Therefore, if the atoms are assumed to be points only, then the atomic scattering factor is first equal to the number of electrons present (i.e., the atomic number of the neutral atom, Z). The atomic scattering factor of an atom falls off with increasing scattering angle or, more precisely with increasing value of $(\sin\theta/\lambda)$. On the other hand, neutrons are scattered by atomic nuclei rather than electrons around the nucleus. Since, the nucleus is very small as compared to the size of the atom (it can be treated as a point atom), the scattering for a non-vibrating nucleus (because it is very heavy) is almost independent of scattering angle.

In order to know the intensity of the X-ray beam by one unit cell in a particular direction where there is a diffraction maximum, it is necessary to sum the waves

© The Author(s), under exclusive license to Springer Nature Singapore Pte Ltd. 2021   299
M. A. Wahab, *Numerical Problems in Crystallography*,
https://doi.org/10.1007/978-981-15-9754-1_8

that arise from all atoms in the unit cell. Mathematically, this involves adding the waves of the same wavelength but with different amplitudes and phases to get the resultant wave from a unit cell is called the structure factor, F, that is,

$$F(hkl) = f_1 . \exp 2\pi i(hx_1 + ky_1 + lz_1) + f_2 . \exp 2\pi i(hx_2 + ky_2 + lz_2) + \cdots$$

where f_1, f_2, etc., are the atomic scattering factors of atoms with fractional coordinates (x_1, y_1, z_1) and (x_2, y_2, z_2), etc., respectively. This can be rewritten in short as

$$F(hkl) = \sum_1^n f_n . \exp 2\pi i(hx_n + ky_n + lz_n)$$

or

$$F(hkl) = \sum_1^n f_n . \cos 2\pi i(hx_n + ky_n + lz_n) + i \sum_1^n f_n . \sin 2\pi i(hx_n + ky_n + lz_n) \quad (8.1)$$

where n is the number of atoms present in the unit cell.

The absolute value of F, that is, |F| gives the amplitude of the resultant wave and can be defined as:

$$|F| = \frac{\text{amplitude of the wave scattered by all atoms in the unit cell}}{\text{amplitude of the X - ray wave scattered by one electron}}$$

The diffracted beam intensity is proportional to the square of the amplitude, that is,

$$I \propto |F(hkl)|^2 \quad (8.2)$$

where,

$$|F(hkl)|^2 = \left(\sum_1^n f_n . \cos 2\pi i(hx_n + ky_n + lz_n) \right)^2 + \left(\sum_1^n f_n . \sin 2\pi i(hx_n + ky_n + lz_n) \right)^2$$

8.2 Determination of Phase Angle, Amplitude and Atomic Structure Factor

Let us look into the problem in a slightly different way. We know that in general a wave can be represented by a complex function $Ae^{\phi i}$, where the phase angle in radians is ϕ and A is the amplitude of the wave as shown in Fig. 8.1. The identity $e^{\phi i}$ is given by

Fig. 8.1 Simple cubic
structure of Po

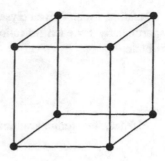

$$e^{\phi i} = \cos\phi + i \sin \phi$$

The structure factor of each atom or ion has two components for each hkl value:

1. Phase angle contains the information only about the position of atom (or ion) in the unit cell; that is, x, y, z, and
2. Amplitude contains information about the scattering of the particular kind of atom (or ion) in the unit cell; for example, a carbon atom.

Phase angle

Phase angle can be determined if the location of the atom in the unit cell is known, while the position of an atom in the unit cell is located by its fractional coordinates x, y, z. For an atom j, the position vector $\vec{r}_j$ is given by

$$\vec{r}_j = x_j\vec{a} + y_j\vec{b} + z_j\vec{c}$$

where $\vec{a}$, $\vec{b}$, and $\vec{c}$ are the basis vectors of the direct lattice. On the other hand, the position of the hkl reflection in the reciprocal lattice is given by

$$H(hkl) = h\,a^* + k\,b^* + l\,c^*$$

where a^*, b^*, and c^* are the basis vectors in the reciprocal lattice. The phase angle of the wave scattered from the atom j in the unit cell is given by the dot product in radians as

$$\phi = 2\pi\,H(hkl).r_j = \left(hx_j + k\,y_j + l\,z_j\right) \tag{8.3}$$

Amplitude

In X-ray diffraction, the amplitude of the scattered wave contains information about the scattering of the particular kind of atom (or ion) in the unit cell; for example, a carbon atom, a potassium atom, etc. When an isolated atom scatters, the atomic scattering factor is a continuous function of $\sin \theta/\lambda$, where λ is the wavelength and θ is any angle between 0 and 90°. The variable $\sin \theta/\lambda$ is conveniently chosen to take the advantage of the situation that the crystal scatters only at the Bragg angle, θ_{hkl}.

Therefore, for any given crystal, the value of the scattering factor must be calculated particularly for each reflection. Rearranging the Bragg's law 2d sin θ = n λ, we obtain

$$\frac{\sin\theta_{hkl}}{\lambda} = \frac{1}{2d_{hkl}}$$

Hence, the atomic scattering factor from the atom j is given by

$$f_j\left(\frac{1}{2d_{hkl}}\right)$$

This is the amplitude of the wave associated with the atom. Finally, the wave scattered from the atom j is given by the complex function as

$$f_j\left(\frac{1}{2d_{hkl}}\right)e^{2\pi iH(hkl).r_j}$$

Atomic Structure Factor

The summation of the atomic scattering factor from all the atoms in the unit cell will provide us the required structure factor for the crystal unit cell. Thus, for each Bragg reflection hkl, the structure factor F(hkl) is given by

$$F(hkl) = \sum_{i=1}^{N} f_j\left(\frac{1}{2d_{hkl}}\right)e^{2\pi iH(hkl).r_j} \tag{8.4}$$

However, it is the D. Cromer and J. Mann (1968) who successfully fitted the atomic scattering factor to a nine-parameter equation and calculated the same as a function (sin θ/λ):

$$F(hkl) = f\left(\frac{\sin\theta_{hkl}}{\lambda}\right) = \sum_{i=1}^{4} a_i\, e^{-b_i\left(\frac{\sin\theta}{\lambda}\right)^2} + c \tag{8.5}$$

where θ is the Bragg angle, λ is the wavelength of the incident X-rays and a_i, b_i (with i = 1 to 4) and c are nine Cromer–Mann coefficients. It is to be noted that sin θ ≤ 1, so that (sin θ/λ) ≤ (1/λ). Table 8.1 provides the Cromer–Mann coefficients for the atomic scattering factors of some elements.

Solved Examples

Example 1 Calculate the phase angle corresponding to atomic position for which hkl = 000.

Solution *Given*: atomic position with hkl = 000, $\phi = ?$
    hkl = 000 represents the undeviated X-ray beam, hence the phase angle

Table 8.1 Cromer–Mann coefficients for the atomic scattering factors of some elements

Element	Z	a_1	a_2	a_3	a_4	b_1	b_2	b_3	b_4	c
O^{2-}	8	4.758	3.637	0	0	7.831	30.05	0	0	1.594
Na^+	11	3.2565	3.9362	1.3998	1.0032	2.6671	6.1153	0.2001	14.039	0.404
$Al^{\beta+}$	13	4.17448	3.3876	1.20296	0.528137	1.93816	4.14553	0.228753	8.28524	0.706786
S	16	6.9053	5.2034	1.4379	1.5863	1.4679	22.2151	0.2536	56.172	0.8669
Cl^-	17	18.2915	7.2084	6.5337	2.3386	0.0066	1.1717	19.5424	60.4486	-16.378
K^+	19	7.9578	7.4917	6.359	1.1915	12.6331	0.7674	-0.002	31.9128	-4.9978
Cr	24	10.6406	7.3537	3.324	1.4922	6.1038	0.392	20.2626	98.7399	1.1832
Cu	29	13.338	7.1676	5.6158	1.6735	3.5828	0.247	11.3966	64.8126	1.191
Zn^{2+}	30	11.9719	7.3862	6.4668	1.394	2.9946	0.2031	7.0826	18.0995	0.7807
Po	84	34.6726	15.4733	13.1138	7.02588	0.700999	3.55078	9.55642	47.0045	13.677

$$\phi = 2\pi\big(H(000).r_j\big) = \Big(0\,x_j + 0\,y_j + 0\,z_j\Big) = 0\,\text{for all atoms.}$$

Example 2 Calculate the phase angle corresponding to the atom at the origin.

Solution *Given*: atomic position with $x_j, y_j, z_j = 0, 0, 0$, (so that $r_j = 0$), $\phi = ?$
The phase angle is given by

$$\phi = 2\pi\big(H(hkl).r_j\big) = (h0 + k0 + l0) = 0\,\text{for jth atom.}$$

Example 3 An atom has the fractional coordinates: 0.4, 0.6, and 0.1. Calculate the phase angle for (212) reflection.

Solution *Given*: atom with the fractional coordinates: $x_j, y_j, z_j = 0.4, 0.6, 0.1$; (hkl) = (212), $\phi = ?$
The phase angle is given by

$$\phi = 2\pi\big(H(hkl).r_j\big) = (2 \times 0.4 + 1 \times 0.6 + 2 \times 0.1)$$
$$= 3.2\,\pi\,\text{radians}$$

Example 4 Determine the general form of structure factor and intensity corresponding to a simple cubic unit cell. Calculate the structure factor for polonium (Po) whose lattice parameter a = 3.359Å and Z = 84 and draw the structure factor curve.

Solution *Given*: Simple cubic unit cell, Po with a = 3.359 Å and Z = 84, $F(hkl) = ?, I = ?$
We know that in a simple cubic system (Fig. 8.1), atoms lie only at the corners of the unit cell. In a cubic unit cell, there are 8 corners and each corner atom contributes 1/8 to the unit cell, therefore, there is only one atom (i.e., $8 \times 1/8 = 1$) per unit cell. This atom is assumed to be present at the origin. Hence, the fractional coordinate of the only atom in the unit cell is (0, 0, 0). Substituting this value in Eq. 8.1 and the resulting value in Eq. 8.2, we obtain

$$F(hkl) = f.[\exp 2\pi i(h.0 + k.0 + l.0)] = f$$
$$\text{and} \qquad I \propto f^2$$

Further, we know that the atomic scattering factor is a function of (sin θ/λ). Therefore, rearranging the Bragg's law 2d sin θ = n λ, we get

$$\frac{\sin\theta_{hkl}}{\lambda} = \frac{1}{2d_{hkl}}$$

Also, for any cubic structure,

$$d_{hkl} = \frac{a}{\left(h^2 + k^2 + l^2\right)^{1/2}}$$

Therefore,

$$\frac{\sin\theta_{hkl}}{\lambda} = \frac{1}{2d_{hkl}} = \frac{\left(h^2 + k^2 + l^2\right)^{1/2}}{2a}$$

and

$$f\left(\frac{\sin\theta_{hkl}}{\lambda}\right) = f\left(\frac{1}{2d_{hkl}}\right) = f\left[\frac{\left(h^2 + k^2 + l^2\right)^{1/2}}{2a}\right]$$

This shows that $\left(\frac{\sin\theta_{hkl}}{\lambda}\right)$ is a function only of hkl values. The structure factor for Po is simplified by substituting $N = 1$ and coordinates of the atom $x_1, y_1, z_1 = 0, 0, 0$, we obtain

$$F(hkl) = \sum_{i=1}^{N} f_{Po}\left(\frac{1}{2d_{hkl}}\right)e^{2\pi iH(hkl)\cdot r_j} = f_{Po}\left(\frac{1}{2d_{hkl}}\right)e^{2\pi i(hx_1 + ky_1 + lz_1)} = f_{Po}\left(\frac{1}{2d_{hkl}}\right)$$

Now, substituting different hkl values along with Cromer–Mann coefficients in Eq. 8.5, the required structure factor can be calculated. The calculation for the first hkl = 001 is

$$F(001) = f(0.1489) = \sum_{i=1}^{4} a_i e^{-b_i(0.1489)^2} + c$$

$$= a_1 e^{-b_1(0.1489)^2} + a_2 e^{-b_2(0.1489)^2} + a_3 e^{-b_3(0.1489)^2} + a_4 e^{-b_4(0.1489)^2} + c$$

$$= 34.6726\,e^{-0.700999(0.1489)^2} + 15.4733\,e^{-3.55078(0.1489)^2} + 13.1138\,e^{-9.55642(0.1489)^2} +$$
$$7.02588\,e^{-47.0045(0.1489)^2} + 13.677$$

$$= 34.13788 + 14.30188 + 10.6099 + 2.47800 + 13.677 = 75.20$$

Calculations for other reflections can be made similarly. They are shown below in tabulated form for the initial 13 hkl values. The atomic scattering curve for Po (z = 84) with 001 and 011 values is shown in Fig. 8.2.

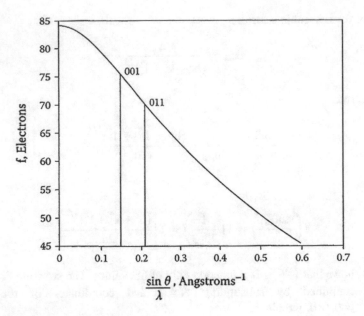

Fig. 8.2 The atomic scattering curve of Po

Structure factor calculation for Po (polonium)

hkl	$\dfrac{(h^2+k^2+l^2)^{1/2}}{2a}$	$F_{Po}(hkl) = f_{Po}, (electron)$
000	0	84
001	0.1489	75.20
011	0.2105	69.97
020	0.2977	63.28
021	0.3328	60.78
111	0.2578	66.24
211	0.3646	58.61
220	0.4210	54.65
221	0.4466	53.49
300	0.4466	53.40
301	0.4707	51.99
311	0.4937	50.69
320	0.5367	48.41
321	0.5570	47.39

Example 5 Determine the general form of structure factor and intensity corresponding to a body-centered cubic unit cell. Calculate the structure factor for Chromium (Cr) whose lattice parameter a = 2.884 Å and Z = 24.

Fig. 8.3 BCC structure of
chromium (Cr)

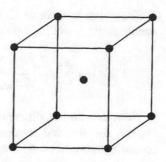

Solution *Given*: Body-centered cubic unit cell, a = 2.884 Å and Z = 24
F(hkl) = ?, I = ?

A simple calculation will give us two atoms in a body-centered cubic unit cell
(Fig. 8.3). The fractional coordinates of atom 1 is (0, 0, 0) and atom 2 is (1/2, 1/2, 1/2), respectively. Substituting these values in Eq. 8.1, we obtain

$$F(hkl) = f.\left[\exp 2\pi i(h.0 + k.0 + l.0) + \exp 2\pi i\left(h.\frac{1}{2} + k.\frac{1}{2} + l.\frac{1}{2}\right)\right]$$

$$= f.[1 + \exp \pi i(h + k + l)]$$

But $\exp \pi i = \cos\pi + i\sin\pi = -1 + 0 = -1$

However in general, $\exp(\pi in) = (-1)^n$, where n is an integer

So that, for a bcc structure

$$F(hkl) = f[1 + \exp \pi i(h + k + l)] = f[1 + 1]$$
$$= 2f, \quad \text{if}(h + k + l) \text{is an even integer}$$

and $I \propto |F(hkl)|^2 = 4f^2$

Similarly,

$$F(hkl) = f[1 - 1] = 0, \quad \text{if } (h + k + l) \text{is an odd integer.}$$

and $I \propto |F(hkl)|^2 = 0$

Thus for a body-centered lattice, the intensities of the diffracted beams coming
from the planes such as (100), (300), (111), (210), (221), etc. whose indices add up
to odd integers are zero, hence absent (this is known as extinction). On the other
hand, the intensities of the diffracted beams coming from the planes such as (110),
(200), (220), (222), etc. whose indices add up to even integers are proportional to
$4f^2$, hence present.

The structure factor for Cr is simplified by substituting N = 2, coordinates of two atoms: $x_1, y_1, z_1 = 0, 0, 0 \cdots$ and $x_2, y_2, z_2 = 1/2, 1/2, 1/2$, we obtain

$$F(hkl) = \sum_{i=1}^{2} f_{Cr}\left(\frac{1}{2d_{hkl}}\right) e^{2\pi i H(hkl) \cdot r_j} = f_{Po}\left(\frac{1}{2d_{hkl}}\right)[1 + \exp\pi i(h+k+l)]$$

Now, substituting different hkl values along with Cromer–Mann coefficients in Eq. 8.5, the required structure factor can be calculated. The calculation for the second value of hkl = 011, when $(h+k+l)$ is an even integer, is

$$F(011) = 2f(0.2452) = 2\sum_{i=1}^{4} a_i e^{-b_i(0.2452)^2} + c$$

$$= 2\left(a_1 e^{-b_1(0.2452)^2} + a_2 e^{-b_2(0.2452)^2} + a_3 e^{-b_3(0.2452)^2} + a_4 e^{-b_4(0.2452)^2} + c\right)$$

$$= 2(10.6406 e^{-6.1038(0.2452)^2} + 7.3537 e^{-0.392(0.2452)^2} + 3.324 e^{-20.2626(0.2452)^2}$$

$$+ 1.4922 e^{-98.7399(0.2452)^2} + 1.1832)$$

$$= 2(7.372065 + 7.18241269 + 0.9830648 + .003941 + 1.1932)$$

$$= 2 \times 16.73468 = 33.46$$

Calculations for other reflections can be made similarly. They are shown below in tabulated form for the initial 12 hkl values.

Structure factor calculation for Cr (Chromium)

hkl	$\dfrac{(h^2 + k^2 + l^2)^{1/2}}{2a}$	$F_{Cr}(hkl) = 2f_{Cr}$, (electron)
000	0	48
001	0.1734	0
011	0.2452	33.46
020	0.3467	27.19
021	0.3877	0
101	0.2452	33.45
111	0.3003	0
120	0.3877	0
121	0.4247	23.32
220	0.4904	20.70
221	0.5201	0
300	0.5201	0
301	0.5482	18/85

Example 6 Determine the general form of structure factor and intensity corresponding to a base-centered unit cell.

Solution *Given*: Base-centered unit cell, $F(hkl) = ?, I = ?$

Fig. 8.4 Base-centered
monoclinic structure

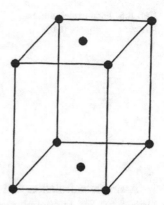

A base-centered unit cell is shown in Fig. 8.4. Each base-centered atom contributes half to the unit cell. Therefore, two base-centered atoms will contribute $(2 \times 1/2 = 1)$ one atom while other atom is contributed from 8 corners. This makes that there are two atoms in the unit cell. The fractional coordinates of atom 1 is (0, 0, 0) and atom 2 is (1/2, 1/2, 0), respectively. Substituting these values in Eq. 8.1, we obtain

$$F(hkl) = f. \left[exp2\pi i(h.0 + k.0 + 1.0) + exp2\pi i \left(h.\frac{1}{2} + k.\frac{1}{2} + 1.0 \right) \right]$$
$$= f.[1 + exp\pi i \ (h+k)]$$

So that, for a base-centered structure

$$F(hkl) = f[1 + exp\pi i(h+k)]$$

It can be seen that if h and k both are odd or both are even integers (i.e., when h and k are unmixed indices), then the sum $(h + k)$ is an even integer. In this case,

$$F(hkl) = 2f \quad and \quad I \propto |F(hkl)|^2 = 4f^2$$

In case of mixed indices (say one is odd and another is even), then the sum $(h + k)$ will be an odd integer. Therefore,

$$F(hkl) = f[1-1] = 0, \ and \ I \ \propto |F(hkl)|^2 = 0$$

Example 7 Determine the general form of structure factor and intensity corresponding to a face-centered cubic unit cell. Calculate the structure factor for Copper (Cu) whose lattice parameter a = 3.615 Å and Z = 29.

Solution *Given*: Face-centered cubic unit cell, $F(hkl) = ?, I = ?$

Fig. 8.5 Face-centered cubic
structure of copper (Cu)

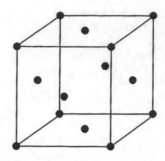

A simple calculation will give us four atoms in a face-centered cubic unit cell
(Fig. 8.5). The fractional coordinates are: (0, 0, 0), (1/2, 1/2, 0), (0, 1/2, 1/2), (1/2,
0, 1/2), respectively. Substituting these values in Eq. 8.1, we obtain

$$F(hkl) = f.\begin{bmatrix} exp2\pi i(h.0 + k.0 + l.0) + exp2\pi i(h.\tfrac{1}{2} + k.\tfrac{1}{2} + l.0) + \\ exp2\pi i(h.0 + k.\tfrac{1}{2} + l.\tfrac{1}{2}) + exp2\pi i(h.\tfrac{1}{2} + k.0 + l.\tfrac{1}{2}) \end{bmatrix}$$

$$= f.[1 + exp\,\pi i(h+k) + exp\,\pi i(k+l) + exp\,\pi i(l+h)]$$

It can be seen that If the reflecting plane indices h, k, l are all odd or all even (i.e.,
unmixed) then the sum (h + k), (k + l) and (l + h) are all even integers and the
structure factor of fcc will be

$$F(hkl) = f[1 + 1 + 1 + 1] = 4f \quad and \quad I \propto |F(hkl)|^2 = 16f^2$$

Further, if indices h, k, l are mixed, it can be seen that

$$F(hkl) = 0, \quad and \quad I \propto |F(hkl)|^2 = 0$$

The structure factor for Cu is simplified by substituting N = 4, coordinates of
four atoms: (0, 0, 0), (1/2, 1/2, 0), (0, 1/2, 1/2), (1/2, 0, 1/2), we obtain

$$F(hkl) = \sum_{i=1}^{4} f_{Cu}\left(\frac{1}{2d_{hkl}}\right)e^{2\pi iH(hkl).r_j}$$

$$= f_{Cu}\left(\frac{1}{2d_{hkl}}\right)[1 + exp\pi i(h+k) + exp\pi i(k+l) + exp\pi i(l+h)]$$

Now, substituting different hkl values along with Cromer–Mann coefficients in
Eq. 8.5, the required structure factor can be calculated. The calculation for the third
value of hkl = 020, when the indices h, k, l are all even, is

$$F(020) = 4f(0.2766) = 4 \sum_{i=1}^{4} a_i e^{-b_i(0.27662)^2} + c$$

$$= 4(a_1 e^{-b_1(0.2766)^2} + a_2 e^{-b_2(0.2766)^2} + a_3 e^{-b_3(0.2766)^2} + a_4 e^{-b_4(0.2766)^2} + c)$$

$$= 4(13.338\, e^{-3.5828(0.2766)^2} + 7.1676\, e^{-0.247(0.2766)^2} + 5.6158\, e^{-11.3966(0.2766)^2}$$

$$+ 1.6735\, e^{-64.8126(0.2766)^2} + 1.191)$$

$$= 4(10.14018 + 7.033423 + 2.34822 + .0117519 + 1.191)$$

$$= 4 \times 20.72457 = 82.89$$

Calculations for other reflections can be made similarly. They are shown below in tabulated form for the initial 13 hkl values.

Structure factor calculation for Cu (Copper)

hkl	$\dfrac{(h^2 + k^2 + l^2)^{1/2}}{2a}$	$F_{Cu}(hkl) = 4f_{Cu}$, (electron)
000	0	116
001	0.1383	0
011	0.1956	0
020	0.2766	82.89
111	0.2396	88.31
120	0.3093	0
121	0.3388	0
220	0.3912	67.13
221	0.4149	0
300	0.4149	0
301	0.4374	0
311	0.4587	59.13
320	0.4987	0
321	0.5175	0

Table 8.2 shows the Miller indices of the planes and the corresponding values of $|F|^2$ for various cubic structures.

Example 8 Cu_3Au is cubic with one Cu_3Au molecule per unit cell. In an ordered form, the atomic positions are: Au: (0, 0, 0), and Cu: (1/2, 1/2, 0), (0, 1/2, 1/2), (1/2, 0, 1/2). In the disordered form, the same positions are occupied at random. Determine the simplified structure factors and the corresponding intensities.

Solution *Given*: Cu_3Au cubic unit cell, $F(hkl) = ?, I = ?$

An ordered form Cu_3Au unit cell is shown in Fig. 8.6. The fractional coordinates are: Au: (0, 0, 0), Cu: (1/2, 1/2, 0), (0, 1/2, 1/2), (1/2, 0, 1/2). Substituting these values in Eq. 8.1, we obtain

Table 8.2 Miller indices of atomic planes and $|F|^2$ for cubic structures

$\sum (h^2 + k^2 + l^2)$	{hkl} planes	$(h^2 + k^2 + l^2)$	Cubic Structures			
			sc	bcc	fcc	dc
1	{100}	$(1^2 + 0^2 + 0^2)$	f^2	–	–	–
2	{110}	$(1^2 + 1^2 + 0^2)$	f^2	$4f^2$	–	–
3	{111}	$(1^2 + 1^2 + 1^2)$	f^2	–	$16f^2$	$64f^2$
4	{200}	$(2^2 + 0^2 + 0^2)$	f^2	$4f^2$	$16f^2$	–
5	{210}	$(2^2 + 1^2 + 0^2)$	f^2	–	–	–
6	{211}	$(2^2 + 1^2 + 1^2)$	f^2	$4f^2$	–	–
7	…					
8	{220}	$(2^2 + 2^2 + 0^2)$	f^2	$4f^2$	$16f^2$	$64f^2$
9	{221}	$(2^2 + 2^2 + 1^2)$	f^2	–	–	–
10	{310}	$(3^2 + 1^2 + 0^2)$	f^2	$4f^2$	–	–
11	{311}	$(3^2 + 1^2 + 1^2)$	f^2	–	$16f^2$	$64f^2$
12	{222}	$(2^2 + 2^2 + 2^2)$	f^2	$4f^2$	$16f^2$	–
13	{320}	$(3^2 + 2^2 + 0^2)$	f^2	–	–	–
14	{321}	$(3^2 + 2^2 + 1^2)$	f^2	$4f^2$	–	–
15	…					
16	{400}	$(4^2 + 0^2 + 0^2)$	f^2	$4f^2$	$16f^2$	$64f^2$
17	{410}	$(4^2 + 1^2 + 0^2)$	f^2	–	–	–
18	{411}	$(4^2 + 1^2 + 1^2)$	f^2	$4f^2$	–	–
19	{331}	$(3^2 + 3^2 + 1^2)$	f^2	–	$16f^2$	$64f^2$
20	{420}	$(4^2 + 2^2 + 0^2)$	f^2	$4f^2$	$16f^2$	–

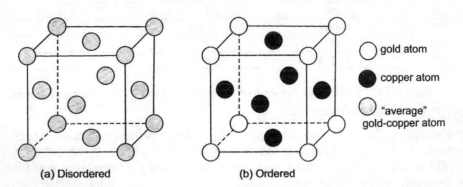

Fig. 8.6 Ordered and disordered form of Cu_3Au cubic structure

$$F(hkl) = f_{Au}.exp2\pi i(h.0 + k.0 + l.0)$$

$$+ f_{Cu}\left[exp2\pi i\left(h.\frac{1}{2} + k.\frac{1}{2} + l.0\right) + exp2\pi i\left(h.0 + k.\frac{1}{2} + l.\frac{1}{2}\right) + exp2\pi i\left(h.\frac{1}{2} + k.0 + l.\frac{1}{2}\right)\right]$$

$$= f_{Au} + f_{Cu}[exp\pi i(h+k) + exp\pi i(k+l) + exp\pi i(l+h)]$$

Exponential terms within the [ ] are identical with the fcc, hence similar criteria will apply. Therefore, when h, k, l are all odd or all even, we obtain

$$F(hkl) = (f_{Au} + 3f_{Cu}) \quad \text{and} \quad I \propto |F(hkl)|^2 = (f_{Au} + 3f_{Cu})^2$$

Similarly, if indices h, k, l are mixed, it can be seen that

$$F(hkl) = (f_{Au} - f_{Cu}), \quad \text{and} \quad I \propto |F(hkl)|^2 = (f_{Au} - f_{Cu})^2$$

In the disordered form, the atomic scattering factor for each atomic site is taken as the weighted average of both the atoms, that is,

$$f_{Av} = \frac{1}{4}(f_{Au} + 3f_{Cu})$$

Hence, the structure factor for h, k, l all odd or all even is given by

$$F(hkl) = 4.\frac{1}{4}(f_{Au} + 3f_{Cu}) = (f_{Au} + 3f_{Cu}) \quad \text{and} \quad I \propto |F(hkl)|^2 = (f_{Au} + f_{Cu})^2$$

and for mixed h, k, l $\quad$ $F(hkl) = 0$

Example 9 Determine the general form of structure factor and intensity corresponding to a diamond cubic unit cell.

Solution *Given*: Diamond cubic unit cell, $F(hkl) = ?, I = ?$

A diamond structure is supposed to be built up from two interpenetrating fcc lattices, which are displaced with respect to one another along the body diagonal of the cube by one-quarter of the length of the diagonal as shown in Fig. 8.7a. Alternatively, it may be considered that each atom in the unit cell to be at the center of a tetrahedron with its four nearest neighbors at the four corners of that tetrahedron as shown in Fig. 8.7b. Accordingly, there are 8 (carbon) atoms in the unit cell. The fractional coordinates are: (0, 0, 0), (1/2, 1/2, 0), (0, 1/2, 1/2), (1/2, 0, 1/2), (1/4, 1/4, 1/4), (3/4, 3/4, 1/4), (1/4, 3/4, 3/4), (3/4, 1/4, 3/4), respectively. Substituting these values in Eq. 8.1, we obtain

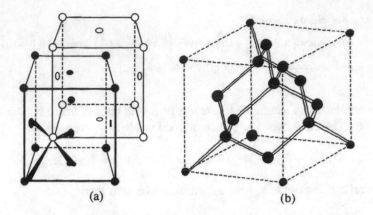

Fig. 8.7 Diamond cubic structure

$$F(hkl) = f_C \cdot \begin{bmatrix} exp2\pi i(h.0+k.0+1.0) + exp2\pi i(h.\frac{1}{2}+k.\frac{1}{2}+1.0) + \\ exp2\pi i(h.0+k.\frac{1}{2}+1.\frac{1}{2}) + exp2\pi i(h.\frac{1}{2}+k.0+1.\frac{1}{2}) + \\ exp2\pi i(h.\frac{1}{4}+k.\frac{1}{4}+1.\frac{1}{4}) + exp2\pi i(h.\frac{3}{4}+k.\frac{3}{4}+1.\frac{1}{4}) + \\ exp2\pi i(h.\frac{1}{4}+k.\frac{3}{4}+1.\frac{3}{4}) + exp2\pi i(h.\frac{3}{4}+k.\frac{1}{4}+1.\frac{3}{4}) \end{bmatrix}$$

$$= f_C \cdot [1 + exp\pi i(h+k) + exp\pi i(k+1) + exp\pi i(1+h)]$$

$$+ exp\frac{\pi i(h+k+1)}{2} f \cdot [1 + exp\pi i(h+k) + exp\pi i(k+1) + exp\pi i(1+h)]$$

$$= f_C \cdot \left(1 + exp\frac{\pi i(h+k+1)}{2}\right) \cdot [1 + exp\pi i(h+k) + exp\pi i(k+1) + exp\pi i(1+h)]$$

The terms within the bracket [ ] are identical as in fcc, hence the same criteria will apply. That is when h, k, l are unmixed indices then the sum $(h + k)$, $(k + l)$ and $(1 + h)$ are all even integers and the structure factor of fcc will be

$$F(hkl) = f_C[1 + 1 + 1 + 1] = 4f_C \quad and \quad I \propto |F(hkl)|^2 = 16f_C^2$$

Further, if indices hkl are mixed, it can be seen that

$$F(hkl) = 0, \quad and \quad I \propto |F(hkl)|^2 = 0$$

Now considering the complete equation for unmixed hkl indices, we can write

$$F(hkl) = 4f_C + \left(1 + exp\frac{\pi i(h+k+1)}{2}\right)$$

Since the exponent contains the term $(h + k + 1)/2$ as a factor of π, which can be a fraction. In order to overcome this difficulty, let us make use of the basic quantum mechanical concept related to the wave function and the amplitude of the wave. Accordingly, for a wave $Ae^{i\varphi}$ and its complex conjugate $Ae^{-i\varphi}$ the intensity (square of the amplitude) is given by

$$A^2 = Ae^{i\varphi} \times Ae^{-i\varphi}$$

Therefore,

$$|F(hkl)|^2 = 4f_C \cdot \left(1 + \exp\frac{\pi i(h+k+1)}{2}\right) \times 4f_C \cdot \left(1 + \exp\frac{-\pi i(h+k+1)}{2}\right)$$

$$= 16f_C^2 \cdot \left\{1 + \exp\frac{\pi i(h+k+1)}{2} + 1 + \exp\frac{-\pi i(h+k+1)}{2}\right\}$$

But, we know that $e^{ix} + e^{-ix} = 2cosx$
Therefore,

$$|F(hkl)|^2 = 16 \, f_C^2 \cdot \left\{2 + 2\cos\frac{\pi(h+k+1)}{2}\right\}$$

$$= 32 \, f_C^2 \cdot \left\{1 + \cos\frac{\pi(h+k+1)}{2}\right\}$$

$$= 32 \, f_C^2 \text{ when} (h+k+1) \text{ is an odd integer}$$

$$= 64 \, f_C^2 \text{ when} (h+k+1) \text{ is an even multiple of } 2$$

$$= 0 \text{ when} (h+k+1) \text{ is an odd multiple of } 2$$

Example 10 Determine the general form of structure factor and intensity corresponding to an hcp (hexagonal close-packed) unit cell.

Solution *Given*: Hexagonal close-packed (hcp) unit cell, $F(hkl) = ?, I = ?$
A simple calculation will give us two atoms in an hcp unit cell (Fig. 8.8). The fractional coordinates of atom 1 is (0, 0, 0) and atom 2 is (1/3, 2/3, 1/2), respectively. Substituting these values in Eq. 8.1, we obtain

$$F(hkl) = f. \left[\exp2\pi i(h.0 + k.0 + 1.0) + \exp2\pi i\left(h.\frac{1}{3} + k.\frac{2}{3} + 1.\frac{1}{2}\right)\right]$$

$$= f. \left[1 + \exp2\pi i\left(h.\frac{1}{3} + k.\frac{2}{3} + 1.\frac{1}{2}\right)\right]$$

In order to solve this, let us put

$$\left[\frac{(h+2k)}{3} + \frac{1}{2}\right] = p$$

Fig. 8.8 Hexagonal closed
packed structure

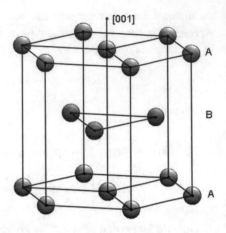

Therefore the structure factor becomes

$$F(hkl) = f.[1 + \exp 2\pi ip]$$

The exponent containing the term p may take fractional values and the expression therefore will remain complex. To overcome this difficulty, let us again make use of the basic quantum mechanical concept related to the wave function and the amplitude of the wave. Accordingly, for a wave $Ae^{i\varphi}$ and its complex conjugate $Ae^{-i\varphi}$ the intensity (square of the amplitude) is given by

$$A^2 = Ae^{i\varphi} \times Ae^{-i\varphi}$$

Therefore,

$$|F(hkl)|^2 = f^2.[1 + \exp(2\pi ip)].[1 + \exp(-2\pi ip)]$$
$$= f^2.[2 + \exp(2\pi ip) + \exp(-2\pi ip)]$$

But, we know that $e^{ix} + e^{-ix} = 2\cos x$

$$|F(hkl)|^2 = f^2.[2 + 2\cos 2\pi p]$$
$$= f^2.[2 + 2(2\cos^2 \pi p - 1)] = f^2.[4\cos^2 \pi p]$$
$$= 4f^2\cos^2 \pi.\left[\frac{(h+2k)}{3} + \frac{1}{2}\right]$$
$$= 0 \text{ when}(h + 2k) \text{ is a multiple of 3 and l is odd.}$$

This indicates that the reflections such as 11.1 (11 $\bar{2}$ 1), 11.3 (11 $\bar{2}$ 3), 22.1 (22 $\bar{4}$ 1), 22.3 (22 $\bar{4}$ 3), etc. are absent in hcp structure. Further, if we assume that (h + 2 k) is a multiple of 3 but l is even, then

$$\left[\frac{(h+2k)}{3} + \frac{1}{2}\right] = n \text{ where n is an integer.}$$

In such a situation,

$$\cos\pi n = \pm 1 \text{ and } \cos^2\pi n = 1$$

Therefore,

$$|F(hkl)|^2 = 4f^2.$$

The results obtained from considering all possible h, k and l values are summarized as follows:

(h + 2k)	l	$|F(hkl)|^2$
3n	Odd	0
3n	Even	$4f^2$
3n ±1	Odd	$3f^2$
3n ±1	Even	f^2

Example 11 Determine the general form of structure factor and intensity corresponding to the unit cell of CsCl structure.

Solution *Given*: Unit cell of CsCl structure, F(hkl) = ?, I = ?

CsCl unit cell contains Cs^+ ions at all the 8 corners of the cube while one Cl^- ion is situated at the body center of the cube (Fig. 8.9). Thus, the structure has one Cs^+ ion and one Cl^- ion per unit cell. The fractional coordinates of Cs ion is (0, 0, 0) and that of the Cl ion is (1/2, 1/2, 1/2), respectively. Substituting these values in Eq. 8.1, we obtain

$$F(hkl) = f_{Cs}.\exp 2\pi i(h.0 + k.0 + 1.0) + f_{Cl}.\exp 2\pi i\left(h.\frac{1}{2} + k.\frac{1}{2} + 1.\frac{1}{2}\right)$$

Fig. 8.9 CsCl crystal structure

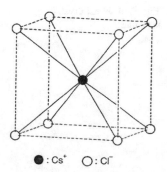

$\bullet$: Cs$^+$ $\bigcirc$: Cl$^-$

where f_{Cs} and f_{Cl} are the scattering factors of Cs^+ and Cl^- ions, respectively. Simplifying the above equation, we obtain

$$F(hkl) = f_{Cs} + f_{Cl}.exp\pi i(h+k+l)$$
$$= f_{Cs} + f_{Cl}, \text{if } (h+k+l) \text{is an even integer,}$$
$$= f_{Cs} - f_{Cl}, \text{if } (h+k+l) \text{is an odd integer}$$

Similarly,

$$|F(hkl)|^2 = \{f_{Cs} + f_{Cl}.\cos\pi(h+k+l)\}^2 + \{f_{Cl}.\sin\pi(h+k+l)\}^2$$
$$= (f_{Cs} + f_{Cl})^2, \text{if } (h+k+l) \text{is an even integer,}$$
$$= (f_{Cs} - f_{Cl})^2, \text{if } (h+k+l) \text{is an odd integer}$$

Although, this structure resembles with bcc, but the results suggest that unlike bcc the reflections in CsCl structure are present (with a diminished intensity) even if (h + k+l) is an odd integer.

Example 12 Determine the general form of structure factor and intensity corresponding to the unit cell of NaCl structure.

Solution *Given*: Unit cell of NaCl structure, $F(hkl) = ?, I = ?$

NaCl structure consists of equal number of sodium and chlorine ions placed at alternate points of a simple cubic lattice as shown in Fig. 8.10. There are four NaCl molecules per unit cell. The fractional coordinates of sodium and chlorine ions are:

$$Na : (0, 0, 0), (1/2, 1/2, 0), (0, 1/2, 1/2), (1/2, 0, 1/2)$$
$$Cl : (1/2, 1/2, 1/2), (0, 0, 1/2), (0, 1/2, 0), (1/2, 0, 0)$$

Now, substituting these values in Eq. 8.1 with the assumption that f_{Na} and f_{Cl} are the scattering factors of Na^+ and Cl^- ions, respectively, we obtain

Fig. 8.10 NaCl crystal structure

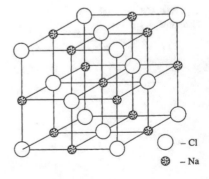

○ – Cl

✳ – Na

$$F(hkl) = f_{Na} \cdot \begin{bmatrix} exp2\pi i(h.0 + k.0 + l.0) + exp2\pi i(h.\frac{1}{2} + k.\frac{1}{2} + l.0) \\ + exp2\pi i(h.0 + k.\frac{1}{2} + l.\frac{1}{2}) + exp2\pi i(h.\frac{1}{2} + k.0 + l.\frac{1}{2}) \end{bmatrix}$$

$$+ f_{Cl} \cdot \begin{bmatrix} exp2\pi i(h.\frac{1}{2} + k.\frac{1}{2} + l.\frac{1}{2}) + exp2\pi i(h.0 + k.0 + l.\frac{1}{2}) \\ + exp2\pi i(h.0 + k.\frac{1}{2} + l.0) + exp2\pi i(h.\frac{1}{2} + k.0 + l.0) \end{bmatrix}$$

$$= f_{Na} \cdot [1 + exp\pi i(h+k) + exp\pi i(k+l) + exp\pi i(l+h)]$$

$$+ exp\pi i(h+k+l) f_{Cl} \cdot [1 + exp - \pi i(h+k) + exp - \pi i(k+l) + exp - \pi i(l+h)]$$

Since $exp(n\pi i) = exp - (n\pi i)$, therefore the above structure factor expression reduces to

$$F(hkl) = (f_{Na} + f_{Cl}.exp\pi i(h+k+l)).[1 + exp\pi i(h+k) + exp\pi i(k+l) + exp\pi i(l+h)]$$

The terms within the bracket [ ] are identical as in fcc, hence the same criteria will apply. That is when h, k, l are unmixed indices then the sum $(h + k)$, $(k + l)$ and $(l + h)$ are all even integers and the structure factor of fcc will be

$$F(hkl) = f[1 + 1 + 1 + 1] = 4f \quad and \quad I \propto |F(hkl)|^2 = 16f^2$$

Further, if indices hkl are mixed, it can be seen that

$$F(hkl) = 0, \quad and \quad I \propto |F(hkl)|^2 = 0$$

Now considering the complete equation for unmixed hkl indices, we can write

$$F(hkl) = 4\{f_{Na} + f_{Cl}.exp\pi i(h+k+l)\}$$
$$= 4(f_{Na} + f_{Cl}), if(h+k+l) is an even integer,$$
$$= 4(f_{Na} - f_{Cl}), if(h+k+l) is an odd integer$$

Similarly,

$$|F(hkl)|^2 = \{f_{Na} + f_{Cl}.cos\pi(h+k+l)\}^2 + \{f_{Cl}.sin\pi(h+k+l)\}^2$$
$$= 16(f_{Na} + f_{Cl})^2, if(h+k+l) is an even integer,$$
$$= 16(f_{Na} - f_{Cl})^2, if (h+k+l) is an odd integer$$

The NaCl structure shows a similar behavior as fcc unit cell. Also, the sodium ions are exactly in phase or exactly out of phase with the chlorine ions. However, when they are out of phase they do not exactly cancel because the atomic scattering factor of chlorine ion is much higher than that of sodium ion. Therefore, the intensities of reflections from some unmixed indices of planes [when $(h + k + l)$ is an even integer] are increased and some others [when $(h + k + l)$ is an odd integer] are decreased.

Example 13 Determine the structure factor and intensity corresponding to the unit cell of ZnS having zinc blende structure.

Solution *Given*: Unit cell of ZnS structure, $F(hkl) = ?$, $I = ?$

When two interpenetrating lattices (as considered for the formation of diamond structure) are of two different elements such as Zn and S, they still produce a tetrahedral arrangement (like carbon atoms in diamond) as shown in Fig. 8.11a. A ZnS (zinc blende) structure has equal number of zinc and sulfur ions distributed on a diamond lattice so that each ion has four opposite ions as the nearest neighbors. The fractional coordinates of Zn and S ions are:

$$Zn : (0,0,0), (1/2,1/2,0), (0,1/2,1/2), (1/2,0,1/2)$$
$$S : (1/4,1/4,1/4), (3/4,3/4,1/4), (1/4,3/4,3/4), (3/4,1/4,3/4)$$

Substituting these values in Eq. 8.1, we obtain structure factor expression similar to that of diamond, that is,

$$
\begin{aligned}
F(hkl) &= f_{Zn}.[1 + \exp\pi i(h+k) + \exp\pi i(k+l) + \exp\pi i(l+h)] \\
&\quad + f_S.\exp\frac{\pi i(h+k+l)}{2}.[1 + \exp\pi i(h+k) + \exp\pi i(k+l) + \exp\pi i(l+h)] \\
&= \left(f_{Zn} + f_S.\exp\frac{\pi i(h+k+l)}{2}\right).[1 + \exp\pi i(h+k) + \exp\pi i(k+l) + \exp\pi i(l+h)]
\end{aligned}
$$

The terms within the bracket [ ] are identical as in fcc and reduces to zero for mixed hkl indices and 4 for unmixed indices. Now considering the complete equation for unmixed hkl indices, we can write

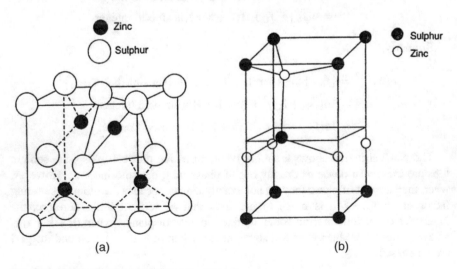

Fig. 8.11 ZnS **a** Zinc blende **b** Wurtzite structure

$$F(hkl) = 4.\left(f_{Zn} + f_S.exp\frac{\pi i(h+k+1)}{2}\right)$$

Now, multiplying the right hand term with its complex conjugate, we can obtain the $|F(hkl)|^2$, that is,

$$|F(hkl)|^2 = 4.\left(f_{Zn} + f_S.exp\frac{\pi i(h+k+1)}{2}\right) \times 4.\left(f_{Zn} + f_S.exp\frac{-\pi i(h+k+1)}{2}\right)$$

On simplification, we obtain

$$|F(hkl)|^2 = 16.\left\{f_{Zn}^2 + f_S^2 + 2f_{Zn}.f_S\cos\frac{\pi(h+k+1)}{2}\right\}$$

$$= 16.\left[f_{Zn}^2 + f_S^2\right] \text{ if } (h+k+1)\text{is an odd integer}$$

$$= 16.\left[f_{Zn}^2 - f_S^2\right]^2 \text{ if } (h+k+1)\text{is an odd multiple of } 2$$

$$= 16.\left[f_{Zn}^2 + f_S^2\right]^2 \text{ if } (h+k+1)\text{is an even multiple of } 2$$

Example 14   A unit cell of tetragonal crystal has its atoms positioned at the following locations:

$$(0, 1/2, 1/4), (1/2, 0, 1/4), (1/2, 0, 3/4), (0, 1/2, 3/4)$$

Determine the corresponding structure factor and intensity.

Solution   *Given*: The fractional coordinates (0, 1/2, 1/4), (1/2, 0, 1/4), (1/2, 0, 3/4), (0, 1/2, 3/4), $F(hkl) = ?, I = ?$

Let f be the scattering factor of the given atom, then the structure factor for the given data is:

$$F(hkl) = f.\begin{bmatrix} exp2\pi i\left(h.0 + k.\frac{1}{2} + 1.\frac{1}{4}\right) + exp2\pi i\left(h.\frac{1}{2} + k.0 + 1.\frac{1}{4}\right) + \\ exp2\pi i\left(h.\frac{1}{2} + k.0 + 1.\frac{3}{4}\right) + exp2\pi i\left(h.0 + k.\frac{1}{2} + 1.\frac{3}{4}\right) \end{bmatrix}$$

$$= f.exp\frac{\pi il}{2}[exp\pi ik + exp\pi ih + exp\pi i(h+1) + exp\pi i(k+1)]$$

Let us consider different cases:

Case I: If h, k, l are mixed indices, such as (110)

$$F(hkl) = f.exp\frac{\pi il}{2}.(exp\pi i + exp\pi i + exp\pi i + exp\pi i)$$

$$= f.(-1 - 1 - 1 - 1) = -4f$$

$$\text{and} \quad |F(hkl)|^2 = 16f^2$$

Case II: If h, k, l are all even unmixed indices, such as (222)

$$F(hkl) = f.exp\pi i.(exp2\pi i + exp2\pi i + exp4\pi i + exp4\pi i)$$
$$= f.(-1)(1 + 1 + 1 + 1) = -4f$$
$$\text{and} \quad |F(hkl)|^2 = 16f^2$$

Case III: If h, k, l are all odd unmixed indices, such as (111)

$$F(hkl) = f.exp\frac{\pi i l}{2}.(exp\pi i + exp\pi i + exp2\pi i + exp2\pi i)$$
$$= f.exp\frac{\pi i l}{2}(-1 - 1 + 1 + 1) = 0$$
$$\text{and} \quad |F(hkl)|^2 = 0$$

Example 15 Determine the structure factor and intensity corresponding to the wurtzite form of ZnS. The crystal is hexagonal and contains two ZnS molecules per unit cell. The Zn and S ions are located at the following positions:

$$Zn : (0, 0, 0), (1/3, 2/3, 1/2,)$$
$$S : (0, 0, 3/8), (1/3, 2/3, 7/8)$$

Solution *Given*: Hexagonal close-packed (wurtzite form of ZnS) unit cell (Fig. 8.11b), 2 ZnS molecules in the unit cell, four fractional coordinates two each for Zn and S, F(hkl) = ?, I = ?

Substituting the given fractional coordinates in Eq. 8.1, we obtain

$$F(hkl) = f_{Zn}.\left[exp2\pi i(h.0 + k.0 + l.0) + exp2\pi i\left(h.\frac{1}{3} + k.\frac{2}{3} + l.\frac{1}{2} \right) \right]$$

$$+ f_S.\left[exp2\pi i\left(h.0 + k.0 + l.\frac{3}{8} \right) + exp2\pi i\left(h.\frac{1}{3} + k.\frac{2}{3} + l.\frac{7}{8} \right) \right]$$

$$= f_{Zn}.\left[1 + exp2\pi i\left\{ \frac{(h + 2k)}{3} \right\}.exp2\pi i.\frac{l}{2} \right]$$

$$+ f_S.\left[exp2\pi i.\frac{3l}{8} + exp2\pi i\left\{ \frac{(h + 2k)}{3} \right\}.exp2\pi i.\frac{7l}{8} \right]$$

$$= f_{Zn}.\left[1 + exp2\pi i\left\{ \frac{(h + 2k)}{3} \right\}.exp2\pi i.\frac{l}{2} \right]$$

$$+ f_S.exp2\pi i.\frac{3l}{8}\left[1 + exp2\pi i\left\{ \frac{(h + 2k)}{3} \right\}.exp2\pi i.\frac{l}{2} \right]$$

$$= 1 + exp2\pi i\left\{ \frac{(h + 2k)}{3} \right\}.exp2\pi i.\frac{l}{2}\left[f_{Zn} + f_S.exp2\pi i.\frac{3l}{8} \right]$$

In order to solve this, let us put

$$\left[\frac{(h+2k)}{3} + \frac{1}{2}\right] = n, \text{where n is an integer including zero}$$

Therefore the structure factor becomes

$$F(hkl) = (1 + \exp2\pi in)\left[f_{Zn} + f_S.\exp2\pi i.\frac{3l}{8}\right]$$

The exponents contain fractional values and the expression therefore will remain complex. To overcome this difficulty, let us simplify this as in case of hcp discussed above, we obtain

$$|F(hkl)|^2 = 4\cos^2\pi n.\left\{f_{Zn}^2 + f_S^2 + 2f_{Zn}.f_S\cos2\pi i.\frac{3l}{8}\right\}$$

$$= 0 \text{ when}(h+2k)\text{is a multiple of 3 and l is odd}$$

Let us further analyze with different values (h + 2 k) and l.

The results obtained from considering all possible h, k and l values are summarized as follows:

(h + 2k)	l	$|F(hkl)|^2$
3n	2p + 1 (as 1, 3, 5, 7 …)	0
3n	8p (as 8, 16, 24…)	$4[f_{Zn} + f_S]^2$
3n	4(2p + 1) (as 4, 12, 20, 28, …)	$4[f_{Zn} - f_S]^2$
3n	2(2p + 1) (as 2, 6, 10, 14, …)	$4[f_{Zn}^2 + f_S^2]$
3n ±1	8p ±1 (as 1, 7, 9, 15, 17…)	$3[f_{Zn}^2 + f_S^2 - \sqrt{2}f_{Zn} f_S]$
3n ±1	4(2p + 1)±1 (as 3, 5, 11, 13, 19, 21 …)	$3[f_{Zn}^2 + f_S^2 + \sqrt{2}f_{Zn}.f_S]$
3n ±1	8p	$[f_{Zn} + f_S]^2$
3n ±1	4(2p + 1)	$[f_{Zn} - f_S]^2$
3n ±1	2(2p + 1)	$[f_{Zn}^2 + f_S^2]$

where p is any integer including zero.

Example 16 Graphite shows hexagonal structure with four atoms per unit cell. Atoms are positioned at the following locations:

$$(0, 0, 0), (1/3, 2/3, 0), (0, 0, 1/2), (2/3, 1/3, 1/2)$$

Show that the structure factor is given by

$$F(hkl) = 4f \cos^2 \pi \left\{ \frac{(h+2k)}{3} \right\} \text{if l is even,}$$

$$= i2f \sin 2\pi \left\{ \frac{(h+2k)}{3} \right\} \text{if l is odd}$$

For what combination h, k, l will the structure factor vanish?

Solution *Given*: Graphite Structure (Fig. 8.12) with fractional coordinates:

$$(0,0,0), (1/3, 2/3, 0), (0, 0, 1/2), (2/3, 1/3, 1/2),$$

Let f be the scattering factor of the carbon atom in graphite, then the structure factor for the given data is:

$$F(hkl) = f. \begin{bmatrix} \exp2\pi i(h.0+k.0+1.0) + \exp2\pi i\left(h.\frac{1}{3}+k.\frac{2}{3}+1.0\right) + \\ \exp2\pi i\left(h.0+k.0+1.\frac{1}{2}\right) + \exp2\pi i\left(h.\frac{2}{3}+k.\frac{1}{3}+1.\frac{1}{2}\right) \end{bmatrix}$$

$$= f. \left[1 + \exp2\pi i \left\{ \frac{(h+2k)}{3} \right\} + \exp2\pi i \frac{1}{2} + \exp2\pi i \left\{ \frac{(2h+k)}{3} \right\} . \exp2\pi i \frac{1}{2} \right]$$

$$= f. \begin{bmatrix} 1 + \exp2\pi i \left\{ \frac{(h+2k)}{3} \right\} + \exp2\pi i \frac{1}{2} + \\ \exp\left\{ -2\pi i \frac{(h+2k)}{3} \right\} . \exp\left\{ 2\pi i \frac{3(h+k)}{3} \right\} . \exp2\pi i \frac{1}{2} \end{bmatrix}$$

Let us consider two different cases:

Case I: when l is even, then the structure factor reduces to

$$F(hkl) = f. \left[1 + \exp\left\{ 2\pi i \frac{(h+2k)}{3} \right\} + 1 + \exp\left\{ -2\pi i \frac{(h+2k)}{3} \right\} . \exp2\pi i(h+k).1 \right]$$

Fig. 8.12 Graphite structure

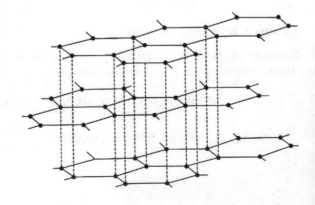

Now, the exponential containing the term (h + k) will become unity for any value h and k. Therefore the structure factor further reduces to

$$F(hkl) = f.\left[2 + \exp 2\pi i\left\{\frac{(h+2k)}{3}\right\} + \exp 2\pi i\left\{-\frac{(h+2k)}{3}\right\}\right]$$

Let us apply the trigonometric function $(e^{ix} + e^{-ix} = 2\cos x)$ to solve the structure factor. Also, consider (h + 2k)/3 = p, then the structure factor reduces to

$$F(hkl) = f.[2 + 2\cos 2\pi p] = 2f\left[1 + 2\cos^2 \pi p - 1\right]$$
$$= 4f\cos^2 \pi p = 4f\cos^2 \pi\left\{\frac{(h+2k)}{3}\right\}$$

Case II: when l is odd, then the structure factor reduces to

$$F(hkl) = f.\left[1 + \exp\left\{2\pi i\frac{(h+2k)}{3}\right\} - 1 + \exp\left\{-2\pi i\frac{(h+2k)}{3}\right\}.\exp 2\pi i(h+k).(-1)\right]$$

Again, the exponential containing the term (h + k) will become unity for any value h and k. Therefore the structure factor further reduces to

$$F(hkl) = f.\left[\exp\left\{2\pi i\frac{(h+2k)}{3}\right\} - \exp\left\{-2\pi i\frac{(h+2k)}{3}\right\}\right]$$

Now, applying the trigonometric function $(e^{ix} - e^{-ix} = 2i\sin x)$, we obtain

$$F(hkl) = i2f\sin 2\pi\left\{\frac{(h+2k)}{3}\right\}$$

Example 17 CaF_2 (Fluorite) is a face-centered cubic with 4 CaF_2 molecules per unit cell. The Ca and F ions are located at the following positions:

$$Ca : (0,0,0), (1/2, 1/2, 0), (0, 1/2, 1/2), (1/2, 0.1/2),$$
$$F : (1/4, 1/4, 1/4), (1/4, 1/4, 3/4), (1/4, 3/4, 1/4), (3/4, 1/4, 1/4),$$
$$(3/4, 3/4, 1/4), (3/4, 1/4, 3/4), (1/4, 3/4, 3/4), (3/4, 3/4, 3/4)$$

Determine the simplified structure factor and evaluate F^2 for (111) and (222) reflections.

Solution *Given*: Unit cell of CaF_2 structure (Fig. 8.13), (hkl) = ?, F^2 for (111) and (222) reflections= ?

Fig. 8.13 CaF$_2$ crystal structure

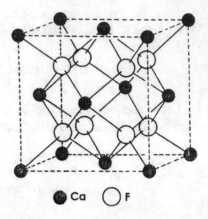

Substituting the values of fractional coordinates in Eq. 8.1, we obtain structure factor expression similar to that of diamond, that is,

$$F(hkl) = f_{Ca}.\left[\begin{array}{l} \exp 2\pi i(h.0 + k.0 + 1.0) + \exp 2\pi i(h.\frac{1}{2} + k.\frac{1}{2} + 1.0) + \\ \exp 2\pi i(h.0 + k.\frac{1}{2} + 1.\frac{1}{2}) + \exp 2\pi i(h.\frac{1}{2} + k.0 + 1.\frac{1}{2}) \end{array}\right]$$

$$+ f_F.\left[\begin{array}{l} \exp 2\pi i(h.\frac{1}{4} + k.\frac{1}{4} + 1.\frac{1}{4}) + \exp 2\pi i(h.\frac{1}{4} + k.\frac{1}{4} + 1.\frac{3}{4}) + \\ \exp 2\pi i(h.\frac{1}{4} + k.\frac{3}{4} + 1.\frac{1}{4}) + \exp 2\pi i(h.\frac{3}{4} + k.\frac{1}{4} + 1.\frac{1}{4}) + \\ \exp 2\pi i(h.\frac{3}{4} + k.\frac{3}{4} + 1.\frac{1}{4}) + \exp 2\pi i(h.\frac{3}{4} + k.\frac{1}{4} + 1.\frac{3}{4}) + \\ \exp 2\pi i(h.\frac{1}{4} + k.\frac{3}{4} + 1.\frac{3}{4}) + \exp 2\pi i(h.\frac{3}{4} + k.\frac{3}{4} + 1.\frac{3}{4}) \end{array}\right]$$

$$= f_{Ca}.[1 + \exp \pi i(h+k) + \exp \pi i(k+1) + \exp \pi i(1+h)]$$

$$+ \exp \frac{\pi i(h+k+1)}{2} f_F.\left[\begin{array}{l} 1 + \exp \pi ih + \exp \pi ik + \exp \pi il] + \\ \exp \frac{3\pi i(h+k+1)}{2} f_F.[1 + \exp \pi i(-h) + \\ \exp \pi i(-k) + \exp \pi i(-l) \end{array}\right]$$

Since, CaF$_2$ (Fluorite) is a face-centered cubic, h, k, and l must be either all odd or all even for an observed reflection; for any other combination, F = 0. Now, let us evaluate F and F^2 for (111) and (222) reflections.

Case I: (111) reflection:
Substituting this value in the final structure factor expression, we have

$$F(111) = 4f_{Ca}$$

$$\text{So that,} \ |F(hkl)|^2 = 16f_{Ca}^2$$

Case II: (222) reflection:
Substituting this value in the final structure factor expression, we have

$$F(222) = 4f_{Ca} - 8f_F$$
$$\text{So that, } |F(hkl)|^2 = 16(f_{Ca} - 2f_F)^2$$

Example 18 Determine the structure factor of a crystal which has a center of inversion and analyze the result.

Solution *Given*: Crystal with center of inversion, $F(hkl) = ?, I = ?$

A crystal with center of inversion can conveniently assumed to be placed at the origin, such that for each atom (x, y, z) in the unit cell, there exists an equivalent atom on the opposite side at (-x, -y,-z). It follows that the structure factor for these two atoms can assume the form:

$$F(hkl) = f.\exp[2\pi i(hx + ky + lz)] + f.\exp[2\pi i(-hx - ky - lz)]$$
$$= f.\exp[2\pi i(hx + ky + lz)] + f.\exp[-2\pi i(hx + ky + lz)]$$

The second term is the complex conjugate of the first, hence the sine terms will get canceled out and the structure factor will reduce to:

$$F(hkl) = 2f\cos 2\pi(hx_i + ky_i + lz_i)$$

It follows from the above that the diffraction pattern from a centrosymmetric crystal is also centrosymmetric. However, even if the crystal does not have a center of symmetry, its diffraction pattern will still be centrosymmetric. This is known as Friedel's law. One can easily show that:

$$I(hkl) = I(\bar{h}\bar{k}\bar{l})$$

We can write the intensity expression for the above simple case as:

$$I(hkl) = F(hkl).F^*(hkl) = f.\exp[2\pi i(hx + ky + lz)].f.\exp[-2\pi i(hx + ky + lz)]$$
$$= f.\exp 2\pi i(hx + ky + lz).f.\exp 2\pi i(\bar{h}x + \bar{k}y + \bar{l}z)$$

Similarly,

$$I(\bar{h}\bar{k}\bar{l}) = F(\bar{h}\bar{k}\bar{l}).F^*(\bar{h}\bar{k}\bar{l}) = f.\exp[2\pi i(\bar{h}x + \bar{k}y + \bar{l}z)].f.\exp[-2\pi i(\bar{h}x + \bar{k}y + \bar{l}z)]$$
$$= f.\exp 2\pi i(\bar{h}x + \bar{k}y + \bar{l}z)f.\exp 2\pi i(hx + ky + lz)$$
$$\Rightarrow F(\bar{h}\bar{k}\bar{l}) = F^*(\bar{h}\bar{k}\bar{l}) \text{ and } F^*(hkl) = F(hkl), \text{ hence } I(hkl) = I(\bar{h}\bar{k}\bar{l})$$

This shows that the intensity contribution from (hkl) and ($\bar{h}\bar{k}\bar{l}$) reflections is the same. This is an important consequence which implies that from a diffraction pattern it is impossible to determine whether or not the crystal has a center of inversion.

Multiple Choice Questions (MCQ)

1. The structure factor corresponding to a simple cubic unit cell is:

   (a) f
   (b) 2f
   (c) 3f
   (d) 4f

2. The intensity corresponding to a simple cubic unit cell is equal to:

   (a) f^2
   (b) $2f^2$
   (c) $3f^2$
   (d) $4f^2$

3. When $(h + k + l)$ is an even integer, the structure factor corresponding to a body-centered cubic unit cell will be:

   (a) f
   (b) 2f
   (c) 3f
   (d) 4f

4. When $(h + k + l)$ is an even integer, the intensity corresponding to a body-centered cubic unit cell is equal to:

   (a) f^2
   (b) $2f^2$
   (c) $3f^2$
   (d) $4f^2$

5. When $(h + k + l)$ is an odd integer, the structure factor corresponding to a body-centered cubic unit cell will be:

   (a) -f
   (b) -2f
   (c) 0
   (d) 2f

6. When $(h + k + l)$ is an odd integer, the intensity corresponding to a body-centered cubic unit cell is equal to:

   (a) $-f^2$
   (b) $2f^2$
   (c) 0
   (d) $4f^2$

7. When $(h + k)$ is an even integer, the structure factor corresponding to a base-centered unit cell will be:

    (a) f
    (b) 2f
    (c) 3f
    (d) 4f

8. When $(h + k)$ is an even integer, the intensity corresponding to a base-centered unit cell is proportional to:

    (a) f^2
    (b) $2f^2$
    (c) $3f^2$
    (d) $4f^2$

9. When the indices h, k, l are all odd or all even, the structure factor corresponding to a face-centered cubic unit cell will be:

    (a) 4f
    (b) 3f
    (c) 2f
    (d) f

10. When the indices h, k, l are all odd or all even, the intensity corresponding to a face-centered cubic unit cell is:

    (a) $20f^2$
    (b) $16f^2$
    (c) $8f^2$
    (d) $4f^2$

11. When the indices h, k, l are mixed, the structure factor corresponding to a face-centered cubic unit cell will be:

    (a) -f
    (b) -2f
    (c) 0
    (d) f

12. When the indices h, k, l are all odd or all even, the intensity corresponding to a face-centered cubic unit cell is:

    (a) $-f^2$
    (b) $-2f^2$
    (c) 0
    (d) $4f^2$

13. Cu_3Au is cubic with one Cu_3Au molecule per unit cell. In an ordered form, the atomic positions are: Au: (0, 0, 0), and Cu: (1/2, 1/2, 0), (0, 1/2, 1/2), (1/2, 0, 1/2). In the disordered form, the same positions are occupied at random. Taking the case of the ordered form and the indices h, k, l are all odd or all even, the simplified structure factor is:

 (a) $(f_{Au} - f_{Cu})$
 (b) $(f_{Au} - 3f_{Cu})$
 (c) $(f_{Au} + f_{Cu})$
 (d) $(f_{Au} + 3f_{Cu})$

14. Cu_3Au is cubic with one Cu_3Au molecule per unit cell. In an ordered form, the atomic positions are: Au: (0, 0, 0), and Cu: (1/2, 1/2, 0), (0, 1/2, 1/2), (1/2, 0, 1/2). In the disordered form, the same positions are occupied at random. Taking the case of the ordered form and the indices h, k, l are mixed, the simplified structure factor is:

 (a) $(f_{Au} - f_{Cu})$
      emContent>
 (b) $(f_{Au} - 3f_{Cu})$
 (c) $(f_{Au} + f_{Cu})$
 (d) $(f_{Au} + 3f_{Cu})$

15. Cu_3Au is cubic with one Cu_3Au molecule per unit cell. In an ordered form, the atomic positions are: Au: (0, 0, 0), and Cu: (1/2, 1/2, 0), (0, 1/2, 1/2), (1/2, 0, 1/2). In the disordered form, the same positions are occupied at random. Taking the case of the ordered form and the indices h, k, l are all odd or all even, the corresponding intensity is:

 (a) $(f_{Au} + f_{Cu})^2$
 (b) $(f_{Au} + 3f_{Cu})^2$
 (c) $(f_{Au} - f_{Cu})^2$
 (d) $(f_{Au} - 3f_{Cu})^2$

16. Cu_3Au is cubic with one Cu_3Au molecule per unit cell. In an ordered form, the atomic positions are: Au: (0, 0, 0), and Cu: (1/2, 1/2, 0), (0, 1/2, 1/2), (1/2, 0, 1/2). In the disordered form, the same positions are occupied at random. Taking the case of the ordered form and the indices h, k, l are mixed, the corresponding intensity is:

 (a) $(f_{Au} + f_{Cu})^2$
 (b) $(f_{Au} + 3f_{Cu})^2$
 (c) $(f_{Au} - f_{Cu})^2$
 (d) $(f_{Au} - 3f_{Cu})^2$

17. Cu_3Au is cubic with one Cu_3Au molecule per unit cell. In an ordered form, the atomic positions are: Au: (0, 0, 0), and Cu: (1/2, 1/2, 0), (0, 1/2, 1/2), (1/2, 0, 1/2). In the disordered form, the same positions are occupied at random. Taking the case of disordered form and the indices h, k, l are all odd or all even, the simplified structure factor is:

   (a) $(f_{Au} - f_{Cu})$
   (b) $(f_{Au} - 3f_{Cu})$
   (c) $(f_{Au} + f_{Cu})$
   (d) $(f_{Au} + 3f_{Cu})$

18. Cu_3Au is cubic with one Cu_3Au molecule per unit cell. In an ordered form, the atomic positions are: Au: (0, 0, 0), and Cu: (1/2, 1/2, 0), (0, 1/2, 1/2), (1/2, 0, 1/2). In the disordered form, the same positions are occupied at random. Taking the case of disordered form and the indices h, k, l are all odd or all even, the corresponding intensity is:

   (a) $(f_{Au} + f_{Cu})^2$
   (b) $(f_{Au} + 3f_{Cu})^2$
   (c) $(f_{Au} - f_{Cu})^2$
   (d) $(f_{Au} - 3f_{Cu})^2$

19. When (h + k), (k + l), (l + h) are all even integers, the structure factor corresponding to a diamond cubic unit cell is:

   (a) $2f_C$
   (b) $4f_C$
   (c) $6f_C$
   (d) $8f_C$

20. When (h + k), (k + l), (l + h) are all even integers, the Intensity corresponding to a diamond cubic unit cell is:

   (a) $2f_C^2$
   (b) $12f_C^2$
   (c) $16f_C^2$
   (d) $20f_C^2$

21. When (h + k + l) is an odd integer, the Intensity corresponding to a diamond cubic unit cell is:

   (a) $8f_C^2$
   (b) $16f_C^2$
   (c) $24f_C^2$
   (d) $32f_C^2$

22. When (h + k + l) is an even multiple of 2, the Intensity corresponding to a diamond cubic unit cell is:

   (a) $64f_C^2$
   (b) $32f_C^2$
   (c) $16f_C^2$
   (d) $8f_C^2$

23. When (h + 2 k) = 3n and l is an even integer, the intensity corresponding to an hcp (hexagonal close packed) unit cell is:

   (a) $2f^2$
   (b) $4f^2$
   (c) $6f^2$
   (d) $8f^2$

24. When (h + 2 k) = 3n ± 1 and l is an odd integer, the intensity corresponding to an hcp (hexagonal close packed) unit cell is:

   (a) $9f^2$
   (b) $6f^2$
   (c) $3f^2$
   (d) f^2

25. When (h + 2 k) = 3n ± 1 and l is an even integer, the intensity corresponding to an hcp (hexagonal close packed) unit cell is:

   (a) $9f^2$
   (b) $6f^2$
   (c) $3f^2$
   (d) f^2

26. When (h + k + l) is an even integer, the structure factor corresponding to the unit cell of CsCl structure is:

   (a) $(f_{Cs} + f_{Cl})$
   (b) $(f_{Cs} - f_{Cl})$
   (c) $(f_{Cs} + 3f_{Cl})$
   (d) $(f_{Cs} + 3f_{Cl})$

27. When (h + k + l) is an odd integer, the structure factor corresponding to the unit cell of CsCl structure is:

   (a) $(f_{Cs} + f_{Cl})$
   (b) $(f_{Cs} + 3f_{Cl})$
   (c) $(f_{Cs} - 3f_{Cl})$
   (d) $(f_{Cs} - 3f_{Cl})$

28. When $(h + k + l)$ is an even integer, the intensity corresponding to the unit cell of CsCl structure is:

    (a) $(f_{Cs} - f_{Cl})^2$
    (b) $(f_{Cs} - 3f_{Cl})^2$
    (c) $(f_{Cs} + 3f_{Cl})^2$
    (d) $(f_{Cs} + 3f_{Cl})^2$

29. When $(h + k + l)$ is an odd integer, the intensity corresponding to the unit cell of CsCl structure is:

    (a) $(f_{Cs} + 3f_{Cl})^2$
    (b) $(f_{Cs} - 3f_{Cl})^2$
    (c) $(f_{Cs} + f_{Cl})^2$
    (d) $(f_{Cs} - f_{Cl})^2$

30. When $(h + k + l)$ is an even integer, the structure factor corresponding to the unit cell of NaCl structure is:

    (a) $4(f_{Na} + f_{Cl})$
    (b) $4(f_{Na} - f_{Cl})$
    (c) $(f_{Na} + f_{Cl})$
    (d) $(f_{Na} - f_{Cl})$

31. When $(h + k + l)$ is an odd integer, the structure factor corresponding to the unit cell of NaCl structure is:

    (a) $4(f_{Na} + f_{Cl})$
    (b) $4(f_{Na} - f_{Cl})$
    (c) $(f_{Na} + f_{Cl})$
    (d) $(f_{Na} - f_{Cl})$

32. When $(h + k + l)$ is an even integer, the intensity corresponding to the unit cell of NaCl structure is:

    (a) $(f_{Na} + f_{Cl})^2$
    (b) $(f_{Na} - f_{Cl})^2$
    (c) $16(f_{Na} + f_{Cl})^2$
    (d) $16(f_{Na} - f_{Cl})^2$

33. When $(h + k + l)$ is an odd integer, the intensity corresponding to the unit cell of NaCl structure is:

    (a) $(f_{Na} + f_{Cl})^2$
    (b) $(f_{Na} - f_{Cl})^2$
    (c) $16(f_{Na} + f_{Cl})^2$
    (d) $16(f_{Na} - f_{Cl})^2$

34. When $(h + k + l)$ is an odd integer, the intensity corresponding to the unit cell of ZnS having zinc blende structure is:

(a) $16[f_{Zn}^2 + f_S^2]$
(b) $16[f_{Zn}^2 - f_S^2]$
(c) $[f_{Zn}^2 + f_S^2]$
(d) $[f_{Zn}^2 - f_S^2]$

35. When $(h + k + l)$ is an odd multiple of 2, the intensity corresponding to the unit cell of ZnS having zinc blende structure is:

(a) $16[f_{Zn}^2 + f_S^2]^2$
(b) $16[f_{Zn}^2 - f_S^2]^2$
(c) $[f_{Zn}^2 + f_S^2]$
(d) $[f_{Zn}^2 - f_S^2]$

36. When $(h + k + l)$ is an even multiple of 2, the intensity corresponding to the unit cell of ZnS having zinc blende structure is:

(a) $[f_{Zn}^2 + f_S^2]$
(b) $[f_{Zn}^2 - f_S^2]$
(c) $16[f_{Zn}^2 + f_S^2]^2$
(d) $16[f_{Zn}^2 - f_S^2]^2$

37. A unit cell of a tetragonal crystal has its atoms positioned at the following locations: (0, 1/2, 1/4), (1/2, 0, 1/4), (1/2, 0, 1/2), (0, 1/2, 3/4). The structure factor corresponding to (110) reflection is:

(a) 4f
(b) 2f
(c) -2f
(d) -4f

38. A unit cell of a tetragonal crystal has its atoms positioned at the following locations: (0, 1/2, 1/4), (1/2, 0, 1/4), (1/2, 0, 1/2), (0, 1/2, 3/4). The intensity corresponding to (110) reflection is:

(a) $20f^2$
(b) $16f^2$
(c) $12f^2$
(d) $8f^2$

39. CaF_2 (Fluorite) is a face-centered cubic with 4 CaF_2 molecules per unit cell. The Ca and F ions are located at the following positions:

$$Ca : (0,0,0), (1/2,1/2,0), (0,1/2,1/2), (1/2,0.1/2),$$
$$F : (1/4,1/4,1/4), (1/4,1/4,3/4), (1/4,3/4,1/4), (3/4,1/4,1/4),$$
$$(3/4,3/4,1/4), (3/4,1/4,3/4), (1/4,3/4,3/4), (3/4,3/4,3/4)$$

The simplified structure factor for (111) reflection is:

(a) f_{Ca}
(b) $2f_{Ca}$
(c) $4f_{Ca}$
(d) $6f_{Ca}$

40. CaF_2 (Fluorite) is a face-centered cubic with 4 CaF_2 molecules per unit cell. The Ca and F ions are located at the following positions:

$$Ca : (0,0,0), (1/2,1/2,0), (0,1/2,1/2), (1/2,0.1/2),$$
$$F : (1/4,1/4,1/4), (1/4,1/4,3/4), (1/4,3/4,1/4), (3/4,1/4,1/4),$$
$$(3/4,3/4,1/4), (3/4,1/4,3/4), (1/4,3/4,3/4), (3/4,3/4,3/4)$$

The intensity corresponding to (111) reflection is:

(a) $4f_{Ca}^2$
(b) $8f_{Ca}^2$
(c) $12f_{Ca}^2$
(d) $16f_{Ca}^2$

41. CaF_2 (Fluorite) is a face-centered cubic with 4 CaF_2 molecules per unit cell. The Ca and F ions are located at the following positions:

$$Ca : (0,0,0), (1/2,1/2,0), (0,1/2,1/2), (1/2,0.1/2),$$
$$F : (1/4,1/4,1/4), (1/4,1/4,3/4), (1/4,3/4,1/4), (3/4,1/4,1/4),$$
$$(3/4,3/4,1/4), (3/4,1/4,3/4), (1/4,3/4,3/4), (3/4,3/4,3/4)$$

The simplified structure factor for (222) reflection is:

(a) $4f_{Ca}$ - $8f_F$
(b) $4f_{Ca}$ - $4f_F$
(c) $8f_{Ca}$ - $8f_F$
(d) $8f_{Ca}$ - $4f_F$

42. CaF_2 (Fluorite) is a face-centered cubic with 4 CaF_2 molecules per unit cell. The Ca and F ions are located at the following positions:

$$Ca : (0,0,0), (1/2,1/2,0), (0,1/2,1/2), (1/2,0.1/2),$$
$$F : (1/4,1/4,1/4), (1/4,1/4,3/4), (1/4,3/4,1/4), (3/4,1/4,1/4),$$
$$(3/4,3/4,1/4), (3/4,1/4,3/4), (1/4,3/4,3/4), (3/4,3/4,3/4)$$

The intensity corresponding to (222) reflection is:

(a) $8(f_{Ca} - 2f_F)^2$
(b) $16(f_{Ca} - 2f_F)^2$
(c) $24(f_{Ca} - 2f_F)^2$
(d) $32(f_{Ca} - 2f_F)^2$

Answers

1. (a)
2. (a)
3. (b)
4. (d)
5. (c)
6. (c)
7. (b)
8. (d)
9. (a)
10. (b)
11. (c)
12. (c)
13. (d)
14. (a)
15. (b)
16. (c)
17. (d)
18. (a)
19. (b)
20. (c)
21. (d)
22. (a)
23. (b)
24. (c)
25. (d)
26. (a)
27. (b)
28. (c)
29. (d)
30. (a)
31. (b)
32. (c)
33. (d)
34. (a)
35. (b)
36. (c)
37. (d)
38. (b)
39. (c)
40. (d)
41. (a)
42. (b)

Chapter 9
Determination of Crystal Structure Parameters

9.1 Steps in Crystal Structure Determinations

Crystal structure determinations in general are not easy and straightforward. They are normally done by trial and error method and taking into account the following three fundamental steps.

1. Indexing (or analysis) of diffraction pattern through the analytical or graphical method to know the nature (shape and size) of the unit cell.
2. Determination of number of atoms in the unit cell from its shape and size, chemical composition and density of the specimen.
3. Determination of atomic positions in the unit cell from suitable analysis of the relative intensities obtained from diffraction experiments.

(a) Interpretation of X-ray Diffraction Data For Cubic Structures

One can use a simple method to identify whether the given (unknown) structure is either bcc or fcc, when X-ray diffraction data is available. We can identify the principal diffracting planes {hkl} and the corresponding 2θ values as given in Table 9.1. Further, for cubic structures the Bragg's law gives us

$$\sin^2\theta = \frac{\lambda^2}{4a^2}\left(h^2 + k^2 + l^2\right)$$

where the lattice parameter, a and the wavelength of the X-ray beam, λ are constants. Eliminating these quantities by taking the ratio of two $\sin^2\theta$ values, we obtain

$$\frac{\sin^2\theta_A}{\sin^2\theta_B} = \frac{h_A^2 + k_A^2 + l_A^2}{h_B^2 + k_B^2 + l_B^2}$$

© The Author(s), under exclusive license to Springer Nature Singapore Pte Ltd. 2021
M. A. Wahab, *Numerical Problems in Crystallography*,
https://doi.org/10.1007/978-981-15-9754-1_9

Table 9.1 Miller indices of diffracting planes for cubic structures

$\sum(h^2+k^2+l^2)$	{hkl} planes	$(h^2+k^2+l^2)$	Diffracting {hkl} planes			
			sc	bcc	fcc	dc
1	{100}	$(1^2+0^2+0^2)$	100	–	–	–
2	{110}	$(1^2+1^2+0^2)$	110	110	–	–
3	{111}	$(1^2+1^2+1^2)$	111	–	111	111
4	{200}	$(2^2+0^2+0^2)$	200	200	200	–
5	{210}	$(2^2+1^2+0^2)$	210	–	–	–
6	{211}	$(2^2+1^2+1^2)$	211	211	–	–
7	...	...	...	...	...	...
8	{220}	$(2^2+2^2+0^2)$	220	220	220	220
9	{221}	$(2^2+2^2+1^2)$	221	–	–	–
10	{310}	$(3^2+1^2+0^2)$	310	310	–	–
11	{311}	$(3^2+1^2+1^2)$	311	–	311	311
12	{222}	$(2^2+2^2+2^2)$	222	222	222	–
13	{320}	$(3^2+2^2+0^2)$	320	–	–	–
14	{321}	$(3^2+2^2+1^2)$	321	321	–	–
15	...	...	...	...	...	...
16	{400}	$(4^2+0^2+0^2)$	400	400	400	400
17	{410}	$(4^2+1^2+0^2)$	410	–	–	–
18	{411}	$(4^2+1^2+1^2)$	411	411	–	–
19	{331}	$(3^2+3^2+1^2)$	331	–	331	331
20	{420}	$(4^2+2^2+0^2)$	420	420	420	–

where θ_A and θ_B are two diffracting angles associated with the principal diffracting planes $\{h_A\,k_A\,l_A\}$ and $\{h_B\,k_B\,l_B\}$, respectively.

Case I From Table 9.1, we observe that the first two principal diffracting planes for bcc structure are (110) and (200). Substituting the indices of these planes in the above equation, we obtain

$$\frac{\sin^2\theta_A}{\sin^2\theta_B} = \frac{1^2+1^2+0^2}{2^2+0^2+0^2} = 0.5$$

Thus, if the crystal structure of the unknown cubic metal is bcc, the ratio of $\sin^2\theta$ values corresponding to the two principal diffracting planes will be 0.5.

Case II From Table 9.1, we observe that the first two principal diffracting planes for fcc structure are (111) and (200). Substituting the indices of these planes in the above equation, we obtain

$$\frac{\sin^2\theta_A}{\sin^2\theta_B} = \frac{1^2 + 1^2 + 1^2}{2^2 + 0^2 + 0^2} = 0.75$$

Thus, if the crystal structure of the unknown cubic metal is fcc, the ratio of $\sin^2\theta$ values corresponding to the two principal diffracting planes will be 0.75.

(b) X-ray Diffraction Methods

(i) Powder Method

The Bragg angle θ and the distance between the pair of arcs S on the powder pattern (Fig. 9.1) are related through

$$\theta = \frac{S}{4}$$

The extinction rules for cubic crystals are provided in Table 9.2.

From these extinction rules, it is possible to derive the ratios of $(h^2 + k^2 + l^2)$ for allowed reflections in the following cubic lattices. They are given as:

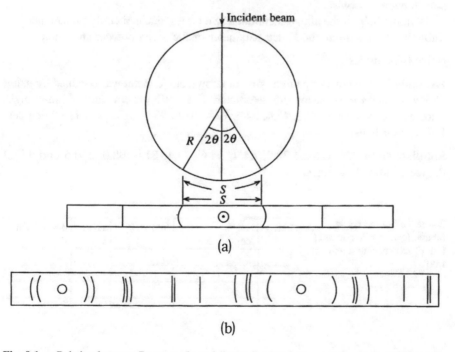

(a)

(b)

Fig. 9.1 a Relation between Bragg angle and linear distance between pair of arcs **b** Schematic powder pattern

Table 9.2 Extinction rules for cubic crystals

Crystals	Allowed reflections
SC	for all values of $(h^2 + k^2 + l^2)$
BCC	for even values of (h + k + l)
FCC	when h, k and l are all odd or all even
DC	when h, k and l are all odd or all even, (h + k +l) should be divisible by four

$$SC \quad 1:2:3:4:5:6:8:9:10:11:12:13:14$$
$$BCC \quad 2:4:6:8:10:12:14:16$$
$$FCC \quad 3:4:8:11:12:16:19:20$$
$$DC \quad 3:8:11:16:19$$

In order to obtain the interplanar spacing/lattice parameter or the Miller indices of possible diffracting planes of the given powder specimen, we need to analyze the corresponding X-ray diffraction data. Depending on the nature of the problem, the datasheets can be prepared in the tabulated form. For the benefit of the readers, a few sample tables are provided below for obtaining interplanar spacing and hkl values, etc., in Tables 9.3, 9.4 and 9.5, respectively. However, problems can also be solved even otherwise.

With the help of the above-mentioned two tables, one can easily prepare another Table 9.5 to determine the lattice parameter of the given powder specimen.

Solved Examples

Example 1 The powder pattern of an unknown cubic metal was obtained by using CuKα radiation in an X-ray diffractometer. The XRD pattern showed initial eight lines with 2θ values of 32.2, 44.4, 64.6, 77.4, 81.6, 98.0, 110.6 and 115.0 degrees. Index these lines.

Solution *Given*: 2θ values: 38.2, 44.4, 64.6, 77.4, 81.6, 98.0, 110.6 and 115.0 degrees. Index these lines.

Table 9.3 Datasheet in tabular form for determining interplanar spacing **d** and (hkl)

Line	Arc length = S	$\theta(°) = \frac{S}{4}$	sin θ	d = λ/2 sin θ
1				
2				
3				
...				

Table 9.4 Datasheet in tabular form for determining hkl values

Line	$\theta(°) = S/4$	$\sin\theta$	$\sin^2\theta$	Common factor (c.f)	$N = \frac{\sin^2\theta}{c.f}$	(hkl)
1						
2						
3						
...						

Table 9.5 Datasheet in tabular form for determining lattice parameter **a**

Line	θ (°)	d(Å)	$N = \frac{\sin^2\theta}{c.f}$	$\sqrt{N}$	$a = d\sqrt{N}$
1					
2					
3					
...					

Preparing the datasheet in tabular form, we have

Line	$2\theta(°)$	$\theta(°)$	$\sin\theta$	$\sin^2\theta$	Common factor (c.f)	$N = \frac{\sin^2\theta}{c.f}$	hkl
1	38.2	19.1	0.327	0.107		3	111
2	44.4	22.2	0.378	0.143		4	200
3	64.6	32.3	0.534	0.285		8	220
4	77.4	38.7	0.625	0.390	0.0356	11	311
5	81.6	40.8	0.653	0.426		12	222
6	98.0	49.0	0.755	0.570		16	400
7	110.6	55.3	0.822	0.676		19	331
8	115.0	57.5	0.843	0.711		20	420

From the values of N, the unknown cubic metal is found to have fcc structure. All hkl values are provided in the last column of the table.

Example 2 Determine the (hkl) and 2θ values of the first four lines on the powder patterns obtained by using CuKα radiation of wavelength 1.54 Å corresponding to a simple cubic structure with a = 4 Å.

Solution *Given*: λ = 1.54 Å, simple cubic structure with a = 4 Å, (hkl) and 2θ values of the first four lines = ?

We know that the Bragg's equation is given by

$$2d\sin\theta = n\lambda \tag{i}$$

For a cubic structure, the relationship between interplanar spacing "d" and lattice parameter "a" is

$$d_{hkl} = \frac{a}{\left(h^2 + k^2 + l^2\right)^{1/2}} \tag{ii}$$

Combining these two equations, we have

$$\sin \theta = \frac{\lambda}{2d_{hkl}} = \frac{\left(h^2 + k^2 + l^2\right)^{1/2}}{2a}$$

From Table 9.1, we know that the first four lines and their corresponding planes for simple cubic structure are: 1(100), 2(110), 3(111) and 4(200).

Therefore, the 2θ angles corresponding to these (hkl) planes are

$$\sin \theta_{100} = \frac{\lambda}{2d_{100}} = \frac{\left(1^2 + 0^2 + 0^2\right)^{1/2}}{2a} = \frac{1 \times 1.54}{2 \times 4}$$

$$\text{or} \quad 2\theta_{100} = 2 \sin^{-1}\left(\frac{\lambda}{2d_{100}}\right) = 2 \sin^{-1}\left(\frac{1 \times 1.54}{2 \times 4}\right) = 22.2°$$

Similarly,

$$\sin \theta_{110} = \frac{\lambda}{2d_{110}} = \frac{\left(1^2 + 1^2 + 0^2\right)^{1/2}}{2a} = \frac{\sqrt{2} \times 1.54}{2 \times 4}$$

$$\text{or} \quad 2\theta_{110} = 2 \sin^{-1}\left(\frac{\lambda}{2d_{110}}\right) = 2 \sin^{-1}\left(\frac{\sqrt{2} \times 1.54}{2 \times 4}\right) = 31.6°$$

$$\sin \theta_{111} = \frac{\lambda}{2d_{111}} = \frac{\left(1^2 + 1^2 + 1^2\right)^{1/2}}{2a} = \frac{\sqrt{3} \times 1.54}{2 \times 4}$$

$$\text{or} \quad 2\theta_{111} = 2\sin^{-1}\left(\frac{\lambda}{2d_{111}}\right) = 2 \sin^{-1}\left(\frac{\sqrt{3} \times 1.54}{2 \times 4}\right) = 39.0°$$

and

$$\sin \theta_{200} = \frac{\lambda}{2d_{200}} = \frac{\left(2^2 + 0^2 + 0^2\right)^{1/2}}{2a} = \frac{2 \times 1.54}{2 \times 4}$$

$$\text{or} \quad 2\theta_{200} = 2 \sin^{-1}\left(\frac{\lambda}{2d_{200}}\right) = 2 \sin^{-1}\left(\frac{2 \times 1.54}{2 \times 4}\right) = 45.3°$$

Example 3 Determine the (hkl) and 2θ values of the first four lines on the powder patterns obtained by using CuKα radiation of wavelength 1.54 Å corresponding to a body-centered cubic structure with a = 4 Å.

Solution *Given*: λ = 1.54 Å, body-centered cubic structure with a = 4 Å, (hkl) and 2θ values of the first four lines = ?

We know that the Bragg's equation is given by

$$2d \sin \theta = n\lambda \qquad \text{(i)}$$

For a cubic structure, the relationship between interplanar spacing "d" and lattice parameter "a" is

$$d_{hkl} = \frac{a}{\left(h^2 + k^2 + l^2\right)^{1/2}} \qquad \text{(ii)}$$

Combining these two equations, we have

$$\sin \theta = \frac{\lambda}{2d_{hkl}} = \frac{\left(h^2 + k^2 + l^2\right)^{1/2}}{2a}$$

From Table 9.1, we know that the first four lines and their corresponding planes for body-centered cubic structure are: 2 (110), 4 (200), 6 (211) and 8 (220). Therefore, the 2θ angles corresponding to these (hkl) planes are

$$\sin \theta_{110} = \frac{\lambda}{2d_{110}} = \frac{\left(1^2 + 1^2 + 0^2\right)^{1/2}}{2a} = \frac{\sqrt{2} \times 1.54}{2 \times 4}$$

$$\text{or} \quad 2\theta_{110} = 2 \sin^{-1}\left(\frac{\lambda}{2d_{110}}\right) = 2 \sin^{-1}\left(\frac{\sqrt{2} \times 1.54}{2 \times 4}\right) = 31.6°$$

Similarly,

$$\sin \theta_{200} = \frac{\lambda}{2d_{200}} = \frac{\left(2^2 + 0^2 + 0^2\right)^{1/2}}{2a} = \frac{2 \times 1.54}{2 \times 4}$$

$$\text{or} \quad 2\theta_{200} = 2 \sin^{-1}\left(\frac{\lambda}{2d_{200}}\right) = 2\sin^{-1}\left(\frac{2 \times 1.54}{2 \times 4}\right) = 45.3°$$

$$\sin \theta_{211} = \frac{\lambda}{2d_{211}} = \frac{\left(2^2 + 1^2 + 1^2\right)^{1/2}}{2a} = \frac{\sqrt{6} \times 1.54}{2 \times 4}$$

$$\text{or} \quad 2\theta_{211} = 2 \sin^{-1}\left(\frac{\lambda}{2d_{211}}\right) = 2 \sin^{-1}\left(\frac{\sqrt{6} \times 1.54}{2 \times 4}\right) = 56.3°$$

and

$$\sin \theta_{220} = \frac{\lambda}{2d_{220}} = \frac{\left(2^2 + 2^2 + 0^2\right)^{1/2}}{2a} = \frac{\sqrt{8} \times 1.54}{2 \times 4}$$

$$\text{or} \quad 2\theta_{220} = 2 \sin^{-1}\left(\frac{\lambda}{2d_{220}}\right) = 2 \sin^{-1}\left(\frac{\sqrt{8} \times 1.54}{2 \times 4}\right) = 66.0°$$

Example 4 Determine the (hkl) and 2θ values of the first four lines on the powder patterns obtained by using $CuK\alpha$ radiation of wavelength 1.54 Å corresponding to a face-centered cubic structure with a = 4 Å.

Solution *Given*: λ = 1.54 Å, face-centered cubic structure with a = 4 Å, (hkl) and 2θ values of the first four lines = ?

We know that the Bragg's equation is given by

$$2d \sin \theta = n\lambda \qquad (i)$$

For a cubic structure, the relationship between interplanar spacing "d" and lattice parameter "a" is

$$d_{hkl} = \frac{a}{\left(h^2 + k^2 + l^2\right)^{1/2}} \qquad (ii)$$

Combining these two equations, we have

$$\sin \theta = \frac{\lambda}{2d_{hkl}} = \frac{\left(h^2 + k^2 + l^2\right)^{1/2}}{2a}$$

From Table 9.1, we know that the first four lines and their corresponding planes for face-centered cubic structure are: 3 (111), 4 (200), 8 (220) and 11 (311).

Therefore, the 2θ angles corresponding to these (hkl) planes are

$$\sin \theta_{111} = \frac{\lambda}{2d_{111}} = \frac{\left(1^2 + 1^2 + 1^2\right)^{1/2}}{2a} = \frac{\sqrt{3} \times 1.54}{2 \times 4}$$

$$\text{or} \quad 2\theta_{111} = 2 \sin^{-1}\left(\frac{\lambda}{2d_{111}}\right) = 2 \sin^{-1}\left(\frac{\sqrt{3} \times 1.54}{2 \times 4}\right) = 39.0°$$

Similarly,

$$\sin \theta_{200} = \frac{\lambda}{2d_{200}} = \frac{\left(2^2 + 0^2 + 0^2\right)^{1/2}}{2a} = \frac{2 \times 1.54}{2 \times 4}$$

$$\text{or} \quad 2\theta_{200} = 2 \sin^{-1}\left(\frac{\lambda}{2d_{200}}\right) = 2 \sin^{-1}\left(\frac{2 \times 1.54}{2 \times 4}\right) = 45.3°$$

$$\sin \theta_{220} = \frac{\lambda}{2d_{220}} = \frac{\left(2^2 + 2^2 + 0^2\right)^{1/2}}{2a} = \frac{\sqrt{8} \times 1.54}{2 \times 4}$$

$$\text{or} \quad 2\theta_{220} = 2 \sin^{-1}\left(\frac{\lambda}{2d_{220}}\right) = 2 \sin^{-1}\left(\frac{\sqrt{8} \times 1.54}{2 \times 4}\right) = 66.0°$$

and

$$\sin \theta_{311} = \frac{\lambda}{2d_{311}} = \frac{(3^2 + 1^2 + 1^2)^{1/2}}{2a} = \frac{\sqrt{11} \times 1.54}{2 \times 4}$$

$$\text{or} \quad 2\theta_{311} = 2\sin^{-1}\left(\frac{\lambda}{2d_{311}}\right) = 2\sin^{-1}\left(\frac{\sqrt{11} \times 1.54}{2 \times 4}\right) = 79.4°$$

Example 5 Determine the (hkl) and 2θ values of the first four lines on the powder patterns obtained by using CuKα radiation of wavelength 1.54 Å corresponding to a simple tetragonal structure with a = b = 4 Å and c = 3 Å.

Solution *Given*: λ = 1.54 Å, simple tetragonal structure with a = b = 4 Å, c = 3 Å, (hkl) and 2θ values of the first four lines = ?

We know that the Bragg's equation is given by

$$2d \sin \theta = n\lambda \tag{i}$$

For a tetragonal structure, the relationship between interplanar spacing 'd' and lattice parameters "a" and "b" is

$$d_{hkl} = \left[\frac{h^2 + k^2}{a^2} + \frac{l^2}{c^2}\right]^{-1/2} \tag{ii}$$

Combining these two equations, we have

$$\sin \theta = \frac{\lambda}{2d_{hkl}} = \frac{\lambda}{2} \cdot \left[\frac{h^2 + k^2}{a^2} + \frac{l^2}{c^2}\right]^{1/2}$$

We know that the structural conditions for tetragonal structure are a = b ≠ c and α = β = γ = 90°, therefore the first four lines and their corresponding planes are: 1 (100). 1 (001). 2 (110) and 2 (011)

Therefore, the 2θ angles corresponding to these (hkl) planes are

$$\sin \theta_{100} = \frac{\lambda}{2d_{100}} = \frac{\lambda}{2} \cdot \left[\frac{1^2 + 0^2}{4^2} + \frac{0^2}{3^2}\right]^{1/2} = \frac{1.54 \times 1}{2 \times 4}$$

$$\text{or} \quad 2\theta_{100} = 2\sin^{-1}\left(\frac{\lambda}{2d_{100}}\right) = 2\sin^{-1}\left(\frac{1.54 \times 1}{2 \times 4}\right) = 22.2°$$

Similarly,

$$\sin \theta_{001} = \frac{\lambda}{2d_{001}} = \frac{\lambda}{2} \cdot \left[\frac{0^2 + 0^2}{4^2} + \frac{1^2}{3^2} \right]^{1/2} = \frac{1.54 \times 1}{2 \times 3}$$

$$\text{or} \quad 2\theta_{001} = 2\sin^{-1}\left(\frac{\lambda}{2d_{001}} \right) = 2\sin^{-1}\left(\frac{1.542}{2 \times 3} \right) = 29.7°$$

$$\sin \theta_{110} = \frac{\lambda}{2d_{110}} = \frac{\lambda}{2} \cdot \left[\frac{1^2 + 1^2}{4^2} + \frac{0^2}{3^2} \right]^{1/2} = \frac{1.54 \times \sqrt{2}}{2 \times 4}$$

$$\text{or} \quad 2\theta_{110} = 2\sin^{-1}\left(\frac{\lambda}{2d_{110}} \right) = 2\sin^{-1}\left(\frac{1.54 \times \sqrt{2}}{2 \times 4} \right) = 31.6°$$

and

$$\sin \theta_{011} = \frac{\lambda}{2d_{011}} = \frac{\lambda}{2} \cdot \left[\frac{0^2 + 1^2}{4^2} + \frac{1^2}{3^2} \right]^{1/2} = \frac{1.54 \times 5}{2 \times 4 \times 3}$$

$$\text{or} \quad 2\theta_{011} = 2\sin^{-1}\left(\frac{\lambda}{2d_{011}} \right) = 2\sin^{-1}\left(\frac{1.54 \times 5}{2 \times 4 \times 3} \right) = 37.5°$$

Example 6 Determine the (hkl) and 2θ values of the first four lines on the powder patterns obtained by using CuKα radiation of wavelength 1.54 Å corresponding to an orthorhombic structure with a = 2 Å, b = 3 Å and c = 4 Å.

Solution *Given*: λ = 1.54 Å, orthorhombic structure with a = 2 Å, b = 3 Å, c = 4 Å, (hkl) and 2θ values of the first four lines = ?
We know that the Bragg's equation is given by

$$2d \sin \theta = n\lambda \tag{i}$$

For an orthorhombic structure, the relationship between interplanar spacing 'd' and lattice parameters "a," "b" and "c" is

$$d_{hkl} = \left[\frac{h^2}{a^2} + \frac{k^2}{b^2} + \frac{l^2}{c^2} \right]^{-1/2} \tag{ii}$$

Combining these two equations, we have

$$\sin \theta = \frac{\lambda}{2d_{hkl}} = \frac{\lambda}{2} \cdot \left[\frac{h^2}{a^2} + \frac{k^2}{b^2} + \frac{l^2}{c^2} \right]^{1/2}$$

We know that the structural conditions for a tetragonal structure are a $\neq$ b $\neq$ c and $\alpha = \beta = \gamma = 90°$, therefore the first four lines and their corresponding planes are: 1 (100), 1 (010). 1 (001) and 2 (110)

Therefore, the 2θ angles corresponding to these (hkl) planes are

$$\sin\theta_{100} = \frac{\lambda}{2d_{100}} = \frac{\lambda}{2} \cdot \left[\frac{1^2}{2^2} + \frac{0^2}{3^2} + \frac{0^2}{4^2}\right]^{1/2} = \frac{1.54 \times 1}{2 \times 2}$$

$$\text{or} \quad 2\theta_{100} = 2\sin^{-1}\left(\frac{\lambda}{2d_{100}}\right) = 2\sin^{-1}\left(\frac{1.54 \times 1}{2 \times 2}\right) = 45.3°$$

Similarly,

$$\sin\theta_{010} = \frac{\lambda}{2d_{010}} = \frac{\lambda}{2} \cdot \left[\frac{0^2}{2^2} + \frac{1^2}{3^2} + \frac{0^2}{4^2}\right]^{1/2} = \frac{1.54 \times 1}{2 \times 3}$$

$$\text{or} \quad 2\theta_{010} = 2\sin^{-1}\left(\frac{\lambda}{2d_{010}}\right) = 2\sin^{-1}\left(\frac{1.54 \times 1}{2 \times 3}\right) = 29.7°$$

$$\sin\theta_{001} = \frac{\lambda}{2d_{001}} = \frac{\lambda}{2} \cdot \left[\frac{0^2}{2^2} + \frac{0^2}{3^2} + \frac{1^2}{4^2}\right]^{1/2} = \frac{1.54 \times 1}{2 \times 4}$$

$$\text{or} \quad 2\theta_{001} = 2\sin^{-1}\left(\frac{\lambda}{2d_{001}}\right) = 2\sin^{-1}\left(\frac{1.54 \times 1}{2 \times 4}\right) = 22.2°$$

and

$$\sin\theta_{110} = \frac{\lambda}{2d_{110}} = \frac{\lambda}{2} \cdot \left[\frac{1^2}{2^2} + \frac{1^2}{3^2} + \frac{0^2}{4^2}\right]^{1/2} = \frac{1.54 \times \sqrt{13}}{2 \times 6}$$

$$\text{or} \quad 2\theta_{110} = 2\sin^{-1}\left(\frac{\lambda}{2d_{110}}\right) = 2\sin^{-1}\left(\frac{1.54 \times \sqrt{13}}{2 \times 6}\right) = 55.1°$$

Example 7 The first three S-values obtained from the powder pattern of a given unknown cubic form of material are 56.8, 94.4 and 112.0 mm, respectively. If the radius of the camera is 57.3 mm and the wavelength of the radiation used is 1.54 Å, determine the crystal structure of the cubic form and its lattice parameter.

Solution *Given*: S-values: $S_1 = 56.8$ mm, $S_2 = 94.4$ mm, $S_3 = 112.0$ mm, $R = 57.3$ mm, $\lambda = 1.54$ Å, crystal structure = ?, a = ?
We know that the angular and linear relationship in a powder pattern is

$$4\theta(\text{degrees}) = S(\text{mm})$$

$$\text{or} \quad \theta = \frac{S}{4}$$

Now, write various S values in tabular form and obtain corresponding θ and d values

Line	Arc length = S	$\theta = S/4$ (°)	$\sin \theta$	$d = \lambda/2 \sin \theta$ (Å)
1	56.8	14.2	0.245	3.14
2	94.4	23.6	0.40	1.92
3	112.0	28.0	0.469	1.64

From the first table, make use of θ. The values of $\sin^2 \theta$, common factor (c.f) are found and added into other columns of the next table. The values of N for allowed reflections give us that the unknown cubic form is the diamond cubic (DC) structure. Finally, we get the value of lattice parameter "a."

Line	θ (°)	$\sin^2 \theta$	Common factor (c.f)	$N = \frac{\sin^2 \theta}{c.f}$	$\sqrt{N}$	$a = d\sqrt{N}$
1	14.2	0.060		3	$\sqrt{3}$	5.44
2	23.6	0.160	0.020	8	$\sqrt{8}$	5.43
3	28.0	0.220		11	$\sqrt{11}$	5.44

Example 8 The first three S-values obtained from the powder pattern of a given unknown cubic form of material are 24.95, 40.9 and 48.05 mm, respectively. If the radius of the camera is 57.3 mm and Molybdenum K_α radiation of wavelength 0.71 Å is used. Determine the crystal structure of the cubic form and its lattice parameter.

Solution *Given*: S-values: $S_1 = 24.95$ mm, $S_2 = 40.9$ mm, $S_3 = 48.05$ mm, R = 57.3 mm, $\lambda = 0.71$ Å, crystal structure = ?, a = ?

Here, we use an alternative method to solve a similar problem as above. We know that the angular and linear relationship in a powder pattern is

$$4\theta(\text{degrees}) = S(\text{mm})$$

$$\text{or} \quad \theta = \frac{S}{4}$$

Therefore,

$$\theta_1 = \frac{S_1}{4} = \frac{24.95}{4} = 6.2375° \text{ and } \sin^2 \theta_1 = 0.0118$$

$$\theta_2 = \frac{S_2}{4} = \frac{40.9}{4} = 10.225° \text{ and } \sin^2 \theta_2 = 0.0315$$

$$\theta_3 = \frac{S_3}{4} = \frac{48.05}{4} = 12.0125° \text{ and } \sin^2 \theta_3 = 0.0433$$

We know that $\sin^2 \theta \propto (h^2 + k^2 + l^2)$ or $\sin^2 \theta \propto N$, then the values of the ratio of $\sin^2 \theta_1 : \sin^2 \theta_2 : \sin^2 \theta_3$ are obtained as 3 : 8 : 11.

According to the ratio of allowed reflections, this gives the diamond cubic (DC) structure. Now, we can obtain the lattice parameter by using the expression

$$\sin^2\theta = \frac{\lambda^2}{4a^2}\left(h^2 + k^2 + l^2\right) = \frac{\lambda^2}{4a^2}N$$

$$\text{or}\quad a_1^2 = \frac{\lambda^2 N_1}{4\sin^2\theta_1} = \frac{(0.71)^2 \times 3}{4 \times 0.0118} = 32.040 \quad \text{or}\quad a_1 = 5.660\,\text{Å}$$

Similarly,

$$\text{or}\quad a_2^2 = \frac{\lambda^2 N_2}{4\sin^2\theta_2} = \frac{(0.71)^2 \times 8}{4 \times 0.0315} = 32.006 \quad \text{or}\quad a_2 = 5.657\,\text{Å}$$

$$\text{or}\quad a_3^2 = \frac{\lambda^2 N_3}{4\sin^2\theta_3} = \frac{(0.71)^2 \times 11}{4 \times 0.0433} = 32.015 \quad \text{or}\quad a_3 = 5.658\,\text{Å}$$

$\Rightarrow$ The lattice parameter a = 5.66 Å

Example 9 In a diffraction experiment, the first reflection from an fcc crystal is observed at $2\theta = 84°$ when the X-ray of wavelength 1.54 Å is used. Determine the indices of possible reflections and the corresponding interplanar spacing.

Solution *Given*: $2\theta = 84°$, that is, $\theta = 42°$, $n = 1$, structure is fcc, $\lambda = 1.54$ Å, $h_1, k_1, l_1 = ?$, $d_1 = ?$, $h_2, k_2, l_2 = ?$, $d_2 = ?$

We know that the ratios of $\left(h^2 + k^2 + l^2\right)$ for allowed reflections in fcc gives are:

$$3 : 4 : 8 : 11 : 12 : 16 : 19 : 20$$

where

3 corresponds to the first reflection from (111) plane
4 corresponds to the second reflection from (200) plane
8 corresponds to the third reflection from (220) plane, and so on

Also, the Bragg's equation for the first-order reflection is

$$2d\sin\theta = n\lambda \qquad\qquad (i)$$

$$\text{or}\quad d_{111} = \frac{\lambda}{2\sin\theta} = \frac{1.54}{2 \times \sin 42} = 1.15\,\text{Å}$$

Further, for cubic crystals, the relationship between d and a is

$$d_{hkl} = \frac{a}{\left(h^2 + k^2 + l^2\right)^{1/2}} \qquad\qquad (ii)$$

From equations (i) and (ii), we have

$$\sin^2 \theta = \frac{\lambda^2}{4a^2} \left(h^2 + k^2 + l^2\right)$$

(iii)

$$\text{or} \quad \sin^2 \theta \propto \left(h^2 + k^2 + l^2\right)$$

Now, for

$$\left(h^2 + k^2 + l^2\right) = 3, \quad \sin^2 \theta \propto 0.45 \quad \text{and } \theta = 42^{\circ}$$
$$\left(h^2 + k^2 + l^2\right) = 4, \quad \sin^2 \theta \propto 0.60 \quad \text{and } \theta = 50.6^{\circ}$$

This gives us

$$d_{200} = \frac{\lambda}{2 \sin \theta} = \frac{1.54}{2 \times \sin 50.6} = 0.996 \,\text{Å}$$

A similar exercise for third reflection (220) will show that d_{220} is greater than 1 because $\sin \theta = \left(\frac{n\lambda}{2d}\right) \leq 1$, which is not a possible reflection. Thus the possible reflections are (111) and (200) and the corresponding interplanar spacing's are 1.15 Å and 0.996 Å, respectively.

Example 10 A powder pattern was obtained for an fcc crystal by using X-ray of wavelength 1.79 Å. The lattice parameter was found to be 3.52 Å. Determine the lowest and highest reflections possible.

Solution *Given*: Crystal is fcc, $\lambda = 1.79$ Å, $a = 3.52$ Å, lowest reflection = ?, highest reflection = ?

We know that the ratios of $\left(h^2 + k^2 + l^2\right)$ for allowed reflections in fcc are:

$$3 : 4 : 8 : 11 : 12 : 16 : 19 : 20$$

They correspond to (111), (200), (220), (113), (222), (400), etc. planes, respectively.

For $\left(h^2 + k^2 + l^2\right) = 3$, we have

$$d_{111} = \frac{a}{\sqrt{3}} \quad \text{and} \quad \sin \theta_{111} = \frac{\lambda}{2d_{111}} = \frac{\sqrt{3} \times 1.79}{2 \times 3.25} = 0.48$$

Similarly, for $\left(h^2 + k^2 + l^2\right) = 4$, we have

$$d_{200} = \frac{a}{2} \quad \text{and} \quad \sin \theta_{200} = \frac{\lambda}{2d_{200}} = \frac{\lambda}{a} = \frac{1.79}{3.25} = 0.51$$

For $(h^2 + k^2 + l^2) = 8$, we have

$$d_{220} = \frac{a}{2\sqrt{2}} \text{ and } \sin\theta_{220} = \frac{\lambda}{2d_{220}} = \frac{\sqrt{2}\lambda}{a} = \frac{\sqrt{2} \times 1.79}{3.25} = 0.72$$

For $(h^2 + k^2 + l^2) = 11$, we have

$$d_{113} = \frac{a}{\sqrt{11}} \text{ and } \sin\theta_{113} = \frac{\lambda}{2d_{113}} = \frac{\sqrt{11}\lambda}{2a} = \frac{\sqrt{11} \times 1.79}{2 \times 3.25} = 0.84$$

For $(h^2 + k^2 + l^2) = 12$, we have

$$d_{222} = \frac{a}{2\sqrt{3}} \text{ and } \sin\theta_{222} = \frac{\lambda}{2d_{222}} = \frac{\sqrt{3}\lambda}{a} = \frac{\sqrt{3} \times 1.79}{3.25} = 0.88$$

A similar calculation for $(h^2 + k^2 + l^2) = 16$, will give the value of d_{400} greater than 1 because $\sin\theta = \left(\frac{n\lambda}{2d}\right) \leq 1$, which is not a possible reflection. Thus the lowest and highest possible reflections for the given λ and a are (111) and (222).

Example 11 In a diffraction experiment of an fcc material only one peak is observed at $2\theta = 121°$ when an X-ray of wavelength 1.54 Å is used. Determine the indices of the plane and the interplanar spacing. Show that the next (higher) peak cannot occur.

Solution *Given*: $\lambda = 1.54$ Å, $2\theta = 121°$, material is fcc, (hkl) = ?, d = ?
For n = 1, the Bragg's equation can be written as

$$d = \frac{\lambda}{2\sin\theta} = \frac{1.54}{2 \times \sin 60.5} - 0.885 \text{ Å}$$

Further, we know that the ratios of $(h^2 + k^2 + l^2)$ for allowed reflections in fcc are:

$$3 : 4 : 8 : 11 : 12 : 16 : 19 : 20$$

For the first peak, $(h^2 + k^2 + l^2) = 3$ and hence (hkl) $\equiv$ (111)
For the second peak, $(h^2 + k^2 + l^2) = 4$ and hence (hkl) $\equiv$ (200)
Also, for the first peak of a cubic crystal, we have

$$d_{111} = \frac{a}{(h^2 + k^2 + l^2)^{1/2}} = \frac{a}{(1^2 + 1^2 + 1^2)^{1/2}} = \frac{a}{\sqrt{3}}$$

or $a = \sqrt{3}d_{111} = \sqrt{3} \times 0.885 = 1.53$ Å

For second peak, we have

$$d_{200} = \frac{a}{\left(h^2 + k^2 + l^2\right)^{1/2}} = \frac{a}{\left(2^2 + 0^2 + 0^2\right)^{1/2}} = \frac{a}{2}$$

or $2\,d_{200} = a = 1.53\,\text{Å}$

However, from Bragg's equation, we have

$$\lambda_{max} \leq 2d$$

For the second peak, this condition is violated and hence cannot occur.

Example 12 In a diffraction experiment, the Bragg's angle corresponding to a reflection for which $\left(h^2 + k^2 + l^2\right) = 8$ is found to be at 14.35° when the X-ray of wavelength 0.71 Å is used. Determine the lattice parameter of the crystal. If there are two other reflections of smaller Bragg's angles, determine the crystal structure.

Solution: *Given*: $\left(h^2 + k^2 + l^2\right) = 8$, $\theta = 14.35°$, $\lambda = 0.71$ Å, a = ?, structure = ?
For the given (hkl), the value of d from Bragg's equation is

$$d = \frac{\lambda}{2\sin\theta} = \frac{0.71}{2 \times \sin 14.35} = 1.43\,\text{Å}$$

Further, for a cubic system, the relationship between a and d is given by

$$a = d\left(h^2 + k^2 + l^2\right)^{1/2} = \frac{\sqrt{8} \times 0.71}{2 \times \sin 14.35} = 4.05\,\text{Å}$$

Since there are two reflections at smaller angles before 14.35°, this indicates that given value 8 lies at third place. This is possible only if the structure is fcc.

Example 13 A monochromatic X-ray beam of wavelength 1.79 Å is used to study the fcc aluminum powder sample with a camera of radius 57.3 mm, determine the first four S-values.

Solution: *Given*: Crystal is fcc, a = 4.05 Å, $\lambda = 1.79$ Å, R = 57.3 mm, first four S-values = ?
We know that the ratios of $\left(h^2 + k^2 + l^2\right)$ for allowed reflections in fcc are:

$$3 : 4 : 8 : 11 : 12 : 16 : 19 : 20$$

They correspond to (111), (200), (220), (113), (222), (400), etc. planes, respectively.

For $(h^2 + k^2 + l^2) = 3$, we have

$$d_{111} = \frac{a}{\sqrt{3}} = \frac{4.05}{\sqrt{3}}$$

Further, from Bragg's law, we have

$$\theta_{111} = \sin^{-1}\left(\frac{\lambda}{2d_{111}}\right) = \sin^{-1}\left(\frac{\sqrt{3} \times 1.79}{2 \times 4.05}\right) = 22.5°$$

Therefore, for 57.3 mm camera radius

$$S_1 = 4\theta_{111} = 4 \times 22.5 = 90.01 \text{ mm}$$

Similarly, for $(h^2 + k^2 + l^2) = 4, 8, 11$ we can obtain

$$d_{200} = \frac{a}{2} = 2.025 \text{ Å}, \quad \theta_{200} = 26.229982 \text{ and } S_2 = 105.0 \text{ mm}$$

$$d_{220} = \frac{a}{\sqrt{8}} = 1.4318912 \text{ Å}, \quad \theta_{220} = 38.685670 \text{ and } S_3 = 154.7 \text{ mm}$$

$$d_{113} = \frac{a}{\sqrt{11}} = 1.2211209 \text{ Å}, \quad \theta_{113} = 47.132856 \text{ and } S_4 = 188.5 \text{ mm}$$

Example 14 First reflection obtained from a sample of copper powder (fcc) has the S-value of 86.7 mm, by using CuKα radiation. Determine the camera radius.

Solution: *Given*: $\lambda_K(\alpha) = 1.54$ Å, S = 86.7 mm, crystal is fcc, so for first reflection, the value of $(h^2 + k^2 + l^2) = 3$, for copper a = 3.61 Å, camera radius = ?
For fcc crystal, we have

$$d_{111} = \frac{a}{(h^2 + k^2 + l^2)^{1/2}} = \frac{a}{(1^2 + 1^2 + 1^2)^{1/2}} = \frac{3.61}{\sqrt{3}}$$

Further, from Bragg's law, we have

$$\theta_{111} = \sin^{-1}\left(\frac{\lambda}{2d_{111}}\right) = \sin^{-1}\left(\frac{\sqrt{3} \times 1.54}{2 \times 3.61}\right) = 21.68°$$

Therefore, the value of S is

$$S_1 = 4\theta_{111} = 4 \times 2168 = 86.72 \text{ mm}$$

This is equal to the given S-value. Since $S = 4\theta$ corresponds to the camera radius 57.3 mm, therefore, 57.3 mm is the required camera radius.

Example 15 The diffraction data of a cubic crystal of an unknown element show peaks at 2θ angles 44.46°, 51.64°, 75.78° and 93.22° when the wavelength of X-ray used is 1.543 Å. Determine the crystal structure, lattice parameter and identify the element.

Solution: *Given*: Crystal is cubic, 2θ angles: 44.46°, 51.64°, 75.78° and 93.22°, $\lambda = 1.543$ Å, crystal structure = ?, a = ?, element = ?

Preparing the datasheet in tabular form, we have

2θ	θ	$\sin\theta$	$\sin^2\theta$	$d = \lambda/2\sin\theta$ (Å)
44.46	22.23	0.3783	0.1431	2.04
51.64	25.82	0.4355	0.1897	1.77
75.78	37.89	0.6141	0.3772	1.26
93.22	46.61	0.7267	0.5281	1.06

From the first table, make use of θ. The values of $\sin^2\theta$, common factor (c.f) are found and added into other columns of the next table. The values of N for allowed reflections give us that the unknown cubic form is the face-centered cubic (fcc) structure. Finally, we get the value of lattice parameter "a."

Line	θ (°)	$\sin^2\theta$	Common factor (c.f)	$N = \frac{\sin^2\theta}{c.f}$	$\sqrt{N}$	$a = d\sqrt{N}$
1	22.23	0.1431		3	$\sqrt{3}$	3.53
2	25.83	0.1897	0.0477	4	$\sqrt{4}$	3.54
3	37.89	0.3772		8	$\sqrt{8}$	3.56
4	46.61	0.5281		11	$\sqrt{11}$	3.52

The average value of "a" from the Table is found to be 3.54 Å. The above data indicates that the element is Nickel.

Example 16 The diffraction data of a cubic crystal of an element show peaks at 2θ angles 38.60°, 55.71°, 69.70°, 82.55°, 95.00° and 107.67° when the wavelength of X-ray used is 1.543 Å. Determine the crystal structure, lattice parameter and identify the element.

Solution: Given: Crystal is cubic, 2θ angles: 38.60°, 55.71°, 69.70°, 82.55°, 95.00° and 107.67°, crystal structure = ?, a = ?, element = ?

Preparing the datasheet in tabular form, we have

2θ	θ	$\sin\theta$	$\sin^2\theta$	$d = \lambda/2\sin\theta$ (Å)
38.60	19.30	0.3305	0.1092	2.334
55.71	27.26	0.4580	0.2097	1.684
69.70	34.85	0.5714	0.3265	1.350
82.55	41.28	0.6597	0.4352	1.169
95.00	47.50	0.7373	0.5436	1.046
107.67	53.84	0.8073	0.6518	0.955

From first table, make use of θ. The values of $\sin^2 \theta$, common factor (c.f) are found and added into other columns of the next table. The values of N for allowed reflections give us that the unknown cubic form is the body-centered cubic (bcc) structure. Finally, we get the value of lattice parameter "a."

Line	θ (°)	$\sin^2 \theta$	Common factor (c.f)	$N = \frac{\sin^2 \theta}{c.f}$	$\sqrt{N}$	$a = d\sqrt{N}$
1	19.30	0.1092		2	$\sqrt{2}$	3.30
2	27.26	0.2183		4	$\sqrt{4}$	3.36
3	34.85	0.3265	0.0546	6	$\sqrt{6}$	3.31
4	41.28	0.4352		8	$\sqrt{8}$	3.31
5	47.50	0.5436		10	$\sqrt{10}$	3.31
6	53.84	0.6518		12	$\sqrt{12}$	3.31

The average value of "a" from the Table is found to be 3.316 Å. The above data indicates that the element is Niobium.

(ii) *Laue Method*

If the crystal to film distance is taken to be D and the distance of the diffraction spot (Laue spot) from the center of the photographic film (Fig. 9.2) is r, then

$$\tan 2\theta = \frac{r}{D}$$

where θ is the Bragg's angle

Solved Examples

Example 1 Bragg's reflection is observed at an angle of 5° in a transmission Laue pattern for KCl crystal. The crystal to film distance is 4 cm; determine the location of the spot with reference to center of the film.

Solution: *Given*: Bragg's angle θ = 5°, crystal to film distance D = 4 cm, distance of reflection from the center of the pattern, r = ?

Fig. 9.2 Relationship among reflecting plane, it is normal and the direction of diffraction beam

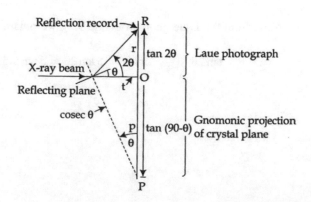

From Laue geometry, we obtain

$$\tan 2\theta = \frac{r}{D}$$

or

$$r = D\tan 2\theta = 4 \times \tan 10 = 4 \times 0.176 = 0.705\,cm$$

Example 2 According to a Laue photograph, the cell parameter of an fcc crystal is 4.50 Å. If the potential difference across the X-ray tube is 50 kV and the crystal to film distance is 5 cm, determine the minimum distance from the center of the pattern at which reflection can occur from the plane of maximum spacing.
Solution: *Given*: Crystal is fcc, a = 4.50 Å, V = 50 kV, crystal to film distance, D = 5 cm, distance of reflection from the center of the pattern, r = ?

Minimum wavelength of the X-ray produced at the electrode potential is given by

$$\lambda_{min} = \frac{hc}{eV} = \frac{6.626 \times 10^{-34} \times 3 \times 10^{8}}{1.6 \times 10^{-19} \times 5 \times 10^{4}}$$
$$= 2.48 \times 10^{-11}m = 0.25\,Å$$

From Bragg's equation, we observe that for λ_{min} the reflection will occur from widely separated planes, that is, (111). Therefore, for fcc

$$d_{111} = \frac{a}{\left(h^2 + k^2 + l^2\right)^{1/2}} = \frac{a}{\left(1^2 + 1^2 + 1^2\right)^{1/2}} = \frac{4.50}{\sqrt{3}}$$

Further, from Bragg's equation, we have

$$\sin\theta_{111} = \frac{\lambda_{min}}{2d_{111}} = \frac{\sqrt{3} \times 0.25}{2 \times 4.50}$$
$$or \quad \theta_{111} = \sin^{-1}\left(\frac{\sqrt{3} \times 0.25}{2 \times 4.50}\right) = 2.76°$$

Now, from the Laue geometry (Fig. 9.2), we obtain

$$\tan 2\theta = \frac{r}{5} \quad or \quad r = 5\tan 2\theta = 5\tan 5.52 = 0.48\,cm$$

Example 3   A transmission Laue pattern of a copper crystal with a = 3.62 Å is obtained on a film kept at a distance of 5 cm. The (100) planes of the crystal make an angle of 5° with the incident beam. Calculate the minimum tube voltage required to produce a 200 reflection. Also, determine the location of the spot with reference to center of the film.

Solution: *Given*: Lattice parameter a = 3.62 Å = d_{100}, D = 5 cm, Bragg's angle $\theta_{100} = 5°$, V_{min} to produce 200 reflection = ?, distance of reflection from the center of the pattern, r = ?

From Bragg's law, we have

$$\lambda = 2d_{200} \sin \theta$$

$$\text{or} \quad d_{200} = \frac{\lambda}{2 \sin \theta} = \frac{d_{100}}{2} = \frac{3.62}{2}$$

This gives us the minimum value of the wavelength, that is,

$$\lambda_{min} = 3.62 \times \sin \theta = 3.62 \times \sin 5° = 0.32 \, \text{Å}$$

Further, we know that

$$\lambda_{min} = \frac{hc}{eV}$$

$$\text{or} \quad V = \frac{hc}{e\lambda} = \frac{6.626 \times 10^{-34} \times 3 \times 10^8}{1.6 \times 10^{-19} \times 0.32 \times 10^{-10}}$$

$$= 38.82 \times 10^3 = 38.82 \, \text{kV}$$

Now, according to the Laue geometry, we have

$$\tan 2\theta = \frac{r}{D}$$

$$\text{or} \quad r = D \tan 2\theta$$

Substituting the values, we obtain

$$r = D \tan 2\theta = 5 \times \tan 10 = 0.88 \, \text{cm}$$

Example 4   A transmission Laue pattern of an iron crystal with a = 2.86 Å is obtained on a film kept at a distance of 5 cm using 30 kV tungsten radiation. How close to the center of the pattern can Laue spots be formed by reflecting planes (110) and (200) assuming that these planes are present in the desired orientation?

Solution: *Given*: Lattice parameter a = 2.86 Å, D = 5 cm, V = 30 kV, distances of reflections from the center of the pattern corresponding to (110) and (200) planes = ?

Minimum wavelength of the X-ray produced at the given electrode potential is given by

$$\lambda_{min} = \frac{hc}{eV} = \frac{6.626 \times 10^{-34} \times 3 \times 10^8}{1.6 \times 10^{-19} \times 3 \times 10^4}$$
$$= 4.14 \times 10^{-11} m = 0.414 \, \text{Å}$$

Case I: Calculations for (110) plane

$$d_{110} = \frac{a}{\left(h^2 + k^2 + l^2\right)^{1/2}} = \frac{a}{\left(1^2 + 1^2 + 0^2\right)^{1/2}} = \frac{2.86}{\sqrt{2}} = 2.02 \, \text{Å}$$

Further, from Bragg's equation, we have

$$\sin \theta_{110} = \frac{\lambda_{min}}{2d_{110}} = \frac{0.414}{2 \times 2.02}$$

$$\text{or} \quad \theta_{110} = \sin^{-1}\left(\frac{0.414}{2 \times 2.02}\right) = 5.88^\circ$$

Now, from the Laue geometry, we obtain

$$\tan 2\theta = \frac{r}{D}$$

$$\text{or} \quad r = D \tan 2\theta = 5 \tan 11.76 = 1.04 \, \text{cm}$$

Case II: Calculations for (200) plane

$$d_{200} = \frac{a}{\left(h^2 + k^2 + l^2\right)^{1/2}} = \frac{a}{\left(2^2 + 0^2 + 0^2\right)^{1/2}} = \frac{2.86}{2} = 1.43 \, \text{Å}$$

Further, from Bragg's equation, we have

$$\sin \theta_{200} = \frac{\lambda_{min}}{2d_{200}} = \frac{0.414}{2 \times 2.86}$$

$$\text{or} \quad \theta_{200} = \sin^{-1}\left(\frac{0.414}{2 \times 2.86}\right) = 8.32^\circ$$

Again, from the Laue geometry, we obtain

$$\tan 2\theta = \frac{r}{D}$$

$$\text{or} \quad r = D \tan 2\theta = 5 \tan 16.64 = 1.49 \, \text{cm}$$

Fig. 9.3 Crystal axes and
X-ray beam

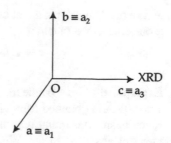

Example 5   A transmission Laue pattern of a cubic crystal with a = 3 Å is obtained
with a horizontal X-ray beam in the following orientation. The [010] axis of the
crystal points along the beam away from the X-ray tube, [001] points vertically
upwards and [100] is horizontal and parallel to the X-ray film. The crystal to film
distance is 5 cm. Determine the wavelengths of the radiation diffracted from the
$(01\bar{2})$ planes and the location of the diffraction spot on the X-ray film.

Solution: *Given*: Three crystallographic axes for beam and film orientation
(Fig. 9.3), a = 3 Å, D = 5 cm, diffraction plane = $(01\bar{2})$, the wavelengths and
location of the diffraction spot on the X-ray film = ?
    For a cubic system, the relationship between the interlayer spacing and the lattice
parameter is

$$d_{01\bar{2}} = \frac{a}{\left(h^2 + k^2 + l^2\right)^{1/2}} = \frac{a}{\left(0^2 + 1^2 + 2^2\right)^{1/2}} = \frac{3}{\sqrt{5}} = 1.34 \, \text{Å}$$

Now from the figure, the angle between the directions [010] and [021] is

$$\cos\theta = \frac{0 + 2 + 0}{1 \cdot (0 + 4 + 1)^{1/2}} = \frac{2}{\sqrt{5}}$$

$$\text{or} \quad \theta = \cos^{-1}\left(\frac{2}{\sqrt{5}}\right) = 26.56$$

From Bragg's law, we have

$$\text{for } n = 1, \quad \lambda = 2d_{01\bar{2}} \sin\theta = 2 \times 1.34 \times \sin 26.56 = 1.20 \, \text{Å}$$

Similarly,

for n = 2, $2\lambda = 2d_{01\bar{2}} \sin\theta = 2 \times 1.34 \times \sin 26.56 = 1.20 \, \text{Å}$, and $\lambda = 0.60 \, \text{Å}$

for n = 3, $3\lambda = 2d_{01\bar{2}} \sin\theta = 2 \times 1.34 \times \sin 26.56 = 1.20 \, \text{Å}$, and $\lambda = 0.40 \, \text{Å}$

for n = 4, $4\lambda = 2d_{01\bar{2}} \sin\theta = 2 \times 1.34 \times \sin 26.56 = 1.20 \, \text{Å}$, and $\lambda = 0.30 \, \text{Å}$

and so on. Thus, the incident beam contains these wavelengths. Now, from the Laue arrangement, we obtain

$$r = D \tan 2\theta = 5 \tan 52.12 = 6.66 \text{ cm}$$

Example 6 A back-reflection Laue pattern of an aluminum crystal (fcc) with a = 4.05 Å is obtained from (111) planes making an angle of 85° with the incident X-ray beam. Assuming that the wavelengths above 2.5 Å are extremely weak and they get absorbed in air. What orders of (111) reflection are present if the tube voltage is 60 kV ?

Solution: *Given*: Lattice parameter a = 4.05 Å, diffraction plane (111), Bragg's angle $\theta_{111} = 85°$, V = 60 kV, orders of (111) reflection possible = ?

Minimum wavelength of the X-ray produced at the given electrode potential is given by

$$\lambda_{min} = \frac{hc}{eV} = \frac{6.626 \times 10^{-34} \times 3 \times 10^8}{1.6 \times 10^{-19} \times 6 \times 10^4}$$
$$= 2.1 \times 10^{-11} \text{m} = 0.21 \text{ Å}$$

For a cubic system, the relationship between the interlayer spacing and the lattice parameter is

$$d_{111} = \frac{a}{\left(h^2 + k^2 + l^2\right)^{1/2}} = \frac{a}{\left(1^2 + 1^2 + 2^2\right)^{1/2}} = \frac{4.05}{\sqrt{3}} = 2.34 \text{ Å}$$

Further, from Bragg's equation, we have

$$\lambda = 2d_{200} \sin \theta$$

for n = 1, $\sin \theta_1 = \dfrac{\lambda_{min}}{2d_{111}} = \dfrac{0.21}{2 \times 2.34}$,   and $\theta_1 = \sin^{-1}\left(\dfrac{0.21}{4.68}\right) = 2.57°$

Similarly,

for n = 10, $\sin \theta_{10} = \dfrac{\lambda_{min}}{2d_{111}} = \dfrac{10 \times 0.21}{2 \times 2.34}$,   and $\theta_1 = \sin^{-1}\left(\dfrac{2.1}{4.68}\right) = 26.66°$

for n = 20, $\sin \theta_{20} = \dfrac{\lambda_{min}}{2d_{111}} = \dfrac{20 \times 0.21}{2 \times 2.34}$,   and $\theta_1 = \sin^{-1}\left(\dfrac{4.2}{4.68}\right) = 63.82°$

for n = 22, $\sin \theta_{22} = \dfrac{\lambda_{min}}{2d_{111}} = \dfrac{22 \times 0.21}{2 \times 2.34}$,   and $\theta_1 = \sin^{-1}\left(\dfrac{4.62}{4.68}\right) = 80.82°$

Fig. 9.4 Formation of layer
line from cones of the
diffracted rays

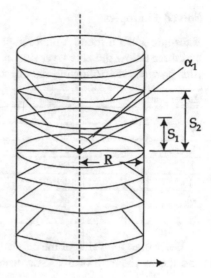

However, for n = 23 onwards will give us sin θ > 1, which is not possible.
Therefore, the possible orders of (111) reflection are up to twenty-second.

(iii)   *Rotation/ Oscillation Method*

For the a-axis rotation/oscillation photograph of the crystal material, the separation
of nth layer line (with respect to zero-layer) and the lattice parameter of the sample
are, respectively, given by (Fig. 9.4)

$$S_n = \frac{(n\lambda/a)}{\sqrt{1 - (n\lambda/a)^2}} R$$

$$\text{and} \quad a = \frac{n\lambda}{S_n}\left(R^2 + S_n^2\right)^{1/2}$$

where R is the radius of the camera used.

Alternatively, for the a-axis (b-axis or c-axis) rotation/oscillation photograph of
the crystal material, the lattice parameter can also be obtained by using the
expression

$$a = \frac{n\lambda}{\sin\left[\tan^{-1}(S_n/R)\right]}$$

Solved Examples

Example 1 In a rotation photograph, seven layer lines are observed, three above and three below the zero layer line. A millimeter scale placed next to the developed film provides the following readings:

Layer line	Height (mm)	Height w.r.t zero layer
3	05.40	34.00
2	22.44	16.96
1	31.84	07.56
0	39.40	–
-1	46.96	07.56
-2	56.35	16.95
-3	73.20	33.80

Determine the value of the cell height, if the diameter of the camera is 57.3 mm and the wavelength of the X-ray used is 1.542 Å.

Solution: Given: mm scale reading of three upper and three lower layer lines, D = 57.3 mm = 5.73 cm, so that R = 2.865 cm, λ = 1.542 Å = 1.542 × 10^{-8} cm.

From the given data, we obtain the height of the layer lines: S_1 = 7.56 mm = 0.756 cm, S_2 = 16.96 mm = 1.696 cm, S_3 = 34 mm = 3.4 cm. Now, to determine the cell parameter along a particular axis, make use of the equation (say)

$$a = \frac{n\lambda}{S_n} \left(R^2 + S_n^2 \right)^{1/2}$$

For n = 1,

$$a_1 = \frac{1 \times 1.542 \times 10^{-8}}{0.756} \left[(2.865)^2 + (0.756)^2 \right]^{1/2}$$
$$= 6.044 \times 10^{-8} \text{cm} = 6.044 \text{ Å}$$

Similar calculations will give us

$$a_2 = 6.054 \text{ Å}$$

Therefore, the average value of a = 6.05 Å.

Example 2 In a rotation photograph, first and second layer lines are observed (w.r. t zero layer line) at 18 mm and 40 mm, respectively. If the wavelength of the MoKα radiation is 0.71 Å and the film diameter 57.3 mm, determine the lattice parameter of the sample.

Solution: *Given*: S_1 = 18 mm   and S_2 = 40 mm,   D = 57.3 mm,   so that R = 28.65 mm, λ = 0.71 Å, a = ?
Case I: We can determine the value of "a" corresponding to the first layer line by using the formula

$$a_1 = \frac{n\lambda}{\sin\left[\tan^{-1}\left(\frac{S_n}{R}\right)\right]} = \frac{1 \times 0.71}{\sin\left[\tan^{-1}\left(\frac{18}{28.65}\right)\right]}$$

$$= \frac{0.71}{\sin[\tan^{-1}(0.628)]}$$

$$= \frac{0.71}{\sin 32.14} = \frac{0.71}{0.53} = 1.33 \, \text{Å}$$

Case II: Similarly, we can determine the value of "a" corresponding to the second layer line

$$a_2 = \frac{n\lambda}{\sin\left[\tan^{-1}\left(\frac{S_n}{R}\right)\right]} = \frac{2 \times 0.71}{\sin\left[\tan^{-1}\left(\frac{40}{28.65}\right)\right]}$$

$$= \frac{1.42}{\sin[\tan^{-1}(1.396)]}$$

$$= \frac{1.42}{\sin 54.39} = \frac{1.42}{0.813} = 1.74 \, \text{Å}$$

Therefore, the average value of a = 1.54 Å.

Example 3 In a rotation photograph six layer lines are observed both above and below the zero layer line. If the heights of these layer lines above (or below) the zero layer are 0.29, 0.59, 0.91, 1.25, 1.65, and 2.2 cm, obtain the cell height of the crystal along the axis of rotation. The radius of the camera is 3 cm and the wavelength of the X-ray is 1.54 Å.

Solution: *Given*: $S_1 = 0.29$ cm, $S_2 = 0.59$ cm, $S_3 = 0.91$ cm, $S_4 = 1.25$ cm, $S_5 = 1.65$ cm, $S_6 = 2.2$ cm, R = 3 cm, $\lambda = 1.54$ Å $= 1.54 \times 10^{-8}$ cm, the unit cell height along the axis of rotation = ?

Let us suppose that the given photograph is the a-axis photograph and then the value of the lattice parameter "a" corresponding to the first layer line is

$$a_1 = \frac{n\lambda}{\sin\left[\tan^{-1}\left(\frac{S_n}{R}\right)\right]} = \frac{1 \times 1.54 \times 10^{-8}}{\sin\left[\tan^{-1}\left(\frac{0.29}{3.0}\right)\right]}$$

$$= \frac{1.54 \times 10^{-8}}{\sin[\tan^{-1}(0.097)]} = \frac{1.54 \times 10^{-8}}{\sin 5.52}$$

$$= \frac{1.54 \times 10^{-8}}{0.096} = 16 \times 10^{-8} \, \text{cm} = 16 \, \text{Å}$$

Making a similar calculation after substituting different values of n and S, we can obtain the same value of the lattice parameter "a." Therefore the average value of the unit cell height of the crystal is 16 Å.

Example 4 In a rotation photograph, the separation of the fourth layer line from the zero layer line is 2.2 cm. If the camera radius is 3 cm and the lattice parameter is 16 Å, determine the wavelength of the X-ray used.

Solution: *Given*: $S_6 = 1.25$ cm, $n = 4$, $R = 3$ cm, lattice parameter $= 16$ Å, $\lambda = ?$
Assuming that the crystal is rotated along the a-axis, therefore using the formula

$$a = \frac{n\lambda}{\sin\left[\tan^{-1}\left(\frac{S_n}{R}\right)\right]}$$

$$\text{or}\quad n\lambda = a \times \sin\left[\tan^{-1}\left(\frac{S_n}{R}\right)\right]$$

Now, substituting different values, we obtain

$$6 \times \lambda = 16 \times \sin\left[\tan^{-1}\left(\frac{1.25}{3}\right)\right]$$

$$= 16 \times \sin 22.62 = 6.154$$

$$\text{or}\quad \lambda = 1.54\,\text{Å}$$

Example 5 A simple cubic crystal is irradiated with X-rays of wavelength 0.9 Å at a glancing angle. The crystal is rotated and the angles for Bragg reflections are measured. Which set of crystal planes will give the smallest angle for first-order reflection? If this angle is 8.9°, determine the spacing between these crystal planes. At what angle will the first-order reflection be obtained from the (110) crystal planes?

Solution: *Given*: $\lambda = 0.9$ Å, set of planes $= ?$, θ (smallest) $= 8.9°$, interlayer spacing $= ?$, $\theta(110) = ?$
For a simple cubic crystal system, the smallest angle corresponds to the largest separation, that is, the {100} planes.
Now, from Bragg's equation (with $n = 1$), we can obtain

$$d_{100} = \frac{\lambda}{2\sin\theta} = \frac{0.9}{2 \times \sin 8.9} = 2.91\,\text{Å}$$

Further, the separation between the (110) planes can be obtained as

$$d_{110} = \frac{a}{\left(h^2 + k^2 + l^2\right)^{1/2}} = \frac{a}{\left(1^2 + 1^2 + 0^2\right)^{1/2}} = \frac{2.91}{\sqrt{2}} = 2.06\,\text{Å}$$

Again using the Bragg's equation (with n = 1), we can obtain

$$\theta_{110} = \sin^{-1}\left(\frac{\lambda}{2 \times d_{110}}\right) = \sin^{-1}\left(\frac{0.9}{2 \times 2.06}\right) = 12.62°$$

Example 6 An oscillation photograph is taken with CuKα radiation the rotation being about an axis of length 7.2 Å. If the camera diameter is 57.3 mm, find the distances of the first three layer lines.

Solution: *Given*: Length of (say) a-axis = 7.2 Å, D = 57.3 mm, so that R = 28.65 mm, λ = 1.54 Å, distances of first three layer lines = ?

Substituting different values, we can obtain the distance of the first layer line

$$S_1 = \frac{(n\lambda/a)}{\sqrt{1-(n\lambda/a)^2}}R = \frac{(1.54/7.2)}{\sqrt{1-(1.54/7.2)^2}} \times 28.65 = \frac{0.214}{\sqrt{1-0.0457}} \times 28.65$$
$$= 6.28 \text{ mm}$$

Making similar calculations by substituting different values, we can obtain

$$S_2 = 13.50 \text{ mm}$$
$$and \quad S_3 = 24.0 \; mm$$

(c) Unit Cell and Density of Crystal Material

Density of a crystal material is defined as

$$\rho = \frac{\text{Mass of the unit cell}}{\text{Volume of the unit cell}} = \frac{Mn}{NV} = \frac{Mn}{Na^3}$$

where,

M = Atomic weight or molecular weight
n = Number of atoms in the unit cell (or No. of formula unit)
N = 6.023×10^{26} (Avogadro's number in SI system)
a = Side of the (cubic) unit cell

Solved Examples

Example 1 A unit cell of NaCl has four formula units. If the side of the unit cell is 5.64 Å, calculate its density.

Solution: *Given*: NaCl with four formula unit, that is, n = 4, a = 5.64 Å = 5.64×10^{-10}m, molecular weight of NaCl = 23 + 35.5 = 58.5, ρ = ?

We know that the density of the crystal material is given by

$$\rho = \frac{\text{Mass}}{\text{Volume}} = \frac{M \times n}{N \times a^3}$$

$$= \frac{4 \times 58.5}{6.023 \times 10^{26} \times (5.64 \times 10^{-10})^3} = 2165 \, \text{kg/m}^3$$

Example 2 The density of NaCl is $2.18 \times 10^3 kg/m^3$, and the atomic weights of sodium and chlorine are 23 and 35.5, respectively. Sodium chloride has fcc structure, calculate its interatomic separation.

Solution: *Given*:   Molecular   weight   of   NaCl = 23 + 35.5 = 58.5, $\rho = 2.18 \times 10^3 \text{kg/m}^3$, structure is fcc, so that n = 4, a = ?, $d = \left(\frac{a}{2}\right) = ?$
 We know that the density of the crystal material is given by

$$\rho = \frac{\text{Mass}}{\text{Volume}} = \frac{n \times M}{N \times a^3}$$

where a is the side of the unit cell.

$$\Rightarrow a^3 = \frac{M \times n}{N \times \rho}$$

or

$$a = \left(\frac{M \times n}{N \times \rho}\right)^{1/3}$$

$$= \left(\frac{4 \times 58.5}{6.023 \times 10^{26} \times 2.18 \times 10^3}\right)^{1/3}$$

$$= 5.63 \times 10^{-10} \, \text{m}$$

Therefore, the interatomic distance

$$d = \left(\frac{a}{2}\right) = \frac{5.63 \times 10^{-10}}{2} = 2.81 \times 10^{-10} \text{m}$$

Example 3 The density of potassium chloride is $1.98 \times 10^3 \text{kg/m}^3$, and its molecular weight is 74.55. Calculate its interatomic separation.

Solution: *Given*:   Since   KCl   has   NaCl   structure,   so   that   n = 4, $\rho = 1.98 \times 10^3 \text{kg/m}^3$, M = 74.55, a = ?, $d = \left(\frac{a}{2}\right) = ?$

We know that the density of the crystal material is given by

$$\rho = \frac{Mass}{Volume} = \frac{n \times M}{N \times a^3}$$

where a is the side of the unit cell.

$$\Rightarrow a^3 = \frac{M \times n}{N \times \rho}$$

or

$$a = \left(\frac{M \times n}{N \times \rho}\right)^{1/3}$$

$$= \left(\frac{4 \times 74.55}{6.023 \times 10^{26} \times 1.98 \times 10^3}\right)^{1/3}$$

$$= 6.30 \times 10^{-10} \, m$$

Therefore, the interatomic distance

$$d = \left(\frac{a}{2}\right) = \frac{6.30 \times 10^{-10}}{2} = 3.15 \times 10^{-10} m$$

Example 4 Silver crystallizes in fcc form and the nearest neighbor distance in silver crystal is 2.87 Å, determine its density.

Solution: *Given*: Crystal structure is fcc, so that n = 4, nearest neighbor distance $= 2.87 Å = 2.87 \times 10^{-10}$m, at. Wt. of silver = 107.68, ρ = ?

We know that the nearest neighbor distance in fcc crystal is $a/\sqrt{2}$

$$\Rightarrow \frac{a}{\sqrt{2}} = 2.87 \times 10^{-10} \quad or \quad a = \sqrt{2} \times 10^{-10} = 4.06 \times 10^{-10} m.$$

Now, the density of crystal material is given by

$$\rho = \frac{Mass}{Volume} = \frac{M \times n}{N \times a^3}$$

$$= \frac{4 \times 107.68}{6.023 \times 10^{26} \times (4.06 \times 10^{-10})^3} = 1.068 \times 10^4 \, kg/m^3$$

Example 5 The crystal structure of α-iron is bcc, density 7860 kg/m³ and the atomic weight 55.85, respectively. Calculate the radius of iron atom.

Solution: *Given*: structure of α-iron is bcc so that n = 2, ρ = 7860 kg/m³, at. wt. = 55.85, r = ?

We know that the density of crystal material is given by

$$\rho = \frac{\text{Mass}}{\text{Volume}} = \frac{M \times n}{N \times a^3}$$

$$\Rightarrow a^3 = \frac{M \times n}{N \times \rho}$$

or

$$a = \left(\frac{M \times n}{N \times \rho}\right)^{1/3}$$

$$= \left(\frac{2 \times 55.85}{6.023 \times 10^{26} \times 7860}\right)^{1/3}$$

$$= 2.86 \times 10^{-10} \text{ m} = 2.86 \text{ Å}$$

Further, in a bcc structure, we have

$$4r = \sqrt{3}a$$

where r is the radius of the atom. Therefore,

$$r = \frac{\sqrt{3} \times 2.86}{4} = 1.24 \text{ Å}$$

Example 6 Zinc has hcp structure. The height of its unit cell is 4.94 Å, nearest neighbor distance 2.7 Å and the atomic weight of zinc 65.38, respectively. Calculate the density of zinc.

Solution: *Given*: $a = b = 2.7 \text{ Å} = 2.7 \times 10^{-10}\text{m}$, $c = 4.94 \text{ Å} = 4.94 \times 10^{-10}\text{m}$, at. wt. of Zn = 65.37, structure is hcp, so that n = 6, $\rho = ?$
    We know that the volume of hcp cell is given by

$$V = \frac{3\sqrt{3}a^2c}{2}$$

$$= \frac{3 \times 1.732 \times (2.7 \times 10^{-10})^2 \times 4.94 \times 10^{-10}}{2}$$

$$= 93.56 \times 10^{-30}\text{m}^3$$

Now, the density of the crystal material is obtained by

$$\rho = \frac{\text{Mass}}{\text{Volume}} = \frac{M \times n}{N \times a^3}$$
$$= \frac{4 \times 65.38}{6.023 \times 10^{26} \times 93.56 \times 10^{-30}}$$
$$= 6961 \text{ kg/m}^3$$

Example 7 Copper crystallizes in cubic form. If its lattice parameter, a = 3.60 Å, at.wt. = 63.6 and density, d = 8960 kg/m³, determine the number of atoms in the unit cell. What is the type of its unit cell?

Solution: *Given*: Copper crystal of cubic form, a = 3.60 Å, at. wt. = 63.6, $\rho = 8960 \text{ kg/m}^3$, n = ?

We know that the density of the crystal material is given by

$$\rho = \frac{\text{Mass}}{\text{Volume}} = \frac{M \times n}{N \times a^3}$$

or

$$n = \frac{a^3 \rho N}{M}$$
$$= \frac{(3.60 \times 10^{-10})^3 \times 8960 \times 6.023 \times 10^{26}}{63.6}$$
$$= 3.959 \cong 4$$

Therefore, there are four atoms in the unit cell. Accordingly, it is fcc.

Example 8 Sodium crystallizes in cubic form. The edge of the unit cell is 4.3 Å. The atomic weight of sodium is 23 and its density is 963 kg/m³. Determine the number of atoms in the unit cell and its type.

Solution: *Given*: Sodium crystal of cubic form, a = 4.3 Å, at. wt. = 23, $\rho = 8960 \text{ kg/m}^3$, n = ?

We know that the density of the crystal material is given by

$$\rho = \frac{\text{Mass}}{\text{Volume}} = \frac{M \times n}{N \times a^3}$$

or

$$n = \frac{a^3 \rho N}{M}$$
$$= \frac{(4.3 \times 10^{-10})^3 \times 963 \times 6.023 \times 10^{26}}{23}$$
$$= 2.005 \cong 2$$

Therefore, there are two atoms in the unit cell. Accordingly, it has bcc form.

Multiple Choice Questions (MCQ)

1. For a bcc crystal structure, the ratio of $\sin^2\theta$ values corresponding to two consecutive diffracting principal planes will be :

    (a) 0.25 Å
    (b) 0.50 Å
    (c) 0.75 Å
    (d) 1.00 Å

2. For an fcc crystal structure, the ratio of $\sin^2\theta$ values corresponding to two consecutive diffracting principal planes will be :

    (a) 0.25 Å
    (b) 0.50 Å
    (c) 0.75 Å
    (d) 1.00 Å

3. The powder pattern of an unknown cubic metal was obtained by using CuKα radiation in an X-ray diffractometer. The XRD pattern showed initial eight lines with 2θ values of 32.2, 44.4, 64.6, 77.4, 81.6, 98.0, 110.6 and 115.0 degrees. The lattice parameter is found to be:

    (a) 4.09 Å
    (b) 5.09 Å
    (c) 6.09 Å
    (d) 7.09 Å

4. The ratios of $(h^2 + k^2 + l^2)$ obtained as 3 : 8 : 11 : 16 : 19 for allowed reflections correspond to:

    (a) SC
    (b) BCC
    (c) FCC
    (d) DC

5. The ratios of $(h^2 + k^2 + l^2)$ obtained as 1 : 2 : 3 : 4 : 5 : 6 : 8 : 9 : 10 for allowed reflections correspond to:

    (a) SC
    (b) BCC
    (c) FCC
    (d) DC

6. The ratios of $(h^2 + k^2 + l^2)$ obtained as 2 : 4 : 6 : 8 : 10 : 12 for allowed reflections correspond to:

   (a) SC
   (b) BCC
   (c) FCC
   (d) DC

7. The ratios of $(h^2 + k^2 + l^2)$ obtained as 3 : 4 : 8 : 11 : 12 : 16 : 19 : 20 for allowed reflections correspond to:

   (a) SC
   (b) BCC
   (c) FCC
   (d) DC

8. The first three S-values obtained from the powder pattern of a given unknown cubic form of material are 56.8, 94.4 and 112.0 mm, respectively. If the radius of the camera is 57.3 mm and the wavelength of the radiation used is 1.54 Å, its lattice parameter is:

   (a) 8.44 Å
   (b) 7.44 Å
   (c) 6.44 Å
   (d) 5.44 Å

9. The first three S-values obtained from the powder pattern of a given unknown cubic form of material are 56.8, 94.4 and 112.0 mm, respectively. If the radius of the camera is 57.3 mm and the wavelength of the radiation used is 1.54 Å, crystal structure of the unknown cubic form is:

   (a) SC
   (b) BCC
   (c) FCC
   (d) DC

10. The first three S-values obtained from the powder pattern of a given unknown cubic form material are 24.95, 40.9 and 48.05 mm, respectively. If the radius of the camera is 57.3 mm and Molybdenum K_α radiation of wavelength used is 0.71 Å. Its lattice parameter is:

   (a) 5.66 Å
   (b) 6.66 Å
   (c) 7.66 Å
   (d) 8.66 Å

11. The first three S-values obtained from the powder pattern of a given unknown cubic form material are 24.95, 40.9 and 48.05 mm, respectively. If the radius of the camera is 57.3 mm and Molybdenum K_α radiation of wavelength 0.71 Å is used. Determine the crystal structure of the cubic form and its lattice parameter:

(a) SC
(b) BCC
(c) FCC
(d) DC

12. In an X-ray diffraction experiment, the first reflection from an fcc crystal is observed at $2\theta = 84°$ when CuKα radiation of wavelength 1.54 Å is used. The indices of possible reflections are:

(a) (111), (200)
(b) (200), (220)
(c) (111), (220)
(d) (200), (311)

13. In an X-ray diffraction experiment, the first reflection from an fcc crystal is observed at $2\theta = 84°$ when CuKα radiation of wavelength 1.54 Å is used. The values of interplanar spacing corresponding to possible reflections are:

(a) 0.15 Å, 0.88 Å
(b) 1.15 Å, 0.99 Å
(c) 2.15 Å, 1.88 Å
(d) 3.15 Å, 1.99 Å

14. A powder pattern is obtained for an fcc crystal with lattice parameter 3.52 Å by using X-ray of wavelength 1.79 Å. The lowest reflection possible is:

(a) (200)
(b) (400)
(c) (111)
(d) (222)

15. A powder pattern is obtained for an fcc crystal with lattice parameter 3.52 Å by using X-ray of wavelength 1.79 Å. The highest reflection possible is:

(a) (200)
(b) (400)
(c) (111)
(d) (222)

16. In a diffraction experiment, only one peak is observed in an fcc material at $2\theta = 121°$ when an X-ray of wavelength 1.54 Å is used. The possible reflection is:

    (a) (111)
    (b) (200)
    (c) (220)
    (d) (222)

17. The Bragg's angle corresponding to a reflection for which $(h^2 + k^2 + l^2) = 8$ is found to be 14.35°. If the X-ray of wavelength 0.71 Å is used, the lattice parameter of the crystal is:

    (a) 3.05 Å
    (b) 4.05 Å
    (c) 5.05 Å
    (d) 6.05 Å

18. The Bragg's angle corresponding to a reflection for which $(h^2 + k^2 + l^2) = 8$ is found to be 14.35°. The X-ray of the wavelength 0.71 Å is used. If there are two other reflections of smaller Bragg's angles, the crystal structure of the sample is:

    (a) SC
    (b) BCC
    (c) FCC
    (d) DC

19. When a monochromatic X-ray of wavelength 1.79 Å is used to study the fcc aluminum powder sample with a camera of radius 57.3 mm, the angle at which the first reflection to occur is:

    (a) 55.5°
    (b) 44.5°
    (c) 33.5°
    (d) 22.5°

20. The first reflection using CuKα radiation from a sample of copper powder (fcc) has an S-value of 86.7 mm. The camera radius is found to be:

    (a) 28.65 mm
    (b) 57.3 mm
    (c) 85.95 mm
    (d) 114.6 mm

21. A diffraction data of a cubic crystal of an unknown element show peaks at 2θ angles 44.46°, 51.64°, 75.78° and 93.22°. If the wavelength of X-ray used is 1.543 Å, the crystal structure is found to be:

    (a) SC
    (b) BCC
    (c) FCC
    (d) DC

22. A diffraction data of a cubic crystal of an unknown element show peaks at 2θ angles 44.46°, 51.64°, 75.78° and 93.22°. If the wavelength of X-ray used is 1.543 Å, the lattice parameter of the element is found to be:

    (a) 3.53 Å
    (b) 4.53 Å
    (c) 5.53 Å
    (d) 6.53 Å

23. A diffraction data of a cubic crystal of an unknown element show peaks at 2θ angles 44.46°, 51.64°, 75.78° and 93.22°. If the wavelength of X-ray used is 1.543 Å, the unknown element is found to be:

    (a) Iron
    (b) Aluminum
    (c) Copper
    (d) Nickel

24. A diffraction data of a cubic crystal of an unknown element show peaks at 2θ angles 38.60°, 55.71°, 69.70°, 82.55°, 95.00° and 107.67°. If the wavelength of X-ray used is 1.543 Å, the crystal structure is found to be:

    (a) SC
    (b) BCC
    (c) FCC
    (d) DC

25. A diffraction data of a cubic crystal of an unknown element show peaks at 2θ angles 38.60°, 55.71°, 69.70°, 82.55°, 95.00° and 107.67°. If the wavelength of X-ray used is 1.543 Å, the lattice parameter of the unknown element is found to be:

    (a) 3.30 Å
    (b) 4.30 Å
    (c) 5.30 Å
    (d) 6.30 Å

26. A diffraction data of a cubic crystal of an unknown element show peaks at 2θ angles 38.60°, 55.71°, 69.70°, 82.55°, 95.00° and 107.67°. If the wavelength of X-ray used is 1.543 Å, identify the element:

   (a) Iron
   (b) Niobium
   (c) Copper
   (d) Nickel

27. Bragg's reflection is observed at an angle of 5° in a transmission Laue pattern for KCl crystal. If the crystal to film distance is 4 cm, the location of the spot with reference to center of the film is:

   (a) 0.505 cm
   (b) 0.605 cm
   (c) 0.705 cm
   (d) 0.805 cm

28. From a Laue photograph of an fcc crystal, the cell parameter is found to be 4.50 Å. If the potential difference across the X-ray tube is 50 kV and the crystal to film distance is 5 cm, the minimum distance from the center of the pattern at which reflection can occur from the plane of maximum spacing is:

   (a) 0.78 cm
   (b) 0.68 cm
   (c) 0.58 cm
   (d) 0.48 cm

29. A transmission Laue pattern of a copper crystal with a = 3.62 Å is obtained on a film kept at a distance of 5 cm. The (100) planes of the crystal make an angle of 5° with the incident beam. The minimum tube voltage required to produce a 200 reflection is found to be:

   (a) 38.82 kV
   (b) 48.82 kV
   (c) 58.82 kV
   (d) 68.82 kV

30. A transmission Laue pattern of a copper crystal with a = 3.62 Å is obtained on a film kept at a distance of 5 cm. The (100) planes of the crystal make an angle of 5° with the incident beam. Location of the spot with reference to center of the film is:

   (a) 0.98 cm
   (b) 0.88 cm
   (c) 0.78 cm
   (d) 0.68 cm

31. A transmission Laue pattern of an iron crystal with a = 2.86 Å is obtained on a film kept at a distance of 5 cm using 30 kV tungsten radiation. Laue spots formed by reflecting planes {110} are obtained. Assuming that these planes are present in the desired orientation, the angle of the orientation is:

    (a) 7.88°
    (b) 6.88°
    (c) 5.88°
    (d) 4.88°

32. A transmission Laue pattern of an iron crystal with a = 2.86 Å is obtained on a film kept at a distance of 5 cm using 30 kV tungsten radiation. Laue spots formed by reflecting planes {200} are obtained. Assuming that these planes are present in the desired orientation, the angle of the orientation is:

    (a) 11.32°
    (b) 10.32°
    (c) 9.32°
    (d) 8.32°

33. A transmission Laue pattern of a cubic crystal with a = 3 Å is obtained with a horizontal X-ray beam in the following orientation. The [010] axis of the crystal points along the beam away from the X-ray tube, [001] points vertically upwards and [100] is horizontal and parallel to the X-ray film. The crystal to film distance is 5 cm. Wavelength of the radiation diffracted from the $(01\bar{2})$ planes for the first-order reflection is:

    (a) 1.20 Å
    (b) .20 Å
    (c) 3.20 Å
    (d) 4.20 Å

34. A transmission Laue pattern of a cubic crystal with a = 3 Å is obtained with a horizontal X-ray beam in the following orientation. The [010] axis of the crystal points along the beam away from the X-ray tube, [001] points vertically upwards and [100] is horizontal and parallel to the X-ray film. The crystal to film distance is 5 cm. Wavelength of the radiation diffracted from the $(01\bar{2})$ planes for the second-order reflection is:

    (a) 0.50 Å
    (b) 0.60 Å
    (c) 0.70 Å
    (d) 0.80 Å

35. A back-reflection Laue pattern of an aluminum crystal (fcc) with a = 4.05 Å is obtained from (111) planes making an angle of 85° with the incident X-ray beam. Assuming that the wavelengths above 2.5 Å are extremely weak and they get absorbed in air. If the tube voltage is 60 kV, the maximum orders of (111) reflection present is:

   (a) 18
   (b) 20
   (c) 22
   (d) 24

36. In a rotation photograph, three layer lines are observed both above and below the zero layer line. A millimeter scale placed next to the developed film gives the following readings:

Layer line	Height (mm)	Height w.r.t zero layer
3	05.40	34.00
2	22.44	16.96
1	31.84	07.56
0	39.40	–
−1	46.96	07.56
−2	56.35	16.95
−3	73.20	33.80

If the diameter of the camera is 57.3 mm and the wavelength of X-ray used is 1.542 Å, the value of the cell height is:

   (a) 9.05 Å
   (b) 8.05 Å
   (c) 7.05 Å
   (d) 6.05 Å

37. In a rotation photograph, first and second layer lines are observed (w.r.t zero layer line) at 18 mm and 40 mm, respectively. If the wavelength of the MoKα radiation is 0.71 Å and the film diameter 57.3 mm, the lattice parameter of the sample is:

   (a) 1.54 Å
   (b) 2.54 Å
   (c) 3.54 Å
   (d) 4.54 Å

38. In a rotation photograph six layer lines are observed both above and below the zero layer line. If the heights of these layer lines above (or below) the zero layer are 0.29, 0.59, 0.91, 1.25, 1.65, and 2.2 cm, respectively. The camera radius is 3 cm and the wavelength of the X-ray is 1.54 Å, the cell height of the crystal along the axis of rotation is:

    (a) 18 Å
    (b) 16 Å
    (c) 14 Å
    (d) 12 Å

39. In a rotation photograph, the separation of the fourth layer line from the zero layer line is 2.2 cm. If the camera radius is 3 cm and the lattice parameter is 16 Å, the wavelength of the X-ray used is:

    (a) 3.54 Å
    (b) 2.54 Å
    (c) 1.54 Å
    (d) 0.54 Å

40. A simple cubic crystal is irradiated with X-rays of wavelength 0.9 Å at a glancing angle. The crystal is rotated and the angles for Bragg reflections are measured. The set of crystal planes will give the smallest angle for first-order reflection is:

    (a) {220}
    (b) {111}
    (c) {110}
    (d) {100}

41. A simple cubic crystal is irradiated with X-rays of wavelength 0.9 Å at a glancing angle. The crystal is rotated and the angles for Bragg reflections are measured. If the glancing angle is 8.9°, angle for the first-order reflection from the (110) crystal planes is:

    (a) 12.62°
    (b) 14.62°
    (c) 16.62°
    (d) 18.62°

42. An oscillation photograph is taken with CuKα radiation the rotation being about an axis of length 7.2 Å. If the camera diameter is 57.3 mm, the distance of the first layer line w.r.t zero layer is:

    (a) 5.28 mm
    (b) 6.28 mm
    (c) 7.28 mm
    (d) 8.28 mm

43. A unit cell of NaCl has four formula units. If the side of the unit cell is 5.64 Å, its density is:

   (a) 1965 kg/m^3
   (b) 2065 kg/m^3
   (c) 2165 kg/m^3
   (d) 2265 kg/m^3

44. Sodium chloride has fcc structure. Its density is $2.18 \times 10^3 \, kg/m^3$, the atomic weight of sodium and chlorine are 23 and 35.5, respectively. The interatomic separation is:

   (a) 5.81×10^{-10}m
   (b) 4.81×10^{-10}m
   (c) 3.81×10^{-10}m
   (d) 2.81×10^{-10}m

45. Potassium chloride has density of $1.98 \times 10^3 \, kg/m^3$ and its molecular weight is 74.55. Its interatomic separation is:

   (a) 7.30×10^{-10}m
   (b) 6.30×10^{-10}m
   (c) 5.30×10^{-10}m
   (d) 4.30×10^{-10}m

46. The nearest neighbor distance in a silver crystal is 2.87 Å. Silver crystallizes in fcc form, its density is:

   (a) 1.068×10^4 kg/m^3
   (b) 2.068×10^4 kg/m^3
   (c) 3.068×10^4 kg/m^3
   (d) 4.068×10^4 kg/m^3

47. α-iron belongs to a bcc structure. The density of α-iron is $7860 \, kg/m^3$ and its atomic weight is 55.85. Radius of the iron atom is:

   (a) 3.24 Å
   (b) 2.24 Å
   (c) 1.24 Å
   (d) 0.24 Å

48. Zinc has hcp structure. The height of the unit cell is 4.94 Å, the nearest neighbor distance is 2.7 Å and the at. Wt. of zinc is 65.38. The density of zinc is:

   (a) 3961 kg/m^3
   (b) 4961 kg/m^3
   (c) 5961 kg/m^3
   (d) 6961 kg/m^3

49. Copper crystallizes in cubic form. If its lattice parameter, a = 3.60 Å, at. wt. = 63.6 and density, $\rho = 8960 \, kg/m^3$, the number of atoms in a unit cell is:

    (a) 4
    (b) 3
    (c) 2
    (d) 1

50. Copper crystallizes in cubic form. If its lattice parameter, a = 3.60 Å, at. wt. = 63.6 and density, $\rho = 8960 \, kg/m^3$, the type of unit cell is:

    (a) SC
    (b) BCC
    (c) FCC
    (d) DC

51. Sodium crystallizes in cubic form. The edge of the unit cell is 4.3 Å. The atomic weight of sodium is 23 and its density is $963 \, kg/m^3$. Determine the number of atoms in the unit cell and its type:

    (a) 1
    (b) 2
    (c) 3
    (d) 4

52. Sodium crystallizes in cubic form. The edge of the unit cell is 4.3 Å. The atomic weight of sodium is 23 and its density is $963 \, kg/m^3$. Determe the number of atoms in the unit cell and its type:

    (a) SC
    (b) BCC
    (c) FCC
    (d) DC

Answers

1. (b)
2. (c)
3. (a)
4. (d)
5. (a)
6. (b)
7. (c)
8. (d)
9. (d)
10. (a)
11. (d)
12. (a)

13. (b)
14. (c)
15. (d)
16. (a)
17. (b)
18. (c)
19. (d)
20. (b)
21. (c)
22. (a)
23. (d)
24. (b)
25. (a)
26. (b)
27. (c)
28. (d)
29. (a)
30. (b)
31. (c)
32. (d)
33. (a)
34. (b)
35. (c)
36. (d)
37. (a)
38. (b)
39. (c)
40. (d)
41. (a)
42. (b)
43. (c)
44. (d)
45. (b)
46. (a)
47. (c)
48. (d)
49. (a)
50. (c)
51. (b)
52. (b)

Bibliography

Ashcroft, N.W., Mermin, N.D.: Solid State Physics. Harcourt Asia PTE Ltd., Singapore (2001)

Cromer, D.T., Mann, J.B.: X-ray scattering factors computed from numerical Hartree-Fock wave function. Acta Crystallograph. A **24**, 321–324 (1968)

Cullity, B.D., Stock, S.R.: Elements of X-Ray Diffraction. Prentice Hall, USA (2001)

Dent Glasser, L.S.: Crystallography and its Applications, Van Nostrand Reinhold Company Limited, USA (1977)

Hammond, C.: The Basics of Crystallography and Diffraction, IUCr, 2nd edn. Oxford University Press, USA (2001)

Kittel, C.: Introduction to Solid State Physics, 5th edn. Wiley Eastern Private Limited, New Delhi (1976)

Hammond, C.: The Basics of Crystallography and Diffraction, IUCr, 2nd edn. Oxford University Press, USA (2001)

Hebbar, K.R.: Basics of X-ray Diffraction and its Applications. I. K. International Publishing House Pvt. Ltd., New Delhi, India (2011)

Wahab, M.A.: Numerical Problems in Solid State Physics, 2nd edn. Narosa Publishing House Pvt. Ltd., New Delhi (2009)

Wahab, M.A.: Essentials of Crystallography, 2nd edn. Narosa Publishing House, New Delhi (2011)

Wahab, M.A.: Solid State Physics: Structure and Properties of Materials, 3rd edn. Narosa Publishing House, New Delhi (2013)

Wahab, M.A.: The mirror: mother of all symmetries in crystals. Adv. Sci. Eng. Med. **12**, 289–313 (2020)

Wahab, M.A., Wahab, K.M.: Resolution of ambiguities and the discovery of two new space lattices. ISST J. Appl. Phys. **6**(1), 1 (2015)

Wahab, M.A., Wahab, K.M.: Genesis of rhombohedral structures and mode of polytype transformations in close packing of identical atoms. Mater. Focus **7**(2), 321 (2018)

© The Editor(s) (if applicable) and The Author(s), under exclusive license to
Springer Nature Singapore Pte Ltd. 2021
M. A. Wahab, *Numerical Problems in Crystallography*,
https://doi.org/10.1007/978-981-15-9754-1

Index

Printed in the United States
by Baker & Taylor Publisher Services